Dr. Euler's Fabulous Formula

4/33

THE MOST REMARKABLE
FORMULA
IN MATH,

$$e^{i\pi} + 1 = 0$$

(FROM SCIENCE HISTORY OF THE UNIVERSE)

DERIVED BE EULER

METHOD. TO FIGURE THIS IT MUST
BE KNOWN THAT

$$e^x = 1 + x + \frac{x^2}{2!} + \frac{x^3}{3!} + \frac{x^4}{4!} + \cdots$$

AND $$\sin x = x - \frac{x^3}{3!} + \frac{x^5}{5!} + \frac{x^7}{7!} + \cdots$$

AND $$\cos x = 1 - \frac{x^2}{2!} + \frac{x^4}{4!} - \frac{x^6}{6!} + \cdots$$

IN THE FIRST LET $x = i u$. and
THIS REDUCES TO:

$$e^{iu} = \left(1 - \frac{u^2}{2!} + \frac{u^4}{4!} \cdots\right) + i\left(u - \frac{u^3}{3!} + \frac{u^5}{5!} \cdots\right)$$

NOW WE SUBSTITUTE AND
FIND

In an entry made in one of his teenage notebooks in April 1933, just before his fifteenth birthday, the future physics Nobel prize winner Richard Feynman (1918–1988) took notice of a major theme of this book. Notice the power series expansions for the exponential, sine, and cosine functions immediately below the "most remarkable result in math." The next line is the start of the standard derivation of Euler's formula (also known as Euler's identity) $e^{iu} = \cos(u) + i\sin(u)$, of which the "remarkable result" is the special case of $u = \pi$. (Feynman's source, *The Science History of the Universe,* was a ten-volume reference set first published in 1909.) Although remembered today as a physicist, Feynman was also a talented mathematician, who wrote in his *The Character of Physical Law* (1965), "To those who do not know mathematics it is difficult to get across a real feeling as to the beauty, the deepest beauty of nature.... If you want to learn about nature, to appreciate nature, it is necessary to understand the language that she speaks in." Feynman would surely have agreed with one of the early working titles to this book: *Complex Numbers Are Real!* (Photograph courtesy of the Archives, California Institute of Technology)

Dr. Euler's Fabulous Formula

CURES MANY MATHEMATICAL ILLS

Paul J. Nahin

PRINCETON UNIVERSITY PRESS PRINCETON AND OXFORD

Copyright © 2006 by Princeton University Press
Published by Princeton University Press, 41 William Street,
Princeton, New Jersey 08540
In the United Kingdom: Princeton University Press, 3 Market Place, Woodstock,
Oxfordshire OX20 1SY

All Rights Reserved

Library of Congress Cataloging-in-Publication Data

Nahin, Paul J.
Dr. Euler's fabulous formula : cures many mathematical ills / Paul J. Nahin.
p.   cm.
Includes bibliographical references and index.
ISBN-13: 978-0-691-11822-2 (cl : acid-free paper)
ISBN-10: 0-691-11822-1 (cl : acid-free paper)
1. Numbers, Complex. 2. Euler's numbers. 3. Mathematics–History. I. Title.

QA255.N339 2006
512.7'88—dc22   2005056550

British Library Cataloging-in-Publication Data is available

This book has been composed in New Baskerville
Printed on acid-free paper. ∞
pup.princeton.edu
Printed in the United States of America

7 9 10 8

For Patricia Ann,
who (like Euler's formula) is both complex and beautiful

May the God who watches over the right use of mathematical symbols, in manuscript, print, and on the blackboard, forgive me [my sins].

—Hermann Weyl, Professor of Mathematics from 1933 to 1952 at the Institute for Advance Study, in his book *The Classical Groups*, Princeton 1946, p. 289

Contents

Chapter 1. Complex Numbers
(an assortment of essays beyond the elementary involving complex numbers)

Chapter 2. Vector Trips
(some complex plane problems in which direction matters)

Chapter 3. The Irrationality of π^2
("higher" math at the sophomore level)

Chapter 4. Fourier Series
(named after Fourier but Euler was there first——but he was, alas, partially WRONG!)

Chapter 5. Fourier Integrals
(what happens as the period of a periodic function becomes infinite, and other neat stuff)

Chapter 6. Electronics and $\sqrt{-1}$
(technological applications of complex numbers that Euler, who was a practical fellow himself, would have loved)

What This Book Is About,
What You Need to Know to Read It,
and WHY You Should Read It

Everything of any importance is founded on mathematics.
—Robert Heinlein, *Starship Troopers* (1959)

Several years ago Princeton University Press published my *An Imaginary Tale: The Story of $\sqrt{-1}$* (1998), which describes the agonizingly long, painful discovery of complex numbers. Historical in spirit, that book still had a lot of mathematics in it. And yet, there was so much I had to leave out or else the book would have been twice its size. This book is much of that "second half" I had to skip by in 1998. While there is some historical discussion here, too, the emphasis is now on more advanced mathematical arguments (but none beyond the skills I mention below), on issues that I think could fairly be called the "sexy part" of complex numbers. There is, of course, some overlap between the two books, but whenever possible I have referred to results derived in *An Imaginary Tale* and have not rederived them here.

To read this book you should have a mathematical background equivalent to what a beginning third year college undergraduate in an engineering or physics program of study would have completed. That is, two years of calculus, a first course in differential equations, and perhaps some preliminary acquittance with matrix algebra and elementary probability. Third year math majors would *certainly* have the required background! These requirements will, admittedly, leave more than a few otherwise educated readers out in the cold. Such people commonly share the attitude of Second World War British Prime Minister Winston Churchill, who wrote the following passage in his 1930 autobiographical work *My Early Years: A Roving Commission*:

I had a feeling once about Mathematics, that I saw it all—Depth beyond depth was revealed to me—the Byss and the Abyss. I saw, as one might see the transit of Venus—or even the Lord Mayor's Show, a quantity passing through infinity and changing its sign from plus to minus. I saw exactly how it happened and why the tergiversation was inevitable: and how the one step involved all the others. It was like politics. But it was after dinner and I let it go!

Churchill was, I'm sure, mostly trying to be funny, but others, equally frank about their lack of mathematical knowledge, seem not to be terribly concerned about it. As an example, consider a review by novelist Joyce Carol Oates of E. L. Doctorow's 2000 novel *City of God* (*New York Review of Books*, March 9, 2000, p. 31). Oates (a professor at Princeton and a winner of the Pulitzer Prize) wrote, "The sciences of the universe are disciplines whose primary language is mathematics, not conventional speech, and it's inaccessible even to the reasonably educated non-mathematician." I disagree. Shouldn't being ignorant of what is taught each year to a *million college freshman and sophomores* (math, at the level of this book) world-wide, the vast majority of whom are *not* math majors, be reason for at least a *little* concern?

Some of Oates's own literary colleagues would also surely disagree with her. As novelist Rebecca Goldstein wrote in her 1993 work *Strange Attractors*, "Mathematics and music *are* God's languages. When you speak them ... you're speaking directly to God." I am also thinking of such past great American poets as Henry Longfellow and Edna St. Vincent Millay. It was Millay, of course, who wrote the often quoted line from her 1923 *The Harp-Weaver*, "Euclid alone has looked on Beauty bare." But it was Longfellow who long ago really laid it on the line, who really put his finger on the knowledge gap that exists without embarrassment in many otherwise educated minds. In the opening passages of Chapter 4 in his 1849 novella *Kavanagh, A Tale*, the dreamy and thoughtful schoolmaster Mr. Churchill and his wife Mary have the following exchange in his study:

> "For my part [says Mary] I do not see how you can make mathematics poetical. There is no poetry in them."
>
> "Ah [exclaims Mr. Churchill], that is a very great mistake! There is something divine in the science of numbers. Like God, it holds

the sea in the hollow of its hand. It measures the earth; it weighs the stars; it illuminates the universe; it is law, it is order, it is beauty. And yet we imagine—that is, most of us—that its highest end and culminating point is book-keeping by double entry. It is our way of teaching it that makes it so prosaic."

You, of course, since you are reading this book, fully appreciate and agree with *this* Mr. Churchill's words!

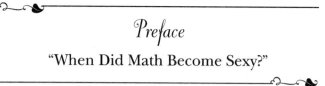

Preface
"When Did Math Become Sexy?"

The above question, from a 2002 editorial[1] in The *Boston Globe,* observes how the concept of beauty in mathematics has moved from the insulated male-dominated world of pipe-smoking, sherry-sipping mathematicians in tweedy coats and corduroy trousers, at the weekly afternoon college seminar, to the "real world" of truck drivers, teenagers, and retired couples looking for a bit of entertainment on a rainy afternoon. You'll see what I mean if you watch the film *Spider-Man 2* (2004); look for Tobey Maguire's casual reference in a Hollywood super-hero adventure flick to Bernoulli's solution to the famous problem of determining the minimum gravitational descent time curve.

In support of its claim, the *Globe* editorial cites three plays and a movie as examples of this remarkable intellectual transition. In the play *Copenhagen* we see a dramatic presentation of a debate between the physicists Niels Bohr and Werner Heisenberg on quantum mechanics. Heisenberg, who gave his very name to the mathematics of inherent uncertainty in nature (discussed in chapter 5), talks at one point of his first understanding of the new quantum theory: it is, he says "A world of pure mathematical structures. I'm too excited to sleep." He then, as the *Globe* put it, "rushes out in the dawn to climb a rock jutting out to sea, the crashing surf all around." It seems a scene we've all seen many times before in films from the 1930s and 1940s, just before (or after) the heroine is bedded. The erotic connection between mathematical insight and sexual orgasm is simply impossible to deny.[2]

The editorial then goes on to discuss the plays *Proof* (in which arcane formulas are presented as "beautiful"), *Q.E.D.* (about the theoretical physicist Richard Feynman, who often talked of the wondrous way mathematics is at the root of any meaningful interpretation of nature), and Ron Howard's Oscar-winning 2001 film *A Beautiful Mind*. While that film was an interpretation of the life of Princeton mathematician John Nash as viewed through a somewhat distorted glass, it did the best a Hollywood movie probably could do in telling a general audience (from teenagers on up) of what Nash's work in game theory was (sort of) all about. Oddly, the *Globe* didn't mention the 1997 film *Good Will Hunting* (I say *oddly* because the stars are Ben Affleck and Matt Damon, two of Boston's own), in which line after line of Fourier integral equations fills the screen during the opening sequences. That film, also an Oscar-winner, also has a mathematical genius—a handsome night-shift custodian at MIT—as its hero. In a powerful emotional reversal of the idea that math is wonderful, Tom Hanks's mob hitman-on-the-run in *The Road to Perdition* (2002) finds common ground with his son when the two discover that they both *hate* math. As poets are so fond of saying, hate and love are two sides of the same coin and so, even in this violent film, math plays an emotional role as male-bonding glue.

Even before the examples mentioned by the *Globe*, mathematics had played a prominent role in a number of mainstream films.[3] Take a look at such movies as *Straw Dogs* (1971), *It's My Turn* (1980), *Stand and Deliver* (1987), *Sneakers* (1992), *The Mirror Has Two Faces* (1996), *Contact* (1997), *Pi* (1998), and *Enigma* (2002), and you'll agree that the *Globe* was right— mathematics (often equated with extreme dorkiness) *has* become sexy! Even television has gotten into the act, with the 2005 series *Numb3rs*, involving an FBI agent with a mathematical genius brother who helps solve crime mysteries (the show's technical adviser is a professor of mathematics at Caltech, where numerous "atmosphere" scenes were shot to achieve the appropriate academic ambiance).

The *Globe* thought this embracing of mathematics by popular culture had happened because "The attraction of math and science is that they require embracing the unknowable." The inclusion of science in this statement is interesting, because many physicists think the most beautiful equations (note the plural) are those of Einstein's theory of gravity. For them it is not the mathematics, itself, that is the source of the beauty, but rather the equations' elegant expression of physical reality. For them the mathematics is the visible flesh, yes, but it is the *physics* that is the soul— and source—of the beauty. The 1933 Physics Nobel prize co-winner Paul Dirac was a notable exponent of that view. Dirac (1902–1984) was famous for his many comments on *technical* beauty[4]; in response to being asked (in Moscow, in 1955) about his philosophy of physics, for example, he wrote on a blackboard "Physical Laws should have mathematical beauty." That blackboard is preserved to this day by admiring Russian physicists.

Of course, as physicists learn more physics their equations change. No one, not even Einstein, is immune to this evolution. Just as Newton's gravitational theory gave way to Einstein's, Einstein's is having to give way to newer ideas that are compatible, as Einstein's equations are not, with quantum mechanics. So, Einstein's physics is, at some fundamentally deep level, "wrong" (or, more graciously, "missing something") and so is only approximately correct. But does that mean the mathematical beauty of the equations in Einstein's theory fades?

I don't think so. In the Introduction I'll discuss a number of views that various writers have put forth on what makes theories (and their equations) beautiful, but one point that isn't mentioned there is perhaps best done so here. The reason I think Einstein's theory is still beautiful, even though we now know it cannot possibly be perfectly correct, is that it is the result of *disciplined reasoning*. Einstein created new physics, yes, but not just willy-nilly. His work was done while satisfying certain severe restrictions. For example, the physical laws of nature must be the same for all observers, no matter what may be their state of motion in the universe. A theory that satisfies such a broad constraint *must*, I think, be beautiful.

Ugly creations, in my opinion, be they theories or paintings, are ones that obey no constraints, that have no discipline in their nature. It is by that criterion, alone, for example, that I place Norman Rockwell *far* above Jackson Pollock as an artist. This will no doubt send most modern art fans into near-fatal convulsions and brand me a cultural Neanderthal (my art historian wife's opinion), but anybody who can observe the result of simply throwing paint on a canvas[5]—what two-year-olds routinely do in ten thousand day-care centers every day (my gosh, what *I* do every time I paint a ceiling!)—and call the outcome art, much less beautiful art, is delusional or at least deeply confused (in my humble opinion). To take my point to the limit, I find the imagery of Jackson Pollock fans exclaiming in awe at the mess formed by paint randomly dripping on the *floor* of the Sistine Chapel, rather than at what Michelangelo *painstakingly*, with *skill* and *discipline*, applied to the ceiling, hilarious. Pollock fans might very well rebut me by saying his works *are* beautiful because he *did* have discipline—the "discipline" never to be hamstrung by discipline! I have heard that argument before from college students, and I must admit I am still trying to come up with a good reply other than rolling my eyes.

In this book the gold standard for mathematical beauty is one of the formulas at the heart of complex number analysis, Euler's formula or identity $e^{i\theta} = \cos(\theta) + i\sin(\theta)$, where $i = \sqrt{-1}$. The special case of $\theta = \pi$ gives $e^{i\pi} = -1$ or, as it is usually written, $e^{i\pi} + 1 = 0$, a compact expression that I think is of exquisite beauty. I think $e^{i\pi} + 1 = 0$ is beautiful because it is true even in the face of *enormous* potential constraint. The equality is precise; the left-hand side is not "almost" or "pretty near" or "just about" zero, but *exactly* zero. That *five* numbers, each with vastly different origins, and each with roles in mathematics that cannot be exaggerated, should be connected by such a simple relationship, is just stunning. *It is beautiful.* And unlike the physics or chemistry or engineering of today, which will almost surely appear archaic to technicians of the far future, Euler's formula will still appear, to the arbitrarily advanced mathematicians ten thousand years hence, to be beautiful and stunning and untarnished by time.

The great German mathematician Hermann Weyl (1885–1955) is famous for declaring, in only a half-joking way, "My work always tried to unite the truth with the beautiful, but when I had to chose one or the other, I usually chose the beautiful." Read on, and I'll try to demonstrate what Weyl meant by showing you some really beautiful ("sexy"?) complex number calculations, many based in part on Euler's formula.

Dr. Euler's Fabulous Formula

Like a Shakespearean sonnet that captures the very essence of
love, or a painting that brings out the beauty of the human
form that is far more than just skin deep, Euler's equation
reaches down into the very depths of existence.
            —Keith Devlin writing of $e^{i\pi} + 1 = 0$[1]

The nineteenth-century Harvard mathematician Benjamin Peirce (1809–
1880) made a tremendous impression on his students. As one of them
wrote many years after Peirce's death, "The appearance of Professor
Benjamin Peirce, whose long gray hair, straggling grizzled beard and
unusually bright eyes sparkling under a soft felt hat, as he walked briskly
but rather ungracefully across the college yard, fitted very well with the
opinion current among us that we were looking upon a real live genius,
who had a touch of the prophet in his make-up."[2] That same former
student went on to recall that during one lecture "he established the
relation connecting π, e, and i, $e^{\pi/2} = \sqrt[i]{i}$, which evidently had a strong
hold on his imagination.[3] He dropped his chalk and rubber (i.e., eraser),
put his hands in his pockets, and after contemplating the formula a few
minutes turned to his class and said very slowly and impressively, 'Gentle-
men, that is surely true, it is absolutely paradoxical, we can't understand
it, and we don't know what it means, but we have proved it, and therefore
we know it must be the truth.' "

Like any good teacher, Peirce was almost certainly striving to be dramatic ("Although we could rarely follow him, we certainly sat up and took
notice"), but with those particular words he reached too far. We certainly
can understand what Peirce always called the "mysterious formula," and
we certainly *do* know what it means. But, yes, it *is* still a wonderful, indeed
beautiful, expression; no amount of "understanding" can ever diminish

its power to awe us. As one limerick (a literary form particularly beloved by mathematicians) puts it,

> *e* raised to the *pi* times *i*,
> And plus 1 leaves you nought but a sigh.
> This fact amazed Euler
> That genius toiler,
> And still gives us pause, bye the bye.

The limerick puts front-and-center several items we need to discuss pretty soon. What are *e*, *pi*, and *i*, and who was Euler? Now, it is hard for me to believe that there are any literate readers in the world who haven't heard of the transcendental numbers $e = 2.71828182\ldots$ and $pi = \pi = 3.14159265\ldots$, and of the imaginary number $i = \sqrt{-1}$. As for Euler, he was surely one of the greatest of all mathematicians. Making lists of the "greatest" is a popular activity these days, and I would wager that the Swiss-born Leonhard Euler (1707–1783) would appear somewhere among the top five mathematicians of all time on the list made by any mathematician in the world today (Archimedes, Newton, and Gauss would give him stiff competition, but what great company *they* are!).

Now, before I launch into the particulars of e, π, and $\sqrt{-1}$, what about the stupefying audacity I displayed in the Preface by declaring $e^{i\pi} + 1 = 0$ to be "an expression of exquisite beauty"? I didn't do that lightly and, indeed, I have "official authority." In the fall 1988 issue of the *Mathematical Intelligencer*, a scholarly quarterly journal of mathematics sponsored by the prestigious publisher of mathematics books and journals, Springer-Verlag, there was the call for a vote on the most beautiful theorem in mathematics. Readers of the *Intelligencer*, consisting almost entirely of academic and industrial mathematicians, were asked to rank twenty-four given theorems on a scale of 0 to 10, with 10 being the most beautiful and 0 the least. The list contained, in addition to $e^{i\pi} + 1 = 0$, such seminal theorems as

(a) The number of primes is infinite;
(b) There is no rational number whose square is 2;

(c) π is transcendental;

(d) A continuous mapping of the closed unit disk into itself has a fixed point.

A distinguished list, indeed.

The results, from a total of 68 responses, were announced in the summer 1990 issue. Receiving the top average score of 7.7 was $e^{i\pi} + 1 = 0$. The scores for the other theorems above, by comparison, were 7.5 for (a), 6.7 for (b), 6.5 for (c), and 6.8 for (d). The lowest ranked theorem (a result in number theory by the Indian genius Ramanujan) received an average score of 3.9. So, it is *official: $e^{i\pi} + 1 = 0$ is* the most beautiful equation in mathematics! (I hope most readers can see my tongue stuck firmly in my cheek as they read these words, and will not send me outraged e-mails to tell me why their favorite expression is so much *more* beautiful.)

Of course, the language used above is pretty sloppy, because $e^{i\pi} + 1 = 0$ is actually *not* an equation. An equation (in a single variable) is a mathematical expression of the form $f(x) = 0$, for example, $x^2 + x - 2 = 0$, which is true only for certain values of the variable, that is, for the *solutions* of the equation. For the just cited quadratic equation, for example, $f(x)$ equals zero for the two values of $x = -2$ and $x = 1$, *only*. There is no x, however, to solve for in $e^{i\pi} + 1 = 0$. So, it isn't an equation. It isn't an identity, either, like Euler's identity $e^{i\theta} = \cos(\theta) + i\sin(\theta)$, where θ is *any* angle, not just π radians. That's what an *identity* (in a single variable) is, of course, a statement that is *identically* true for *any* value of the variable. There isn't any variable at all, anywhere, in $e^{i\pi} + 1 = 0$: just five constants. (Euler's identity is at the heart of this book and it will be established in Chapter 1.) So, $e^{i\pi} + 1 = 0$ isn't an equation and it isn't an identity. Well, then, what *is* it? It is a *formula* or a *theorem*.

More to the point for us, here, isn't semantics but rather the issue I first raised in the Preface, that of *beauty*. What could it possibly mean to say a mathematical statement is "beautiful"? To that I reply, what does it mean to say a kitten asleep, or an eagle in flight, or a horse in full gallop, or a laughing baby, or ... is beautiful? An easy answer is that it is all simply in the eye of the beholder (the ultimate "explanation," I suppose, for the popularity of Jackson Pollock's drip paintings), but I think (at least in the mathematical case) that there are deeper possibilities. The author of the *Intelligencer* poll (David Wells, the writer of a number of popular mathematical works), for example, offered several good suggestions as to what makes a mathematical expression *beautiful.*

To be beautiful, Wells writes, a mathematical statement must be simple, brief, important, and, obvious when it is stated but perhaps easy to overlook otherwise, *surprising.* (A similar list was given earlier by H. E. Huntley in his 1970 book *The Divine Proportion.*) I think Euler's identity (and its offspring $e^{i\pi} + 1 = 0$) scores high on all four counts, and I believe you will think so too by the end of this book. Not everyone agrees, however, which should be no surprise—there is *always* someone who doesn't agree with *any* statement! For example, in his interesting essay "Beauty in Mathematics," the French mathematician François Le Lionnais (1901–1984) starts off with high praise, writing of $e^{i\pi} + 1 = 0$ that it

> [E]stablishes what appeared in its time to be a fantastic connection between the most important numbers in mathematics, 1, π, and e [for some reason 0 and i are ignored by Le Lionnais]. It was generally considered "the most important formula of mathematics."[4]

But then comes the tomato surprise, with a very big splat in the face: "Today the intrinsic reason for this compatibility has become so obvious[!] that the same formula seems, if not insipid, at least entirely natural."

Well, good for François and his fabulous powers of insight (or is it hindsight?), but such a statement is rightfully greeted with the same skepticism that most mathematicians give to claims from those who say they can "see geometrical shapes in the fourth dimension." Such people only *think* they do. They are certainly "seeing things," all right, but I doubt very much it's the true geometry of hyperspace. When you are finished here, $e^{i\pi} + 1 = 0$ *will* be "obvious," but borderline *insipid?* Never!

At this point, for completeness, I should mention that the great English mathematician G. H. Hardy (1887–1947) had a very odd view of what constitutes beauty in mathematics: to be beautiful, mathematics must be *useless*! That wasn't a sufficient condition but, for the ultra-pure Hardy, it was a necessary one. He made this outrageous assertion in his famous 1940 book *A Mathematician's Apology*, and I can't believe there is a mathematician today (no matter how pure) who would subscribe to Hardy's conceit. Indeed, I think Hardy's well-known interest and expertise in Fourier series and integrals, mathematics *impossible by 1940 for practical, grease-under-the-fingernails electrical engineers to do without* (as you'll see in chapters 5 and 6), is proof enough that his assertion was nonsense even as he wrote it. To further illustrate how peculiar was Hardy's thinking on this issue, he called the physicists James Clerk Maxwell (1831–1879), and Dirac, "real" mathematicians. That is comical because Maxwell's equations for the electromagnetic field are what make the oh-so-*useful* gadgets of radios and cell-phones possible, and Dirac always gave much credit to his undergraduate training in electrical engineering as being the inspiration behind his very nonrigorous development of the impulse "function" in quantum mechanics![5]

As a counterpoint to mathematical beauty, it may be useful to mention, just briefly, an example of mathematical *ugliness*. Consider the 1976 "proof" of the four-color theorem for planar maps. The theorem says that four colors are both sufficient *and necessary* to color all possible planar maps so that countries that share a border can have different colors.[6] This problem, which dates from 1852, defied mathematical attack until two mathematicians at the University of Illinois programmed a computer to automatically "check" many hundreds of specific special cases. The details are unimportant here—my point is simply that this particular "proof" is almost always what mathematicians think of when asked "What is an example of ugly mathematics?" If this seems a harsh word to use, let me assure you that I am not the first to do so. The two Illinois programmers themselves have told the story of the reaction of a mathematician friend when informed they had used a computer:[7] the friend "exclaimed in horror, 'God would never permit the best proof of such a beautiful theorem to be so ugly.'"

Although nearly all mathematicians believe the result, nearly all dislike how it was arrived at because the computer calculations hide from view so much of the so-called "solution." As the English mathematician who first started the four-color problem on its way into history, Augustus

De Morgan (1806–1871), wrote in his book *Budget of Paradoxes*, "Proof requires a **person** who can give and a **person** who can receive" (my emphasis). There is no mention here of an automatic machine performing hundreds of millions of intermediate calculations (requiring *weeks* of central processor time on a supercomputer) that not even a single person has ever completely waded through.[8]

Before leaving computer proofs, I should admit that there is one way such an approach *could* result in beautiful mathematics. Imagine that, unlike in the four color problem, the computer discovered one or more counter-examples to a proposed theorem. Those specific counter-examples could then be verified in the traditional manner by as many independent minds as cared to do it. An example of this, involving Euler, dates from 1769.[9] The disproof of a statement, by presenting a specific counter-example, is perhaps the most convincing of all methods (the counter-example's origin in a computer analysis is irrelevant once we have the counter-example in-hand), and is generally thought by mathematicians to be a beautiful technique.

There are, of course, lots of beautiful mathematical statements that I think might give $e^{i\pi} + 1 = 0$ a run for its money but weren't on the original *Intelligencer* list. Just to give you a couple of examples, consider first the infinite series

$$S = \sum_{n=1}^{\infty} \frac{1}{n} = 1 + \frac{1}{2} + \frac{1}{3} + \frac{1}{4} \cdots.$$

This series is called the *harmonic series,* and the question is whether the sum S is finite or infinite, that is, does the series *converge* or does it *diverge*? Nearly everyone who sees this for the first time thinks S should be finite (a mathematician would say *S exists*) because each new term is smaller than the previous term. Indeed, the terms are tending toward zero, which *is, indeed,* a *necessary* condition for the series to converge to a finite sum—but it isn't a *sufficient* condition. For a series to converge, the terms must not only go to zero, they must go to zero *fast enough* and, in the case of the harmonic series, they do not. (If the signs of the harmonic series alternate *then* the sum *is* finite: ln(2).) Thus, we have the beautiful, *surprising* statement that,

$$\lim_{k \to \infty} \sum_{n=1}^{k} \frac{1}{n} = \infty,$$

which has been known since about 1350. This theorem should, I think, have been on the original *Intelligencer* list.[10]

By the way, the proof of this beautiful theorem is an example of a beautiful mathematical *argument*. The following is not the original proof (which is pretty slick, too, but is more widely known and so I won't repeat it here[11]). We'll start with the *assumption* that the harmonic series converges, i.e., that its sum S is some finite number. Then,

$$S = 1 + \frac{1}{2} + \frac{1}{3} + \frac{1}{4} + \cdots$$
$$= \left(1 + \frac{1}{3} + \frac{1}{5} + \frac{1}{7} + \cdots\right) + \left(\frac{1}{2} + \frac{1}{4} + \frac{1}{6} + \frac{1}{8} + \cdots\right)$$
$$= \left(1 + \frac{1}{3} + \frac{1}{5} + \frac{1}{7} + \cdots\right) + \frac{1}{2}\left(1 + \frac{1}{2} + \frac{1}{3} + \frac{1}{4} + \cdots\right)$$
$$= \left(1 + \frac{1}{3} + \frac{1}{5} + \frac{1}{7} + \cdots\right) + \frac{1}{2}S.$$

So,

$$\frac{1}{2}S = 1 + \frac{1}{3} + \frac{1}{5} + \frac{1}{7} + \cdots.$$

That is, the sum of just the *odd* terms alone is one-half of the total sum. Thus, the sum of just the *even* terms alone must be the other half of S. Therefore, the *assumption* that S exists has led us to the conclusion that

$$1 + \frac{1}{3} + \frac{1}{5} + \frac{1}{7} + \cdots = \frac{1}{2} + \frac{1}{4} + \frac{1}{6} + \frac{1}{8} + \cdots.$$

But this equality is clearly not true since, term by term, the left-hand side is larger than the right-hand side ($1 > \frac{1}{2}, \frac{1}{3} > \frac{1}{4}, \frac{1}{5} > \frac{1}{6}, \cdots$). So, our initial assumption that S exists must be wrong, that is, S does *not* exist, and the harmonic series must diverge. This beautiful argument is called a *proof by contradiction.*

The most famous proof by contradiction is Euclid's proof of theorem (a) on the original *Intelligencer* list. I remember when I first saw (while still in high school) his demonstration of the infinity of the primes; I was thrilled by its elegance and its *beauty.* For me, a proof by contradiction became one of the signatures of a beautiful mathematical demonstration. When Andrew Wiles (1953–) finally cracked the famous problem of Fermat's last theorem in 1995, it was with a proof "by contradiction." And the proof I'll show you in chapter 3 of the irrationality of π^2, using Euler's formula, is a proof "by contradiction."

Celebrity intellectual Marilyn vos Savant ("world's highest IQ") is not impressed by this line of reasoning, however, as she rejects *any* proof by contradiction. As she wrote in her now infamous (and famously embarrassing) book on Wiles's proof,

> But how can one ever really prove anything by contradiction? Imaginary numbers are one example. The square root of $+1$ is a real number because $+1 x + 1 = +1$; however, the square root of -1 is imaginary because -1 times -1 would also equal $+1$, instead of -1. This appears to be a contradiction. [The "contradiction" escapes me, and I have no absolutely idea why she says this.] Yet it is accepted, and imaginary numbers are used routinely. But how can we justify using them to *prove* a contradiction?

This is, of course, as two reviewers of her book put it, an example of "inane reasoning" (the word *drivel* was also used to describe her book),[12] and so let me assure you that proof by contradiction *is* most certainly a valid technique.

So, imagine my surprise when I read two highly respected mathematicians call such a demonstration "a wiseguy argument!" They obviously meant that in a humorous way, but the phrase still brought me up short. I won't give Euclid's proof of the infinity of the primes here (you can find it in any book on number theory), but rather let me repeat what Philip Davis and Reuben Hersh said in their acclaimed 1981 book, *The Mathematical Experience*, about the traditional proof by contradiction of theorem (b) in the *Intelligencer* list. To prove that $\sqrt{2}$ is not rational, let's assume that it *is*. That is, *assume* (as did Pythagoras some time in the sixth century B.C.) that there are two integers m and n such that

$$\sqrt{2} = \frac{m}{n}.$$

We can further assume that m and n have no common factors, because if they *do* then simply cancel them and then call what remains m and n.

So, squaring, $2n^2 = m^2$, and thus m^2 is even. But that means that m itself is even, because you can't get an even m^2 by squaring an odd integer (any odd integer has the form $2k + 1$ for k some integer, and $(2k + 1)^2 = 4k^2 + 4k + 1$, which is odd). But, since m is even, then there must be an integer r such that $m = 2r$. So, $2n^2 = 4r^2$ or, $n^2 = 2r^2$, which means n^2 is even. And so n is even. To summarize, we have deduced that m and n are *both* even integers *if* the integers m and n exist. But, we started by assuming that m and n have no common factors (in particular, two even numbers have the common factor 2), and so our assumption that m and n exist has led to a logical contradiction. Thus, m and n do *not* exist! What a beautiful proof—it uses only the concept that the integers can be divided into two disjoint sets, the evens and the odds.

Davis and Hersh don't share my opinion, however, and besides the "wiseguy" characterization they suggest that the proof also has a problem "with its emphasis on logical inexorableness that seems heavy and plodding." Well, to that all I can say is all proofs should have such a problem! But what is very surprising to me is what they put forth as a better proof. They start as before, until they reach $2n^2 = m^2$. Then, they say, imagine that whatever m and n are, we factor them into their prime components. Thus, for m^2 we would have a sequence of *paired* primes (because $m^2 = m \cdot m$), and similarly for n^2. I now quote their dénouement:

> But (aha!) in $2n^2$ there is a 2 that has no partner.
> Contradiction.

Huh? *Why* a contradiction? Well, because they are invoking (although they never say so explicitly) a theorem called "the fundamental theorem of arithmetic," which says the factorization of any integer (in the realm of the ordinary integers) into a product of primes is *unique*. They do sort of admit this point, saying "Actually, we have elided some formal details." Yes, I think so!

Davis and Hersh claim their proof would be preferred to the Pythagorean one by "nine professional mathematicians out of ten [because it] exhibits a higher level of aesthetic delight." I suppose they might well be right, but *I* think the unstated unique factorization result a pretty big hole to jump over. It isn't hard to prove for the ordinary integers, but it isn't either trivial or, I think, even obvious — indeed, it isn't difficult to create different realms of real integers in which it isn't even true![13] I therefore have a very big problem with that "aha!" It certainly is a huge step beyond just the concept of evenness and oddness, which is all that the Pythagorean proof uses.

For my second example of a beautiful mathematical expression, this one due to Euler, consider the infinite product expansion of the sine function:

$$\sin(x) = x \prod_{n=1}^{\infty} \left(1 - \frac{x^2}{n^2}\right).$$

You don't have to know much mathematics to "know" that this is a pretty amazing statement (which may account for why many think it is "pretty," too). I think a high school student who has studied just algebra and trigonometry would appreciate this. As an illustration of this statement's significance, it is only a few easy steps from it to the conclusion that

$$\sum_{n=1}^{\infty} \frac{1}{n^2} = \frac{1}{1^2} + \frac{1}{2^2} + \frac{1}{3^2} + \cdots = \frac{\pi^2}{6} = 1.644934\cdots,$$

a beautiful line of mathematics in its own right (I'll show you a way to do it, different from Euler's derivation, later in the book).[14] Failed attempts to evaluate $\sum_{n=1}^{\infty}(1/n^2)$ had been frustrating mathematicians ever since the Italian Pietro Mengoli (1625–1686) first formally posed the problem in 1650, although many mathematicians must have thought of this next obvious extension beyond the harmonic series long before Mengoli; Euler finally cracked it in 1734.[15] All this *is* beautiful stuff but, in the end, I still think $e^{i\pi} + 1 = 0$ is the best. This is, in part, because you can *derive* the infinite product expression for sin(x) via the intimate link between sin(x) and $i = \sqrt{-1}$, as provided by Euler's identity (see note 14 again).

Let me end this little essay with the admission that perhaps mathematical beauty *is* all in the eye of the beholder, just like a Jackson Pollock painting. At the end of his 1935 presidential address to the London Mathematical Society, for example, the English mathematician G. N. Watson stated that a particular formula gave him "a thrill which is indistinguishable from the thrill which I feel when I enter the Sagrestia Nuova of the Capelle Medicee and see before me the austere beauty of the four statues representing Day, Night, Evening, and Dawn which Michelangelo has set over the tombs of Guiliano de'Medici and Lorenzo de'Medici."[16] Wow, that's quite a thrill!

In a series of lectures he gave to general audiences at the Science Museum in Paris in the early 1980s, the Yale mathematician Serge Lang tried to convey what *he* thought beautiful mathematics is, using somewhat less dramatic imagery than Watson's.[17] He never gave a formal definition, but several times he said that, whatever it was, he knew it when he saw it because it would give him a "chill in the spine." Lang's phrase reminds me of Supreme Court Justice Potter Stewart, who, in a 1964 decision dealing with pornography, wrote his famous comment that while he couldn't define it he "knew it when he saw it." Perhaps it is the same at the other end of the intellectual spectrum, as well, with beautiful mathematics.

Being able to appreciate beautiful mathematics is a privilege, and many otherwise educated people who can't sadly understand that they are "missing out" on something precious. In autobiographical recollections that he wrote in 1876 for his children, Charles Darwin expressed his feelings on this as follows:

> During the three years which I spent at Cambridge my time was wasted, as far as academical studies were concerned . . . I attempted mathematics, and even went during the summer of 1828 with a private tutor . . . but I got on very slowly. The work was repugnant to me, chiefly from my not being able to see any meaning in the early steps in algebra. This impatience was very foolish, and in after years I have deeply regretted that I did not proceed far enough at least to understand something of the great leading principles of mathematics, *for men thus endowed seem to have an extra sense* [my emphasis].[18]

I started this section with a limerick, so let me end with one. I think, if you read this book all the way through, then, contrary to Professor Peirce, you'll agree with the following (although I suspect the first two lines don't really apply to *you!*):

> I used to think math was no fun,
> 'Cause I couldn't see how it was done.
> Now Euler's my hero
> For I now see why zero,
> Equals $e^{pi\ i} + 1$.

Okay, enough with the bad poetry. Let's begin the good stuff. Let's do some "complex" mathematics.

Chapter 1
Complex Numbers

1.1 The "mystery" of $\sqrt{-1}$.

Many years ago a distinguished mathematician wrote the following words, words that may strike some readers as somewhat surprising:

> I met a man recently who told me that, so far from believing in the square root of minus one, he did not even believe in minus one. This is at any rate a consistent attitude. There are certainly many people who regard $\sqrt{2}$ as something perfectly obvious, but jib at $\sqrt{-1}$. This is because they think they can visualize the former as something in physical space, but not the latter. Actually $\sqrt{-1}$ is a much simpler concept.[1]

I say these words are "somewhat surprising" because I spent a fair amount of space in *An Imaginary Tale* documenting the *confusion* about $\sqrt{-1}$ that was common among many very intelligent thinkers from past centuries.

It isn't hard to appreciate what bothered the pioneer thinkers on the question of $\sqrt{-1}$. In the realm of the ordinary real numbers, every positive number has two real square roots (and zero has one). A *negative* real number, however, has *no* real square roots. To have a solution for the equation $x^2 + 1 = 0$, for example, we have to "go outside" the realm of the real numbers and into the expanded realm of the complex numbers. It was the need for this expansion that was the intellectual roadblock, for so long, to understanding what it means to say $i = \sqrt{-1}$ "solves" $x^2 + 1 = 0$. We can completely sidestep this expansion,[2] however, if we approach the problem from an entirely new (indeed, an unobvious) direction.

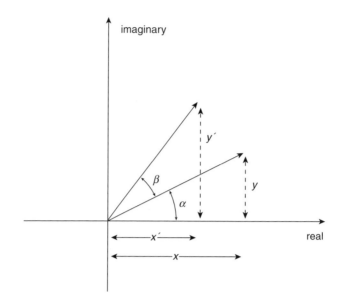

Figure 1.1.1. A rotated vector

A branch of mathematics called *matrix theory,* developed since 1850, formally illustrates (I think) what the above writer may have had in mind. In figure 1.1.1 we see the *vector* of the complex number $x + iy$, which makes angle α with the positive real axis, *rotated* counterclockwise through the additional angle of β to give the vector of the complex number $x' + iy'$. Both vectors have the same length r, of course, and so $r = \sqrt{x^2 + y^2} = \sqrt{x'^2 + y'^2}$. From the figure we can immediately write $x = r\cos(\alpha)$ and $y = r\sin(\alpha)$, and so, using the addition formulas for the sine and cosine

$$x' = r\cos(\alpha + \beta) = r\left[\cos(\alpha)\cos(\beta) - \sin(\alpha)\sin(\beta)\right],$$
$$y' = r\sin(\alpha + \beta) = r\left[\sin(\alpha)\cos(\beta) + \cos(\alpha)\sin(\beta)\right].$$

Now, focus on the x', y' equations and replace $r\cos(\alpha)$ and $r\sin(\alpha)$ with x and y, respectively. Then,

$$x' = x\cos(\beta) - y\sin(\beta),$$
$$y' = y\cos(\beta) + x\sin(\beta) = x\sin(\beta) + y\cos(\beta).$$

Writing this pair of equations in what is called *column vector/matrix* notation, we have

$$\begin{bmatrix} x' \\ y' \end{bmatrix} = \begin{bmatrix} \cos(\beta) & -\sin(\beta) \\ \sin(\beta) & \cos(\beta) \end{bmatrix} \begin{bmatrix} x \\ y \end{bmatrix} = \mathbf{R}(\beta) \begin{bmatrix} x \\ y \end{bmatrix},$$

where $\mathbf{R}(\beta)$ is the so-called *two-dimensional matrix rotation operator* (we'll encounter a different sort of operator—the *differentiation operator*—in chapter 3 when we prove the irrationality of π^2). That is, the column vector $\begin{bmatrix} x \\ y \end{bmatrix}$, when operated on (i.e., when multiplied[3]) by $\mathbf{R}(\beta)$, is rotated counterclockwise through the angle β into the column vector $\begin{bmatrix} x' \\ y' \end{bmatrix}$.

Since $\beta = 90°$ is the CCW rotation that results from multiplying $x + iy$ by i, this would seem to say that $i = \sqrt{-1}$ can be associated with the 2×2 matrix $\mathbf{R}(90°)$

$$\begin{bmatrix} \cos(90°) & -\sin(90°) \\ \sin(90°) & \cos(90°) \end{bmatrix} = \begin{bmatrix} 0 & -1 \\ 1 & 0 \end{bmatrix}.$$

Does this mean that we might, with merit, call this the *imaginary matrix?* To see that this actually makes sense, indeed that it makes a *lot* of sense, recall the 2×2 identity matrix

$$\mathbf{I} = \begin{bmatrix} 1 & 0 \\ 0 & 1 \end{bmatrix},$$

which has the property that, if $\mathbf{A}$ is any 2×2 matrix, then $\mathbf{AI} = \mathbf{IA} = \mathbf{A}$. That is, $\mathbf{I}$ plays the same role in matrix arithmetic as does 1 in the arithmetic of the realm of the ordinary real numbers. In that realm, of course, $i^2 = -1$, and the "mystery" of $\sqrt{-1}$ is that it itself is *not* (as I mentioned earlier) in the real of the ordinary real numbers. In the realm of 2×2 matrices, however, there is no such "mystery" because the square of the "imaginary matrix" (a perfectly respectable 2×2 matrix) is

$$\begin{bmatrix} 0 & -1 \\ 1 & 0 \end{bmatrix} \begin{bmatrix} 0 & -1 \\ 1 & 0 \end{bmatrix} = \begin{bmatrix} -1 & 0 \\ 0 & -1 \end{bmatrix} = -\mathbf{I}.$$

That is, unlike the ordinary real numbers, the realm of 2×2 matrices *does* have a member whose square is equal to the negative of the 2×2 matrix that plays the role of unity.

To carry the analogy with the ordinary real numbers just a bit further, the *zero* 2×2 matrix is $\mathbf{0} = \begin{bmatrix} 0 & 0 \\ 0 & 0 \end{bmatrix}$, since any 2×2 matrix multiplied by $\mathbf{0}$ gives $\mathbf{0}$. In addition, just as $(1/a) \cdot a = 1$ for any real number a different from zero ($1/a$ is the *inverse* of a), we'll call $\mathbf{A}^{-1}$ the *inverse matrix* of $\mathbf{A}$ if $\mathbf{AA}^{-1} = \mathbf{A}^{-1}\mathbf{A} = \mathbf{I}$. (This notation immediately suggests $\mathbf{A}^0 = \mathbf{I}$.) One has to be careful not to carry the analogy of 2×2 matrices to the ordinary real numbers too far, however. There *are* profound differences. For example, the number 1 has two square roots, -1 and 1. And its matrix counterpart $\mathbf{I}$ does indeed also have $-\mathbf{I}$ and $\mathbf{I}$ as square roots (we call a matrix $\mathbf{S}$ the square root of A if $\mathbf{S}^2 = \mathbf{SS} = \mathbf{A}$). But $\mathbf{I}$ also has an *infinity more* of square roots! You can see this by writing $\mathbf{S}$ as

$$\mathbf{S} = \begin{bmatrix} a & b \\ c & -a \end{bmatrix}$$

and then setting $\mathbf{S}^2 = \mathbf{I}$, that is,

$$\begin{bmatrix} a & b \\ c & -a \end{bmatrix}\begin{bmatrix} a & b \\ c & -a \end{bmatrix} = \begin{bmatrix} a^2 + bc & 0 \\ 0 & cb + a^2 \end{bmatrix}$$
$$= \begin{bmatrix} 1 & 0 \\ 0 & 1 \end{bmatrix}.$$

Thus, a, b, and c can be *any* three real numbers that satisfy the condition $a^2 + bc = 1$. In fact, they don't even have to be real. If $a = \sqrt{2}$, then $bc = -1$, which is satisfied by $b = c = i$. Thus,

$$\mathbf{S} = \begin{bmatrix} \sqrt{2} & i \\ i & \sqrt{2} \end{bmatrix}$$

is a square root of $\mathbf{I}$, too. Even more astonishing than an infinity of square roots, perhaps, is the fact that it is possible to have two *nonzero* matrices whose product *is* zero! Nothing like that happens in the ordinary real numbers. As an example of this remarkable possibility, I'll let you confirm that if

$$\mathbf{A} = \begin{bmatrix} 4 & 0 \\ -1 & 0 \end{bmatrix} \text{ and } \mathbf{B} = \begin{bmatrix} 0 & 0 \\ 7 & 1 \end{bmatrix},$$

then $\mathbf{AB} = \mathbf{0}$ even though $\mathbf{A} \neq \mathbf{0}$ and $\mathbf{B} \neq \mathbf{0}$. Matrices are *not* "just like" numbers at all.

This may all seem mostly a pretty game in symbol pushing, but it is actually much more than that. Here's why. Suppose we apply *two* successive rotations to the arbitrary vector $\begin{bmatrix} x \\ y \end{bmatrix}$, first with a rotation angle of β and then with a rotation angle of α. That is, suppose

$$\begin{bmatrix} x' \\ y' \end{bmatrix} = \mathbf{R}(\alpha)\mathbf{R}(\beta) \begin{bmatrix} x \\ y \end{bmatrix}.$$

The result should be equivalent to a *single* angular rotation through the angle $\alpha + \beta$ and, further, which rotation is actually the first one shouldn't matter. Thus, the rotation operator $\mathbf{R}$ should have the following two properties:

$$\mathbf{R}(\alpha)\mathbf{R}(\beta) = \mathbf{R}(\alpha + \beta),$$
$$\mathbf{R}(\alpha)\mathbf{R}(\beta) = \mathbf{R}(\beta)\mathbf{R}(\alpha).$$

The second statement says that $\mathbf{R}$ has very special *commutation* properties (see note 3 again). It is the first relation, however, that really has a nice result tucked away in it. It tells us that if we apply n successive, equal rotations of angle β, then the result is the same as for a single rotation of angle of $n\beta$. That is, for n *any* integer (positive, negative, or zero),

$$\boxed{\mathbf{R}^n(\beta) = \mathbf{R}(n\beta)},$$

which is the *matrix operator* form of De Moivre's theorem (I'll explain this comment in just a moment). That is,

$$\begin{bmatrix} \cos(\beta) & -\sin(\beta) \\ \sin(\beta) & \cos(\beta) \end{bmatrix}^n = \begin{bmatrix} \cos(n\beta) & -\sin(n\beta) \\ \sin(n\beta) & \cos(n\beta) \end{bmatrix}.$$

An interesting special case is that of $n = -1$, which says

$$\mathbf{R}^{-1}(\beta) = \mathbf{R}(-\beta).$$

That is, $\mathbf{R}(-\beta)$ is the inverse of $\mathbf{R}(\beta)$. Inverse matrices are, in general, not trivial to calculate, but the rotation matrix is an exception to that. After all, the inverse of a CCW rotation through angle β is just the CW rotation through angle β (which is equivalent to the CCW rotation through angle $-\beta$). Thus,

$$\mathbf{R}^{-1}(\beta) = \begin{bmatrix} \cos(-\beta) & -\sin(-\beta) \\ \sin(-\beta) & \cos(-\beta) \end{bmatrix} = \begin{bmatrix} \cos(\beta) & \sin(\beta) \\ -\sin(\beta) & \cos(\beta) \end{bmatrix}.$$

We can verify that this matrix has the proper mathematical behavior by direct calculation:

$$\mathbf{R}^{-1}(\beta)\mathbf{R}(\beta) = \begin{bmatrix} \cos(\beta) & \sin(\beta \\ -\sin(\beta) & \cos(\beta) \end{bmatrix} \begin{bmatrix} \cos(\beta) & -\sin(\beta) \\ \sin(\beta) & \cos(\beta) \end{bmatrix}$$

$$= \begin{bmatrix} \cos^2(\beta) + \sin^2(\beta) & -\cos(\beta)\sin(\beta) + \sin(\beta)\cos(\beta) \\ -\sin(\beta)\cos(\beta) + \cos(\beta)\sin(\beta) & \sin^2(\beta) + \cos^2(\beta) \end{bmatrix}$$

$$= \begin{bmatrix} 1 & 0 \\ 0 & 1 \end{bmatrix} = \mathbf{I}.$$

I'll let *you* show that $\mathbf{R}(\beta)\mathbf{R}^{-1}(\beta) = \mathbf{I}$, too.

The matrix $\begin{bmatrix} \cos(\beta) & -\sin(\beta) \\ \sin(\beta) & \cos(\beta) \end{bmatrix}$ rotates the vector $\begin{bmatrix} x \\ y \end{bmatrix}$ through the CCW angle β. So does multiplication of the vector by $e^{i\beta} = \cos(\beta) + i\sin(\beta)$. Thus, the matrix operator expression in the above box simply says $(e^{i\beta})^n = e^{in\beta}$ (which is hardly surprising!), and Euler's formula turns this into

$$[\cos(\beta) + i\sin(\beta)]^n = \cos(n\beta) + i\sin(n\beta),$$

which is the well-known De Moivre's theorem. So, in this sense, it appears that the introduction of matrix notation hasn't told us anything we didn't already know. That isn't really true (and I'll explain that remark in just a moment); for now, however, these results do present us with a quite interesting mathematical question—can we show, without any reference to the *physical* concept of rotation, the *mathematical* truth of

$$\begin{bmatrix} \cos(\beta) & -\sin(\beta) \\ \sin(\beta) & \cos(\beta) \end{bmatrix}^n = \begin{bmatrix} \cos(n\beta) & -\sin(n\beta) \\ \sin(n\beta) & \cos(n\beta) \end{bmatrix}?$$

The answer is *yes*, at least for the case of n any positive integer, although it will still require us to take notice of complex numbers. The cases of $n = 0$ and 1 are of course trivially obvious by inspection.

The pure mathematical argument I'll show you in the next section is based on what is perhaps the most famous theorem in matrix algebra, the *Cayley-Hamilton theorem*, named after the English mathematician Arthur Cayley (1821–1895) and the Irish mathematician William Rowan Hamilton (1805–1865). Of Cayley in particular, who was so pure a mathematician that he would have been positively *repelled* by mere physical arguments, the great physicist James Clerk Maxwell (of "Maxwell's equations for the electromagnetic field" fame) wrote "Whose soul too large for vulgar space, in n dimensions flourished."

1.2 The Cayley-Hamilton and De Moivre theorems.

The Cayley-Hamilton theorem is deceptively simple to state. It applies to all square matrices of any size ($n \times n$, n any positive integer), but we will need to treat it here only for the $n = 2$ case. To start, the *determinant* of any 2×2 matrix **A** is defined as

$$\det \mathbf{A} = \det \begin{bmatrix} a_{11} & a_{12} \\ a_{21} & a_{22} \end{bmatrix} = a_{11}a_{22} - a_{21}a_{12},$$

that is, the determinant of **A** is the difference of the two diagonal products of **A**.

Next, we define what is called the *characteristic polynomial* of **A** as the polynomial in the scalar parameter λ given by

$$p(\lambda) = \det(\mathbf{A} - \lambda \mathbf{I}).$$

The equation $p(\lambda) = 0$ is called the *characteristic equation* of **A**. For example, if $\mathbf{A} = \begin{bmatrix} 1 & 4 \\ 2 & 3 \end{bmatrix}$, then we have

$$\mathbf{A} - \lambda \mathbf{I} = \begin{bmatrix} 1 & 4 \\ 2 & 3 \end{bmatrix} - \begin{bmatrix} \lambda & 0 \\ 0 & \lambda \end{bmatrix} = \begin{bmatrix} 1 - \lambda & 4 \\ 2 & 3 - \lambda \end{bmatrix},$$

and so the characteristic polynomial of **A** is

$$p(\lambda) = \det \begin{bmatrix} 1 - \lambda & 4 \\ 2 & 3 - \lambda \end{bmatrix} = (1 - \lambda)(3 - \lambda) - 8,$$

that is, $p(\lambda) = \lambda^2 - 4\lambda - 5$. The solutions to the characteristic equation $p(\lambda) = 0$ are called the *characteristic values* of **A**. In this example, since $\lambda^2 - 4\lambda - 5 = (\lambda - 5)(\lambda + 1)$, we see that **A** has two characteristic values: $\lambda = 5$ and $\lambda = -1$.

With all of this preliminary discussion done, you can now understand the statement of the Cayley-Hamilton theorem: any square matrix **A** satisfies its own characteristic equation, that is, substituting **A** for λ in $p(\lambda) = 0$ gives $p(\mathbf{A}) = \mathbf{0}$. In this example, the Cayley-Hamilton theorem says $\mathbf{A}^2 - 4\mathbf{A} - 5\mathbf{I} = \mathbf{0}$, which you can easily verify for yourself. Indeed, for the 2×2 case, it is just as easy to work through the algebra *in general* to show that the characteristic equation

$$\det \begin{bmatrix} a_{11} - \lambda & a_{12} \\ a_{21} & a_{22} - \lambda \end{bmatrix} = 0$$

is indeed identically satisfied when **A** is substituted in for λ. Indeed, that is all Cayley himself did. That is, he didn't prove the theorem for all positive integer n, but only (in 1858) for the $n = 2$ and $n = 3$ cases by working through the detailed algebra (the $n = 3$ case is significantly grubbier than the $n = 2$ case). Hamilton's name is attached to the theorem because he did the even more intricate algebra for the $n = 4$ case. "Working through the algebra" for arbitrary n, however, clearly isn't the path to follow for larger values of n. The *general* proof, for arbitrary n, requires more mathematical machinery than I care to get into here (all we need is the 2×2 case, anyway), and so I'll simply refer you to any good book on linear algebra and matrix theory for that.

Now, what's the relevance of the Cayley-Hamilton theorem for us? One of its major uses by applied mathematicians, engineers, and physicists is in the calculation of high powers of a matrix. Given $\mathbf{A}$, it is easy to directly calculate $\mathbf{A}^2$, $\mathbf{A}^3$, or $\mathbf{A}^4$ (if $\mathbf{A}$ isn't very large), but even for a mere 2×2 matrix the calculation of $\mathbf{A}^{3,973}$ gives one reason to pause. And such calculations *do* occur; two examples are in the probabilistic theory of Markov chains and in the engineering subject of control theory.[4] With modern computers, of course, the numerical calculation of high matrix powers is easy and fast (MATLAB, the language used to generate all of the computer plots in this book, can blast through $\mathbf{A}^{3,973}$ in the snap of a finger for a 2×2 matrix). Our interest here, however, is in a *mathematical* solution (which we'll then use to establish the matrix form of De Moivre's theorem for integer $n \geq 0$).

For a 2×2 matrix the characteristic polynomial will be quadratic in λ, and so we can write the characteristic equation $p(\lambda) = 0$ in the general form of $\lambda^2 + \alpha_1 \lambda + \alpha_2 = 0$, where α_1 and α_2 are constants. Thus, by the Cayley-Hamilton theorem, $\mathbf{A}^2 + \alpha_1 \mathbf{A} + \alpha_2 \mathbf{I} = \mathbf{0}$. Now, suppose we divide λ^n by $\lambda^2 + \alpha_1 \lambda + \alpha_2$. The most general result is a polynomial of degree $n-2$ *and a remainder of at most degree one* (this argument obviously doesn't apply for $n < 2$, and this restriction means our method here, while mathematically pure, is *not* as broad as is the physical rotation idea in the previous section which has no such restriction), that is,

$$\frac{\lambda^n}{\lambda^2 + \alpha_1 \lambda + \alpha_2} = q(\lambda) + \frac{r(\lambda)}{\lambda^2 + \alpha_1 \lambda + \alpha_2},$$

where $r(\lambda) = \beta_2 \lambda + \beta_1$, with β_1 and β_2 constants. That is,

$$\boxed{\lambda^n = (\lambda^2 + \alpha_1 \lambda + \alpha_2) q(\lambda) + \beta_2 \lambda + \beta_1}.$$

This is a polynomial identity in λ, and it is another theorem in matrix algebra that the replacement of λ with $\mathbf{A}$ in such an identity results in a valid matrix identity. (This is not hard to establish, I'll simply take it as plausible here, and refer you to any good book on linear algebra for a formal proof.) Thus,

$$\mathbf{A}^n = (\mathbf{A}^2 + \alpha_1\mathbf{A} + \alpha_2\mathbf{I})q(\mathbf{A}) + \beta_2\mathbf{A} + \beta_1\mathbf{I}.$$

But since $\mathbf{A}^2 + \alpha_1\mathbf{A} + \alpha_2\mathbf{I} = \mathbf{0}$ by the Cayley-Hamilton theorem,

$$\mathbf{A}^n = \beta_1\mathbf{I} + \beta_2\mathbf{A}.$$

All we have left to do is to determine the constants β_1 and β_2, and that is a straightforward task. Here's how to do it.

Returning to the above boxed equation, which holds for *all* λ, insert the two characteristic values λ_1 and λ_2 (for which, by definition, $\lambda^2 + \alpha_1\lambda + \alpha_2 = 0$). Then,

$$\lambda_1^n = \beta_2\lambda_1 + \beta_1,$$
$$\lambda_2^n = \beta_2\lambda_2 + \beta_1.$$

These equations are easy to solve for β_1 and β_2 in terms of λ_1 and λ_2, and I'll let you do the algebra to confirm that

$$\beta_1 = \frac{\lambda_2\lambda_1^n - \lambda_1\lambda_2^n}{\lambda_2 - \lambda_1},$$

$$\beta_2 = \frac{\lambda_2^n - \lambda_1^n}{\lambda_2 - \lambda_1}.$$

So, our general result for $\mathbf{A}^n$ is, for the 2×2 case,

$$\mathbf{A}^n = \frac{\lambda_2^n - \lambda_1^n}{\lambda_2 - \lambda_1}\mathbf{A} + \frac{\lambda_2\lambda_1^n - \lambda_1\lambda_2^n}{\lambda_2 - \lambda_1}\mathbf{I}.$$

One might wonder what happens if $\lambda_1 = \lambda_2$ (giving a division by zero in the formula), but that is actually easily handled by writing $\lambda_1 = \lambda_2 + \varepsilon$ and then letting $\epsilon \to 0$. In any case, for our use here in establishing De Moivre's theorem, you'll next see that $\lambda_1 \neq \lambda_2$ and so we actually don't have to worry about that particular problem.

For us, we have

$$\mathbf{A} = \begin{bmatrix} \cos(\beta) & -\sin(\beta) \\ \sin(\beta) & \cos(\beta) \end{bmatrix}$$

and so

$$\mathbf{A} - \lambda\mathbf{I} = \begin{bmatrix} \cos(\beta) - \lambda & -\sin(\beta) \\ \sin(\beta) & \cos(\beta) - \lambda \end{bmatrix},$$

so the characteristic equation $\det(\mathbf{A} - \lambda\mathbf{I}) = 0$ is simply $[\cos(\beta) - \lambda]^2 + \sin^2(\beta) = 0$, which reduces to $\lambda^2 - 2\lambda\cos(\beta) + 1 = 0$. The quadratic formula then gives us the characteristic values of $\lambda_1 = \cos(\beta) - i\sin(\beta)$ and $\lambda_2 = \cos(\beta) + i\sin(\beta)$. Euler's formula then immediately tells us that we can write λ_1 and λ_2 as $\lambda_1 = e^{-i\beta}$ and $\lambda_2 = e^{i\beta}$. Plugging these complex exponentials into the general formula for $\mathbf{A}^n$, we have

$$\mathbf{A}^n = \begin{bmatrix} \cos(\beta) & -\sin(\beta) \\ \sin(\beta) & \cos(\beta) \end{bmatrix}^n$$

$$= \frac{e^{i\beta}e^{-in\beta} - e^{-i\beta}e^{in\beta}}{e^{i\beta} - e^{-i\beta}} \begin{bmatrix} 1 & 0 \\ 0 & 1 \end{bmatrix} + \frac{e^{in\beta} - e^{-in\beta}}{e^{i\beta} - e^{-i\beta}} \begin{bmatrix} \cos(\beta) & -\sin(\beta) \\ \sin(\beta) & \cos(\beta) \end{bmatrix}$$

$$= \frac{e^{-i(n-1)\beta} - e^{i(n-1)\beta}}{2i\sin(\beta)} \begin{bmatrix} 1 & 0 \\ 0 & 1 \end{bmatrix} + \frac{2i\sin(n\beta)}{2i\sin(\beta)} \begin{bmatrix} \cos(\beta) & -\sin(\beta) \\ \sin(\beta) & \cos(\beta) \end{bmatrix}$$

$$= \frac{-2i\sin\{(n-1)\beta\}}{2i\sin(\beta)} \begin{bmatrix} 1 & 0 \\ 0 & 1 \end{bmatrix} + \frac{\sin(n\beta)}{\sin(\beta)} \begin{bmatrix} \cos(\beta) & -\sin(\beta) \\ \sin(\beta) & \cos(\beta) \end{bmatrix}$$

$$= \begin{bmatrix} -\dfrac{\sin\{(n-1)\beta\}}{\sin(\beta)} & 0 \\ 0 & -\dfrac{\sin\{(n-1)\beta\}}{\sin(\beta)} \end{bmatrix}$$

$$+ \begin{bmatrix} \dfrac{\sin(n\beta)\cos(\beta)}{\sin(\beta)} & -\sin(n\beta) \\ \sin(n\beta) & \dfrac{\sin(n\beta)\cos(\beta)}{\sin(\beta)} \end{bmatrix}$$

$$= \begin{bmatrix} \dfrac{\sin(n\beta)\cos(\beta) - \sin\{(n-1)\beta\}}{\sin(\beta)} & -\sin(n\beta) \\ \sin(n\beta) & \dfrac{\sin(n\beta)\cos(\beta) - \sin\{(n-1)\beta\}}{\sin(\beta)} \end{bmatrix}.$$

Now, $\sin\{(n-1)\beta\} = \sin(n\beta)\cos(\beta) - \cos(n\beta)\sin(\beta)$. Thus,

$$\frac{\sin(n\beta)\cos(\beta) - \sin\{(n-1)\beta\}}{\sin(\beta)}$$

$$= \frac{\sin(n\beta)\cos(\beta) - \sin(n\beta)\cos(\beta) + \cos(n\beta)\sin(\beta)}{\sin(\beta)}$$

$$= \cos(n\beta).$$

So,

$$\begin{bmatrix} \cos(\beta) & -\sin(\beta) \\ \sin(\beta) & \cos(\beta) \end{bmatrix}^n = \begin{bmatrix} \cos(n\beta) & -\sin(n\beta) \\ \sin(n\beta) & \cos(n\beta) \end{bmatrix}.$$

Thus, at last, we have De Moivre's theorem in matrix form for n a nonnegative integer *without any mention of physical rotations.*

It is actually pretty straightforward to extend De Moivre's theorem from just nonnegative integers to all of the integers, using a purely mathematical argument. (Remember, the *automatic* inclusion of negative powers is a most attractive feature of the rotation argument.) Here's one way to do it. We first write

$$\{\cos(\beta) + i\sin(\beta)\}^{-1} = \frac{1}{\cos(\beta) + i\sin(\beta)}$$

$$= \frac{\cos(\beta) - i\sin(\beta)}{\{\cos(\beta) + i\sin(\beta)\}\{\cos(\beta) - i\sin(\beta)\}}$$

$$= \frac{\cos(\beta) - i\sin(\beta)}{\cos^2(\beta) + \sin^2(\beta)} = \cos(\beta) - i\sin(\beta).$$

Since $\cos(-\beta) = \cos(\beta)$ and $\sin(-\beta) = -\sin(\beta)$ (the cosine and sine are said to be *even* and *odd* functions, respectively),

$$\boxed{\{\cos(\beta) + i\sin(\beta)\}^{-1} = \cos(-\beta) + i\sin(-\beta)}.$$

Now, using the fact that $k = (-1)(-k)$ for any k, we have

$$\{\cos(\beta) + i\sin(\beta)\}^k = \left[\{\cos(\beta) + i\sin(\beta)\}^{-1}\right]^{-k};$$

substituting the boxed expression into the right-hand side gives

$$\{\cos(\beta) + i\sin(\beta)\}^k = [\cos(-\beta) + i\sin(-\beta)]^{-k},$$

and this is true for *any* integer k (positive, zero, *and* negative). In particular, if k is negative, then $-k$ is positive and the right-hand side becomes (from De Moivre's theorem, which we have already established for positive powers) $\cos(k\beta) + i\sin(k\beta)$. That is, for $k < 0$ (and so now for *all* integer k),

$$\{\cos(\beta) + i\sin(\beta)\}^k = \cos(k\beta) + i\sin(k\beta)$$

and we are done.

De Moivre's theorem is a powerful analytical tool, and let me show you one application of it right now. Applying the binomial theorem to the left-hand side of De Moivre's theorem, we have

$$\{\cos(\beta) + i\sin(\beta)\}^k = \sum_{j=0}^{k} \binom{k}{j} \cos^{k-j}(\beta)\{i\sin(\beta)\}^j,$$

and so, by the right-hand side of De Moivre's theorem,

$$\begin{aligned}
\cos(k\beta) + i\sin(k\beta) = {} & \cos^k(\beta) + \binom{k}{1}\cos^{k-1}(\beta)i\sin(\beta) \\
& + \binom{k}{2}\cos^{k-2}(\beta)i^2\sin^2(\beta) + \binom{k}{3}\cos^{k-3}(\beta)i^3\sin^3(\beta) \\
& + \binom{k}{4}\cos^{k-4}(\beta)i^4\sin^4(\beta) + \cdots.
\end{aligned}$$

Equating the real and imaginary parts of both sides of this identity, we have

$$\begin{aligned}
\cos(k\beta) = {} & \cos^k(\beta) - \binom{k}{2}\cos^{k-2}(\beta)\sin^2(\beta) \\
& + \binom{k}{4}\cos^{k-4}(\beta)\sin^4(\beta) + \cdots
\end{aligned}$$

and

$$\sin(k\beta) = \binom{k}{1}\cos^{k-1}(\beta)\sin(\beta) - \binom{k}{3}\cos^{k-3}(\beta)\sin^3(\beta) + \cdots.$$

Thus,

$$\tan(k\beta) = \frac{\sin(k\beta)}{\cos(k\beta)}$$

$$= \frac{\binom{k}{1}\cos^{k-1}(\beta)\sin(\beta) - \binom{k}{3}\cos^{k-3}(\beta)\sin^3(\beta) + \cdots}{\cos^k(\beta) - \binom{k}{2}\cos^{k-2}(\beta)\sin^2(\beta) + \binom{k}{4}\cos^{k-4}(\beta)\sin^4(\beta) - \cdots}$$

$$= \frac{\binom{k}{1}[\cos^k(\beta)\sin(\beta)]/\cos(\beta) - \binom{k}{3}\cos^k(\beta)\sin^3(\beta)/\cos^3(\beta) + \cdots}{\cos^k(\beta) - \binom{k}{2}\cos^k(\beta)\sin^2(\beta)/\cos^2(\beta) + \binom{k}{4}\cos^k(\beta)\sin^4(\beta)/\cos^4(\beta) - \cdots}$$

or,

$$\tan(k\beta) = \frac{\binom{k}{1}\tan(\beta) - \binom{k}{3}\tan^3(\beta) + \cdots}{1 - \binom{k}{2}\tan^2(\beta) + \binom{k}{4}\tan^4(\beta) - \cdots}.$$

This result says we can write $\tan(k\beta)$ as the ratio of two polynomials in $\tan(\beta)$, with integer coefficients. Notice, since $\binom{k}{j} = 0$ for any positive integers j and k such that $j > k$, that this expression does indeed reduce to the obviously correct $\tan(\beta) = \tan(\beta)$ for the special case of $k = 1$. And for the next case of $k = 2$ we get

$$\tan(2\beta) = \frac{2\tan(\beta)}{1 - \tan^2(\beta)},$$

which agrees with the double-angle formula for $\sin(2\beta)$ and $\cos(2\beta)$, that is,

$$\tan(2\beta) = \frac{\sin(2\beta)}{\cos(2\beta)} = \frac{2\sin(\beta)\cos(\beta)}{\cos^2(\beta) - \sin^2(\beta)} = \frac{2\cos(\beta)/\sin(\beta)}{[\cos^2(\beta)]/\sin^2(\beta) - 1}$$

$$= \frac{2/\tan(\beta)}{[1/\tan^2(\beta)] - 1} = \frac{2\tan(\beta)}{1 - \tan^2(\beta)}.$$

For $k = 5$, however (for example), we get the not so obvious formula

$$\tan(5\beta) = \frac{5\tan(\beta) - 10\tan^3(\beta) + \tan^5(\beta)}{1 - 10\tan^2(\beta) + 5\tan^4(\beta)},$$

a result easily confirmed by direct calculation for any value of β you wish.

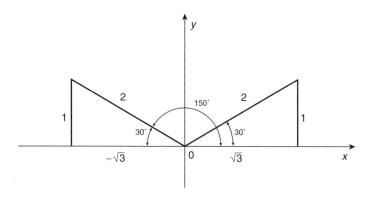

Figure 1.2.1. The 30°–60°–90° triangle ("the side opposite the 30° angle is half the hypotenuse")

For example, from high school geometry's good old 30°–60°–90° triangle we know that $\tan(30°) = 1/\sqrt{3} = \sqrt{3}/3$ (see figure 1.2.1) and that $\tan(150°) = -\tan(30°) = -\sqrt{3}/3$. Our formula agrees with this because it says that $\tan(5 \cdot 30°)$ is

$$\frac{5(\sqrt{3}/3) - 10(\sqrt{3}/3)^3 + (\sqrt{3}/3)^5}{1 - 10(\sqrt{3}/3)^2 + 5(\sqrt{3}/3)^4} = \frac{5\sqrt{3}/3 - 30\sqrt{3}/3^3 + 9\sqrt{3}/3^5}{1 - 30/3^2 + 45/3^4}$$

$$= \frac{5 \cdot 3^4\sqrt{3} - 30 \cdot 3^2\sqrt{3} + 9\sqrt{3}}{3^5 - 30 \cdot 3^3 + 45 \cdot 3}$$

$$= \sqrt{3}\frac{405 - 270 + 9}{243 - 810 + 135}$$

$$= \sqrt{3}\frac{144}{-432} = -\frac{\sqrt{3}}{3}.$$

This sort of success should really make us feel that we have a very powerful tool in hand with De Moivre's theorem in particular, and with complex numbers in general. The rest of this chapter is devoted to a number of very different problems that I hope will enhance that feeling.

1.3 Ramanujan sums a series.

In 1912 the self-taught Indian genius Srinivasa Ramanujan (1887–1920) used complex numbers and Euler's formula to solve an "interesting" problem that had been posed the previous year in the *Journal of the Indian*

Mathematical Society. What makes any problem "interesting" is often a matter of taste, but for us *this* problem will be interesting because at one time it "interested" a genius.[5] The problem was to express the function

$$P(x) = \sum_{m=1}^{\infty} \frac{(-1)^m \cos(mx)}{(m+1)(m+2)}$$

in closed form. Here's how Ramanujan did it.

If we make the perhaps "obvious" auxiliary definition

$$Q(x) = \sum_{m=1}^{\infty} \frac{(-1)^m \sin(mx)}{(m+1)(m+2)},$$

then we can use Euler's formula to write

$$P(x) + iQ(x) = \sum_{m=1}^{\infty} \frac{(-1)^m e^{imx}}{(m+1)(m+2)} = \sum_{m=1}^{\infty} \frac{(-1)^m z^m}{(m+1)(m+2)},$$

where $z = e^{ix}$. We can write this last sum as

$$\sum_{m=1}^{\infty} \frac{(-1)^m z^m}{(m+1)(m+2)} = \sum_{m=1}^{\infty} (-1)^m \left\{ \frac{1}{m+1} - \frac{1}{m+2} \right\} z^m$$

$$= \sum_{m=1}^{\infty} (-z)^m \frac{1}{m+1} - \sum_{m=1}^{\infty} (-z)^m \frac{1}{m+2}.$$

Now, as shown in freshman calculus, the Maclaurin power series expansion for $\ln(1+x)$ is

$$\ln(1+x) = x - \frac{1}{2}x^2 + \frac{1}{3}x^3 - \frac{1}{4}x^4 + \frac{1}{5}x^5 - \cdots = -\sum_{n=1}^{\infty} \frac{(-1)^n}{n} x^n.$$

That derivation assumed x is real, but if we now assume that the expansion holds even for our complex-valued z (at the end of this section I'll explore this assumption just a bit), then we see that

$$\sum_{m=1}^{\infty} (-z)^m \frac{1}{m+1} = -\frac{z}{2} + \frac{z^2}{3} - \frac{z^3}{4} + \frac{z^4}{5} - \frac{z^5}{6} + \cdots,$$

and so

$$z \sum_{m=1}^{\infty} (-z)^m \frac{1}{m+1} = -\frac{z^2}{2} + \frac{z^3}{3} - \frac{z^4}{4} + \frac{z^5}{5} - \cdots = \ln(1+z) - z,$$

or,

$$\sum_{m=1}^{\infty} (-z)^m \frac{1}{m+1} = \frac{\ln(1+z)}{z} - 1.$$

Also,

$$\sum_{m=1}^{\infty} (-z)^m \frac{1}{m+2} = -\frac{z}{3} + \frac{z^2}{4} - \frac{z^3}{5} + \frac{z^4}{6} - \cdots,$$

and so

$$z^2 \sum_{m=1}^{\infty} (-z)^m \frac{1}{m+2} = -\frac{z^3}{3} + \frac{z^4}{4} - \frac{z^5}{5} + \frac{z^6}{6} - \cdots,$$

$$= -\ln(1+z) + z - \frac{1}{2}z^2.$$

or

$$\sum_{m=1}^{\infty} (-z)^m \frac{1}{m+2} = -\frac{\ln(1+z)}{z^2} + \frac{1}{z} - \frac{1}{2}.$$

Thus,

$$\sum_{m=1}^{\infty} (-z)^m \frac{1}{m+1} - \sum_{m=1}^{\infty} (-z)^m \frac{1}{m+2}$$

$$= \frac{\ln(1+z)}{z} - 1 + \frac{\ln(1+z)}{z^2} - \frac{1}{z} + \frac{1}{2}$$

$$= \ln(1+z) \left\{ \frac{1}{z} + \frac{1}{z^2} \right\} - \frac{1}{z} - \frac{1}{2}.$$

Since $z = e^{ix}$, $1/z = e^{-ix}$ and $1/z^2 = e^{-i2x}$, and we therefore have

$$P(x) + iQ(x) = \ln(1 + e^{ix}) \left\{ e^{-ix} + e^{-i2x} \right\} - e^{-ix} - \frac{1}{2}.$$

Now,

$$\ln(1 + e^{ix}) = \ln\left\{e^{ix/2}(e^{-ix/2} + e^{ix/2})\right\} = \ln\left\{e^{ix/2}2\cos\left(\frac{x}{2}\right)\right\}$$

$$= \ln(e^{ix/2}) + \ln\left\{2\cos\left(\frac{x}{2}\right)\right\} = \ln\left\{2\cos\left(\frac{x}{2}\right)\right\} + i\frac{x}{2}.$$

Thus,

$$P(x) + iQ(x) = \left[\ln\left\{2\cos\left(\frac{x}{2}\right)\right\} + i\frac{x}{2}\right]\left\{e^{-ix} + e^{-i2x}\right\} - e^{-ix} - \frac{1}{2}$$

$$= \left[\ln\left\{2\cos\left(\frac{x}{2}\right)\right\} + i\frac{x}{2}\right]$$

$$\times \left[\cos(x) + \cos(2x) - i\left\{\sin(x) + \sin(2x)\right\}\right] - \cos(x)$$

$$+ i\sin(x) - \frac{1}{2}$$

$$= \ln\left\{2\cos\left(\frac{x}{2}\right)\right\}[\cos(x) + \cos(2x)]$$

$$+ \frac{x}{2}\left\{\sin(x) + \sin(2x)\right\} - \cos(x) - \frac{1}{2}$$

$$+ i\left[\frac{x}{2}\left\{\cos(x) + \cos(2x)\right\} - \ln\left\{2\cos\left(\frac{x}{2}\right)\right\}\right.$$

$$\left.\times \left\{\sin(x) + \sin(2x)\right\} + \sin(x)\right],$$

and so, equating real parts, we get Ramanujan's answer:

$$P(x) = \sum_{m=1}^{\infty} \frac{(-1)^m \cos(mx)}{(m+1)(m+2)}$$

$$= \ln\left\{2\cos\left(\frac{x}{2}\right)\right\}[\cos(x) + \cos(2x)]$$

$$+ \frac{x}{2}\left\{\sin(x) + \sin(2x)\right\} - \cos(x) - \frac{1}{2}.$$

(Equating imaginary parts would, of course, give us the sum for $Q(x)$, as well.) We must impose a constraint on this result, however; it is valid only in the interval $-\pi < x < \pi$, which keeps the argument of the log function nonnegative.

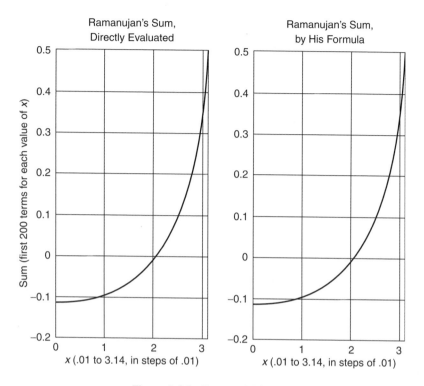

Figure 1.3.1. Ramanujan's sum

The central question now, of course, is whether Ramanujan's result "correct"? By "correct" I mean in the sense of plugging the same value for x into each side of the above formula and getting the same answer. After all, we started with a formula derived for *real* values of x, but then used *complex* values. Did that perhaps cause a problem? This is easy to check by computer, and figure 1.3.1 shows the behavior of both sides of the formula. The left-hand plot directly evaluates the sum of the first 200 terms of the original series, for each of 314 values of x (from 0.01 to 3.14, in steps of 0.01). The right-hand plot shows Ramanujan's expression for the same values of x. The two plots are virtually identical—an overlay of one plot on the other reveals no discrepancy visible to the eye. This illustration isn't a *proof*, of course,[6] but only the most rigid purist would remain unconvinced! This is, I think, a quite powerful demonstration of the utility of complex numbers, and of Euler's formula.

This use of Euler's identity to sum a series is more than a mere trick. Here's another example of its use to sum a famous series.[7] We start with

$$\ln\left[2\cos(u)\right] = \ln\left[2\frac{e^{iu}+e^{-iu}}{2}\right]$$

$$= \ln\left[e^{iu}(1+e^{-i2u})\right]$$

$$= \ln(e^{iu}) + \ln(1+e^{-i2u}) = iu + \ln(1+e^{-i2u}).$$

Thus,

$$\int_0^{\pi/2} \ln\left[2\cos(u)\right] du = \int_0^{\pi/2} \left\{ iu + \ln(1+e^{-i2u}) \right\} du.$$

If we again suppose the power series for $\ln(1+x)$ holds even for complex values (an assumption we should feel a little more confidence in making, given our success with Ramanujan's problem), then with $x = e^{-i2u}$ we have

$$\int_0^{\pi/2} \ln\left[2\cos(u)\right] du = i\left(\frac{u^2}{2}\Big|_0^{\pi/2} - \int_0^{\pi/2}\sum_{n=1}^{\infty}\frac{(-1)^n}{n}e^{-i2un}\,du\right)$$

$$= i\frac{\pi^2}{8} - \sum_{n=1}^{\infty}\frac{(-1)^n}{n}\int_0^{\pi/2}e^{-i2un}\,du.$$

Considering for now just the integral, we have

$$\int_0^{\pi/2} e^{-i2un}\,du = i\left(\frac{e^{-i2un}}{2n}\Big|_0^{\pi/2}\right) = i\frac{e^{-i\pi n}-1}{2n}.$$

Now, from Euler's formula we have $e^{-i\pi n} = \cos(\pi n) - i\sin(\pi n) = \cos(\pi n)$, as $\sin(\pi n) = 0$ for all integer n. Since $\cos(\pi n) = (-1)^n$ for all integer n, then

$$\int_0^{\pi/2} e^{-i2un}\,du = i\frac{(-1)^n - 1}{2n} = \begin{cases} -i\frac{1}{n} & \text{for } n \text{ odd,} \\ 0 & \text{for } n \text{ even.} \end{cases}$$

Substituting this into our last expression for $\int_0^{\pi/2} \ln\left[2\cos(u)\right] du$, we have

$$\int_0^{\pi/2} \ln\left[2\cos(u)\right] du = i\frac{\pi^2}{8} - \sum_{\text{odd } n>0} \frac{(-1)^n}{n}\left[-i\frac{1}{n}\right]$$

$$= i\frac{\pi^2}{8} - i\sum_{\text{odd } n>0} \frac{1}{n^2} = i\left[\frac{\pi^2}{8} - \sum_{k=0}^{\infty} \frac{1}{(2k+1)^2}\right].$$

Now, if there is only one thing we can say about $\int_0^{\pi/2} \ln\left[2\cos(u)\right] du$ it is that it is *real*, since the integrand is real over the entire interval of integration. The only way this fact is compatible with the above result, which says that the integral is *pure imaginary*, is that the integral is *zero.* Thus,

$$\sum_{k=0}^{\infty} \frac{1}{(2k+1)^2} = \frac{\pi^2}{8},$$

a result first found by Euler himself (by different means), and in yet a different way in section 4.3. As a little extra payoff, since we now know that

$$\int_0^{\pi/2} \ln\left[2\cos(u)\right] du = 0 = \int_0^{\pi/2} \{\ln(2) + \ln\left[\cos(u)\right]\} du$$

$$= \frac{\pi}{2}\ln(2) + \int_0^{\pi/2} \ln\left[\cos(u)\right] du,$$

we also have the pretty result

$$\int_0^{\pi/2} \ln\left[\cos(u)\right] du = -\frac{\pi}{2}\ln(2).$$

1.4 Rotating vectors and negative frequencies.

Euler's formula allows us to write the real-valued time functions $\cos(\omega t)$ and $\sin(\omega t)$ in terms of complex exponentials (for now the ω is an arbitrary constant, but it does have a simple physical interpretation, as you'll soon see). That is,

$$\cos(\omega t) = \frac{e^{i\omega t} + e^{-i\omega t}}{2},$$

$$\sin(\omega t) = \frac{e^{i\omega t} - e^{-i\omega t}}{2i}.$$

There is a very nice geometrical interpretation to these analytical expressions, which as they stand, admittedly, do look pretty abstract. $e^{i\omega t}$ is a vector in the complex plane with constant length one, making angle ωt with the positive real axis. So, at time $t = 0$ the $e^{i\omega t}$ vector has angle zero and thus lies directly along the positive real axis, pointing from 0 to 1. As time increases the angle ωt increases, and so the vector *rotates* (or *spins*) around the origin in the complex plane, in the counter-clockwise sense. It makes one complete rotation when $\omega t = 2\pi$ radians. If that occurs when $t = T$, then $\omega = 2\pi/T$, which has the units of *radians per second*. Accordingly, ω is called the *angular frequency* of the vector's rotation.

If we measure frequency in units of rotations (or *cycles*) per second, which is probably the more natural measure, and if we denote that number by ν, then $\nu = 1/T$ (T is called the *period* of the rotation) and $w = 2\pi\nu$. In the world of mechanical technology, ν is typically in the range of values from zero to several thousands or tens of thousands,[8] while in electronic technology ν can vary from zero up to the frequency of gamma rays (10^{20} hertz[9]). For example, ordinary AM and FM radio use frequencies in the 10^6 hertz and 10^8 hertz range, respectively.

The only difference between $e^{i\omega t}$ and $e^{-i\omega t}$ is that the angle of $e^{-i\omega t}$ is $-\omega t$, that is, $e^{-i\omega t}$ rotates in the opposite direction, that is, in the *clock*wise sense, as time increases, and this finally lets us see what writing $\cos(\omega t)$ in terms of complex exponentials *means*. Figure 1.4.1a shows $e^{i\omega t}$ and $e^{-i\omega t}$ at some arbitrary time t. The imaginary components of the two vectors are equal in magnitude and *opposite* in direction (that is, straight up and straight down), and so the imaginary components cancel at every instant of time. The real components of the two vectors, however, are equal in magnitude and in the *same* direction at every instant of time (that is, they always *add*). Thus, the two exponentials add

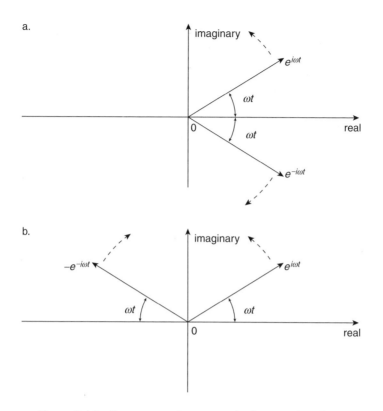

Figure 1.4.1. Counter-rotating vectors in the complex plane

vectorially to give an oscillation that lies completely and always along the real axis. The division of this sum by 2 to get $\cos(\omega t)$ is done because the real component of each individual rotating vector is $\cos(\omega t)$, by itself. The same imagery holds for the complex exponential form of $\sin(\omega t)$, except that now the vectorial addition of $e^{i\omega t}$ and $-e^{-i\omega t}$ is as shown in figure 1.4.1b, which gives rise to an oscillation that lies completely and always along the imaginary axis, so the division of the sum by $2i$ gives the *real-valued* $\sin(\omega t)$. (Note that the $-e^{-i\omega t}$ vector is indeed as shown, since $-e^{-i\omega t} = -\cos(\omega t) + i\sin(\omega t)$, that is, its real component is the *negative* of the real component of $e^{i\omega t}$ and its imaginary component *equals* the imaginary component of $e^{i\omega t}$.)

Rather than speaking of both $e^{i\omega t}$ and $e^{-i\omega t}$ as rotating with the positive angular frequency ω and then using the additional words counterclockwise and clockwise to distinguish between the two vectors engineers and mathematicians often simply write $e^{-i\omega t} = e^{i(-\omega)t}$ and say that if $e^{i\omega t}$ is a vector rotating at positive frequency ω; then $e^{-i\omega t}$ is a vector rotating at the *negative frequency* $-\omega$. This usually strikes beginning students as mysterious ("How can something vary at a frequency *less* than zero?"), and of course *real* time functions cannot. But *complex* time functions can.

Writing $\cos(\omega t)$ and $\sin(\omega t)$ in terms of complex exponentials can be the key to solving otherwise difficult problems, and I'll return to this point at great length in Chapter 4. But for now, let me show you one quick but impressive example of what can be done with this approach. Let's evaluate the following integral (it was a challenge problem in *An Imaginary Tale*, presented without solution there, p. 70), and I'll tell you why it is of historical interest when we are done):

$$\int_0^\pi \sin^{2n}(\theta)\,d\theta, \, n = 0, 1, 2, 3, \cdots.$$

Since

$$\sin(\theta) = \frac{e^{i\theta} - e^{-i\theta}}{2i},$$

we have

$$\sin^{2n}(\theta) = \frac{(e^{i\theta} - e^{-i\theta})^{2n}}{2^{2n}(i)^{2n}}.$$

And since $(i)^{2n} = (\sqrt{-1})^{2n} = ((\sqrt{-1})^2)^n = (-1)^n$ and $2^{2n} = 4^n$, we have

$$\sin^{2n}(\theta) = \frac{(e^{i\theta} - e^{-i\theta})^{2n}}{(-1)^n 4^n}.$$

From the binomial theorem we have

$$(e^{i\theta} - e^{-i\theta})^{2n} = \sum_{k=0}^{2n} \binom{2n}{k} e^{ki\theta}(-e^{-i\theta})^{2n-k}$$

$$= \sum_{k=0}^{2n} \binom{2n}{k} e^{ik\theta} \frac{(-e^{-i\theta})^{2n}}{(-e^{-i\theta})^k} = \sum_{k=0}^{2n} \binom{2n}{k} e^{ik\theta} \frac{(-1)^{2n} e^{-i2n\theta}}{(-1)^k e^{-ik\theta}}.$$

Since $(-1)^{2n} = 1$, we have

$$(e^{i\theta} - e^{-i\theta})^{2n} = \sum_{k=0}^{2n} \binom{2n}{k} e^{i2k\theta} \frac{e^{-i2n\theta}}{(-1)^k} = \sum_{k=0}^{2n} \binom{2n}{k} \frac{e^{i2(k-n)\theta}}{(-1)^k},$$

and so

$$\int_0^\pi \sin^{2n}(\theta) \, d\theta = \frac{1}{(-1)^n 4^n} \sum_{k=0}^{2n} \frac{\binom{2n}{k}}{(-1)^k} \int_0^\pi e^{i2(k-n)\theta} \, d\theta.$$

Now, concentrate for the moment on the integral at the right. If $k \neq n$, then

$$\int_0^\pi e^{i2(k-n)\theta} \, d\theta = \left(\frac{e^{i2(k-n)\theta}}{i2(k-n)} \right)\Bigg|_0^\pi = 0$$

because, at the lower limit of $\theta = 0$, $e^{i2(k-n)\theta} = e^0 = 1$ and, at the upper limit of $\theta = \pi$, $e^{i2(k-n)\theta} = e^{i(k-n)2\pi} = e^{\text{integer multiple of } 2\pi i} = 1$, too. But if $k = n$ we can't use this integration because that gives a division by zero. So, set $k = n$ *before* doing the integral. Thus, for $k = n$ we have the integrand reducing to one and so our integral is just

$$\int_0^\pi d\theta = \pi.$$

That is,

$$\boxed{\int_0^\pi e^{i2(k-n)} \, d\theta = \begin{cases} 0 & \text{if } k \neq n \\ \pi & \text{if } k = n \end{cases}}.$$

Thus,

$$\int_0^\pi \sin^{2n}(\theta) \, d\theta = \frac{1}{(-1)^n 4^n} \frac{\binom{2n}{n}}{(-1)^n} \pi = \frac{\pi}{4^n} \binom{2n}{n}.$$

Because of the symmetry of $\sin^{2n}(\theta)$ over the interval 0 to π, it is clear that reducing the interval of integration to 0 to $\pi/2$ cuts the value of the integral in half, and so

$$\int_0^{\pi/2} \sin^{2n}(\theta)\, d\theta = \frac{\pi/2}{4^n}\binom{2n}{n},$$

which is often called *Wallis's integral,* after the English mathematician John Wallis (1616–1703). This is a curious naming, since Wallis did *not* evaluate this integral! His name is nonetheless attached to the integral because integrals of the form $\int_0^{\pi/2}\sin^n(\theta)\,d\theta$ can be used to derive Wallis's famous product formula for π:

$$\frac{\pi}{2} = \frac{2\cdot 2}{1\cdot 3}\cdot\frac{4\cdot 4}{3\cdot 5}\cdot\frac{6\cdot 6}{5\cdot 7}\cdot\frac{8\cdot 8}{7\cdot 9}\cdot\frac{10\cdot 10}{9\cdot 11}\cdots,$$

a formula which is, in fact, much more easily derived from Euler's infinite product formula for the sine function.[10]

1.5 The Cauchy-Schwarz inequality and falling rocks.

We can use an argument based on both analytic geometry and complex numbers to derive one of the most useful tools in analysis—the so-called Cauchy-Schwarz inequality. As one writer put it many years ago, "This [the inequality] is an exceptionally potent weapon. There are many occasions on which persons who know and think of using this formula can shine while their less fortunate brethren flounder."[11] At the end of the section, once we have the inequality in our hands, I'll show you an amusing example of what that writer had in mind, of how the inequality can squeeze a solution out of what seems to be nothing more than a vacuum. We'll use the Cauchy-Schwarz inequality again in chapter 5.

The derivation of the inequality is short and sweet. If $f(t)$ and $g(t)$ are *any* two real-valued functions of the real variable t, then it is certainly true, for λ *any* real constant, and U and L two more constants (either or both of which may be arbitrarily large, i.e., infinite), that

$$\int_L^U \{f(t)+\lambda g(t)\}^2\, dt \geq 0.$$

This is so because, as something real and *squared*, the integrand is nowhere negative. We can expand the integral to read

$$\lambda^2 \int_L^U g^2(t)\,dt + 2\lambda \int_L^U f(t)g(t)\,dt + \int_L^U f^2(t)\,dt \geq 0.$$

And since these three definite integrals are constants (call their values a, b, and c) we have the left-hand side as simply a quadratic in λ,

$$h(\lambda) = a\lambda^2 + 2b\lambda + c \geq 0,$$

where

$$a = \int_L^U g^2(t)\,dt, \, b = \int_L^U f(t)g(t)\,dt, \, c = \int_L^U f^2(t)\,dt.$$

The inequality has a simple geometric interpretation: a plot of $h(\lambda)$ versus λ cannot *cross* the λ-axis. At most that plot (a parabolic curve) may just touch the λ-axis, allowing the "greater-than-or-equal to" condition to collapse to the special case of strict equality, that is, to $h(\lambda) = 0$, in which the λ-axis is the horizontal tangent to the parabola. This, in turn, means that there can be no real solutions (other than a double root) to $a\lambda^2 + 2b\lambda + c = 0$, because a real solution *is* the location of a λ-axis crossing. That is, the two solutions to the quadratic must be the *complex* conjugate pair

$$\lambda = \frac{-2b \pm \sqrt{4b^2 - 4ac}}{2a} = \frac{-b \pm \sqrt{b^2 - ac}}{a},$$

where, of course, $b^2 \leq ac$ is the condition that gives complex values to λ (or a real double root if $b^2 - ac = 0$). Thus, our inequality $h(\lambda) \geq 0$ requires that

$$\left\{ \int_L^U f(t)g(t)\,dt \right\}^2 \leq \left\{ \int_L^U f^2(t)\,dt \right\} \left\{ \int_L^U g^2(t)\,dt \right\}$$

and we have the Cauchy-Schwarz inequality.[12]

As a simple example of the power of this result, consider the following. Suppose I climb to the top of a 400-foot-high tower and then drop a rock. Suppose further that we agree to ignore such complicating "little" details as air drag. Then, as Galileo discovered centuries ago, if we call $y(t)$ the distance the rock has fallen downward toward the ground (from my hand), where $t = 0$ is the instant I let go of the rock, and if g is the acceleration of gravity (about 32 feet/second^2), then the speed of the falling rock is $v(t) = dy/dt = gt = 32t$ feet/second (I am measuring time in seconds). So (with s a dummy variable of integration)

$$y(t) = \int_0^t v(s)\,ds = \int_0^t 32s\,ds = \left(16s^2\Big|_0^t\right) = 16t^2.$$

If the rock hits the ground at time $t = T$, then $400 = 16T^2$ or, $T = 5$ seconds.

With this done, it now seems to be a trivial task to answer the following question: What was the rock's *average* speed during its fall? Since it took five seconds to fall 400 feet, then I suspect $9,999$ out of $10,000$ people would reply "Simple. It's 400 feet divided by 5 seconds, or 80 feet/second." But what would that ten-thousandth person say, you may wonder? Just this: what we just calculated is called the *time average*, that is,

$$V_{\text{time}} = \frac{1}{T}\int_0^T v(t)\,dt.$$

That is, the integral is the total area under the $v(t)$ curve from $t = 0$ to $t = T$, and if we imagine a *constant* speed V_{time} from $t = 0$ to $t = T$ bounding the same area (as a rectangle, with V_{time} as the constant height), then we have the above expression. And, indeed, since $v(t) = 32t$, we have

$$V_{\text{time}} = \frac{1}{5}\int_0^5 32t\,dt = \frac{32}{5}\left\{\frac{1}{2}t^2\Big|_0^5\right\} = \frac{32 \cdot 25}{10} = 80 \text{ feet/sec},$$

just as we originally found.

And then, as we all nod in agreement with this sophisticated way of demonstrating what was "obviously so" from the very start, our odd-man-out butts in to say "But there is *another* way to calculate an average speed for the rock, and it gives a *different* result!" Instead of looking at the falling rock in the *time* domain, he says, let's watch it in the *space* domain. That is, rather than speaking of $v(t)$, the speed of the rock as a function of the *time* it has fallen, let's speak of $v(y)$, the speed of the rock as a function of the *distance* it has fallen. In either case, of course, the units are feet/second. If the total distance of the fall is L (400 feet in our example), then we can talk of a *spatial average*,

$$V_{\text{space}} = \frac{1}{L} \int_0^L v(y)\,dy.$$

Since $t = \sqrt{y/16} = \frac{1}{4}\sqrt{y}$, then $v(y) = 32(\frac{1}{4}\sqrt{y}) = 8\sqrt{y}$. When $y = 400$ feet, for example, this says that $v(400 \text{ feet}) = 8\sqrt{400} = 160$ feet/sec at the end of the fall, which agrees with our earlier calculation of $v(5 \text{ seconds}) = 160$ feet/sec. Thus, for our falling rock,

$$V_{\text{space}} = \frac{1}{400} \int_0^{400} 8\sqrt{y}\,dy = \frac{1}{50} \left\{ \frac{2}{3} y^{3/2} \right\}\Big|_0^{400}$$

$$= \frac{1}{75} \cdot 400^{3/2} = \frac{400\sqrt{400}}{75} \approx 107 \text{ feet/sec}.$$

We could discuss, probably for quite a while, just what this result means, but all I want to do here is to point out that $V_{\text{time}} \leq V_{\text{space}}$. In fact, although we have so far analyzed only a very specific problem, it can be shown that *no matter how* $v(t)$ varies with t (*no matter how* $v(y)$ varies with y), even if we take into account the air drag we neglected before, $V_{\text{time}} \leq V_{\text{space}}$. This is a very general claim, of course, and there is a beautiful way to show its truth using the Cauchy-Schwarz inequality (which, of course, is the whole point of this section!) Here's how it goes.

Let $g(t) = v(t)/T$ and $f(t) = 1$. Then, the Cauchy-Schwarz inequality says

$$\left(\frac{1}{T}\int_0^T v(t)\,dt\right)^2 \leq \left\{\int_0^T dt\right\}\left\{\frac{1}{T^2}\int_0^T v^2(t)\,dt\right\}$$

$$= \frac{1}{T}\int_0^T v^2(t)\,dt = \frac{1}{T}\int_0^T \left(\frac{dy}{dt}\right)^2 dt.$$

Now, taking advantage of the suggestive nature of differential notation,

$$\left(\frac{dy}{dt}\right)^2 dt = \frac{dy}{dt}\cdot\frac{dy}{dt}\cdot dt = \frac{dy}{dt}\cdot dy.$$

If we insert this into the last integral above, then we must change the limits on the integral to be consistent with the variable of integration, that is, with y. Since $y = 0$ when $t = 0$, and $y = L$ when $t = T$, we have

$$\left(\frac{1}{T}\int_0^T v(t)\,dt\right)^2 \leq \frac{1}{T}\int_0^L \frac{dy}{dt}\,dy = \frac{1}{T}\int_0^L v(y)\,dy.$$

Now, $L = \int_0^T v(t)\,dt$, and so

$$\left\{\frac{1}{T}\int_0^T v(t)\,dt\right\}\left\{\frac{1}{T}\int_0^T v(t)\,dt\right\} = \left(\frac{1}{T}\int_0^T v(t)\,dt\right)\frac{L}{T}$$

$$\leq \frac{1}{T}\int_0^L v(y)\,dy.$$

Thus,

$$\frac{1}{T}\int_0^T v(t)\,dt \leq \frac{1}{L}\int_0^L v(y)\,dy,$$

and so $V_{\text{time}} \leq V_{\text{space}}$ as claimed, and we are done.

At no place in the above analysis did I make any assumptions about the details of either $v(t)$ or $v(y)$, and so our result is completely general. This result would be substantially more difficult to derive without the Cauchy-Schwarz inequality, and so don't forget how we got *it*—with arguments that depend in a central way on the concept of complex numbers.

1.6 Regular *n*-gons and primes.

In *An Imaginary Tale* I included some discussion of the equation $z^n - 1 = 0$; *one* solution to it is obviously $z = 1$, which means that $(z - 1)$ must be a factor of $z^n - 1$. In fact, as I mentioned in *AIT*, it is easy to verify by direct multiplication that

$$(z - 1)(z^{n-1} + z^{n-2} + \cdots + z + 1) = z^n - 1.$$

The second factor is called a *cyclotomic polynomial,* and I also rather casually mentioned in *AIT* that the polynomial gets its name from its association with the construction of regular *n*-gons. I did not pursue that comment in *AIT*, however; here I will, with a discussion of the connection cyclotomic polynomials have with one of the great triumphs of mathematics, as well as another connection they have with an impressive *mistake*. To set the stage for all that, let me very quickly review a few elementary matters concerning $z^n - 1 = 0$.

To formally "solve" $z^n - 1 = 0$ is "easy." The solutions are $z = 1^{1/n}$, one of which (as mentioned above) is obviously 1. Since there are n solutions to an *n*th-degree equation, then there must be $n - 1$ others, and Euler's formula tells us, quickly and easily, what they are. Since $1 = e^{i2\pi k}$, where k is *any* integer, we can write

$$z = 1^{1/n} = (e^{i2\pi k})^{1/n} = e^{i2\pi k/n} = e^{i360° k/n},$$

where k is any integer ($k = 0$ gives us the solution of 1). For $k = 1$, $2, \ldots, n - 1$ we get the other $n - 1$ solutions, which are, in general, complex (unless n is even, and then $k = n/2$ also gives a real solution, but more on that later). Any other integer value for k simply repeats one of the n solutions; for example, $k = n$ gives the same solution as does $k = 0$.

We can now see several interesting properties that these solutions possess. First, all have an absolute value of one, that is, $|e^{i(\text{real})}| = 1$. Therefore, *all* the solutions to $z^n - 1 = 0$ lie on the circumference of the circle with radius one centered on the origin. Second, the solutions are evenly spaced around the circle, separated by the constant angular displacement of $2\pi/n$ radians, with each step around the circle from one solution to the next caused by incrementing k by one. Let me now quote to you from Felix Klein's 1895 book *Famous Problems in Elementary Geometry*. (Klein (1849–1925) is perhaps most famous today for his discovery of the *single*-sided *closed* surface, called the *Klein bottle*, which exists only in spaces with dimension greater than three and, consequently, also in many science fiction stories.) In his book Klein wrote

> Let us trace in the z-plane ($z = x + iy$) a circle of radius 1. To divide this circle into n equal parts, beginning at $z = 1$, is the same as to solve the equation
>
> $$z^n - 1 = 0.$$
>
> This equation admits the root $z = 1$; let us suppress this root by dividing by $z - 1$, which is the same geometrically as to disregard the initial point of the division. We thus obtain the equation
>
> $$z^{n-1} + z^{n-2} + \cdots + z + 1 = 0,$$
>
> which may be called the *cyclotomic equation*.

The even spacing of the roots of $z^n - 1 = 0$ around the unit circle explains why the word *cyclotomy* is used when studying this equation—it comes from *cycl* (circle) + *tom* (divide).

Notice, too, that not only is $z = 1$ a solution for *any* integer n, so is $z = -1$ *if* n is even. This is most obvious from just $(-1)^{\text{even}} - 1 = 1 - 1 = 0$, but our general formula for z tells us this, too. If n is even then there is an integer m such that $n = 2m$. Thus,

$$z = e^{i360°k/n} = e^{i360°k/2m} = e^{i180°k/m},$$

and so when $k = m = \frac{1}{2}n$ we have

$$z = e^{i180°} = \cos(180°) + i\sin(180°) = -1.$$

So, for n even, there are two real solutions to $z^n - 1 = 0$ (± 1) and $n - 2$ complex ones. For n odd, there are one real solution ($+1$) and $n - 1$ complex ones. In any case, therefore, there are always an *even* number of complex solutions. This leads us to our third general observation about the solutions to $z^n - 1 = 0$.

Since there are always an even number of complex roots, and since those roots are evenly spaced around a circle centered on the origin, then by symmetry half the complex roots are on the top half of the circle, and half are on the bottom half. Also by symmetry, we can conclude that each root on the top half is the conjugate of a root on the bottom half (and vice versa, of course). For example, suppose $n = 5$ and we are talking about the roots of $z^5 - 1 = 0$. Since n is odd, the only real root is $+1$, and the other four roots are complex. Those four roots are

$$z_1 = e^{i360° \cdot 1/5} = e^{i72°},$$
$$z_2 = e^{i360° \cdot 2/5} = e^{i144°},$$
$$z_3 = e^{i360° \cdot 3/5} = e^{i216°},$$
$$z_4 = e^{i360° \cdot 4/5} = e^{i288°}.$$

The first two roots are on the top half of the circle, and the last two are on the bottom half. Furthermore, z_1 and z_4 are a conjugate pair, as are z_2 and z_3 (all these assertions are obvious when you look at figure 1.6.1).

Now we are ready for the first example of this section. Let me set the stage with a little history. On March 30, 1796, the German genius Carl Friedrich Gauss (1777–1855), still a month short of his nineteenth birthday, made a wonderful discovery. This discovery, so stunning it is said to have been the event that pursuaded him to pick mathematics as his career path, was that it is possible to construct the regular heptadecagon (the regular n-gon, with $n = 17$) using compass-and-straightedge alone. The reason this result was immediately seen as "important" was that, until Gauss, there had been no progress on the constructability (or not) of regular n-gons for 2,000 years. That is, there had been nothing new since Euclid!

In Euclid we can find compass-and-straightedge constructions of the equilateral triangle (3-gon), the square (4-gon), the pentagon (5-gon), and the pentadecagon (15-gon); it is nearly trivial, too, to see that once

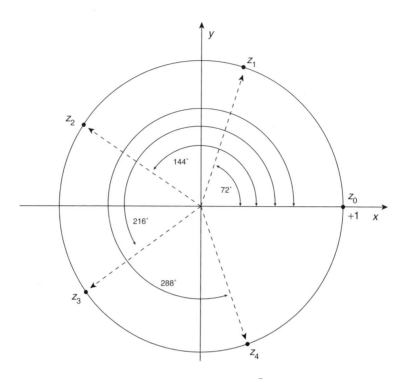

Figure 1.6.1. The solutions to $z^5 - 1 = 0$

we have constructed *any* regular n-gon then we can construct a regular $2n$-gon by simply bisecting the central angles formed by joining the center of the regular n-gon to its vertices. For example, once you have a 3- or 4- or 5- or 15-gon, it is easy to get a 6-gon (hexagon), a 8-gon (octagon), a 10-gon (decagon), a 12-gon (dodecagon), a 20-gon (icosagon), a 30-gon (triacontagon), a 40-gon (tetracontagon), a $10,000$-gon (myriagon), etc., etc., etc. These are all *even*-gons, however, and until Gauss nobody knew *anything* about the constructability of *odd*-gons other than what Euclid had shown twenty centuries before.

Now Gauss didn't actually *construct* a 17-gon. What he did was to show that it is *possible* to construct it. What does that mean? Take a look at figure 1.6.2, which shows the vertices of a regular 17-gon spaced (equally, of course) around the circumference of the unit circle centered on the origin (point O). Two adjacent vertices are labeled P and S, and their central angle is θ. The point T on the radius OS is determined by

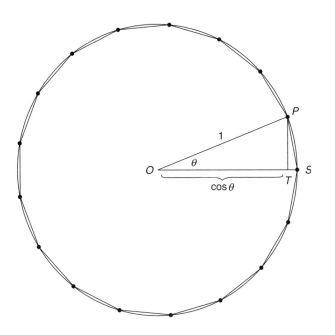

Figure 1.6.2. Gauss's regular 17-gon

dropping a perpendicular from P to OS. It is clear that $\cos(\theta) = T$, that is, since $\theta = 2\pi/17$ radians, $T = \cos(2\pi/17)$. So here's what Gauss did. He showed that T, i.e., that $\cos(2\pi/17)$, is constructable by compass-and-straightedge. Thus, with T located, one only has to erect a perpendicular at T to find P. That, in turn, defines the length of each of the 17-gon's sides and so it is a simple matter to swing a compass around the circle to find *all* of the vertices of the 17-gon. That's it!

But what does it mean to say "$\cos(2\pi/17)$ is constructable"? Simply that it can be written as a finite expression including only the operations of addition, subtraction, bisection, and square roots, all of which can be done by compass and straightedge.[13] Specifically, Gauss showed that

$$\cos\left(\frac{2\pi}{17}\right) = -\frac{1}{16} + \frac{1}{16}\sqrt{17} + \frac{1}{16}\sqrt{34 - 2\sqrt{17}}$$

$$+ \frac{1}{8}\sqrt{17 + 3\sqrt{17} - \sqrt{34 - 2\sqrt{17}} - 2\sqrt{34 + 2\sqrt{17}}}.$$

You can easily check that each side does indeed evaluate to the same result (to sixteen decimal places, at least) of 0.93247222940436. This impressive expression also gives us a step-by-step process by which the regular 17-gon can actually be constructed by compass-and-straightedge. Starting with the unit length (a radius of the circle), it is easy to extend it to length 17, take its square root, etc., etc., etc. The division by 8 is simply three consecutive bisections, and the division by 16 is the result of one more bisection. It isn't elegant, but it does the job of locating T.[14]

With Gauss's demonstration of the 17-gon's constructability came the fact that the 34-gon, the 68-gon, the 136-gon, etc., could all be constructed. In his 1801 *Disquisitiones Arithmeticae*, Gauss stated that a necessary and sufficient condition for a regular n-gon to be constructable is that n must have the form

$$n = 2^k F_i F_j F_k \cdots,$$

where k is any nonnegative integer ($k = 0, 1, 2, 3, \ldots$) and the Fs are *distinct Fermat primes*, primes named after the French genius Pierre de Fermat (1601–1665). A *Fermat number* is of the form $F_p = 2^{2^p} + 1$, where p is a nonnegative integer. If F_p happens to be prime, then we have a Fermat prime. Fermat himself showed that $p = 0, 1, 2, 3$, and 4 do result in primes, that is,

$$F_0 = 2^1 + 1 = 3,$$
$$F_1 = 2^2 + 1 = 5,$$
$$F_2 = 2^4 + 1 = 17,$$
$$F_3 = 2^8 + 1 = 257,$$
$$F_4 = 2^{16} + 1 = 65,537$$

are all primes. From 1640 on, Fermat made it plain that he *thought* F_p always to be prime, and toward the end of his life he claimed that he could prove it. But he couldn't have actually had such a proof because the very next case, F_5, is *not* a prime: in 1732 Euler found the factors[15] of $F_5 = 2^{32} + 1$.

Euclid's ancient constructions for $n = 3$ and 5 obviously follow Gauss's formula in the above box for $k = 0$ with F_0, and for $k = 0$ with F_1, respectively. Gauss's own 17-gon follows for $k = 0$ with F_2. (It is only for $k = 0$, of course, that we get *odd* values for n.) Although Gauss *said* his above formula for constructable n-gons is both necessary and sufficient, and he did indeed prove sufficiency, he did *not* prove the necessity part. That wasn't done until 1837, when the French mathematician and engineer Pierre Wantzel (1814–1848) proved that if the odd prime factors of n are *not* distinct Fermat primes then the regular n-gon can *not* be constructed. So, for example, we now see that it is futile to search for a compass-and-straightedge construction of the regular 7-gon (heptagon). Somewhat more subtle is the regular 9-gon (nonagon or, alternatively, enneagon) because for $k = 0$ we can write $9 = 2^0 \cdot 3 \cdot 3 = 2^0 \cdot F_0 \cdot F_0$. But this fails Gauss's condition that the F-factors be *distinct*. So, the regular 9-gon is not constructable either.

After Gauss's 17-gon, the next smallest *prime*-gon[16] that is constructable is the 257-gon, which was finally constructed in 1832 by the German mathematician F. J. Richelot (1808–1875). You might think that only a crazy person would try to construct a bigger n-gon, but you'd be wrong. As Professor Klein stated in his *Famous Problems* book, "To the regular polygon of 65,537 sides Professor Hermes of Lingen devoted ten years of his life, examining all the roots furnished by Gauss' method. His MSS. are preserved in the mathematical seminary in Göttingen." The man Klein was referring to was the German schoolteacher Johann Gustav Hermes (1846–1912), and his 1894 manuscript is indeed still stored to this day at the University of Göttingen. In a note on the Web (December 5, 2002), University of San Francisco mathematician John Stillwell, a well-known historian of the subject, posted the following fascinating follow-up to the story of Hermes's manuscript:

> When I visited Göttingen in July this year it was still there, and in pretty good condition. They keep it in the library of the mathematics institute, in a suitcase made specially for it called the "Koffer." The manuscript is around 200 large sheets, about the size of tabloid newspaper pages, and bound into a book. Evidently, Hermes worked out his results on scratch paper first, then entered the results very meticulously into the book. His penmanship is

incredibly neat (by today's standards) and in some places so small it can barely be read with the naked eye. He also made an enormously detailed ruler and compass construction, whose purpose is not clear to me, of a 15-gon elaborated with scores of extra circles. I could not see what Hermes is really driving at in the manuscript, but it is not sheer nonsense. According to colleagues at Göttingen, he knew some algebraic number theory, and seems to use it in his investigations, but he doesn't reach a clear conclusion. You won't find the 65537*th* root of 1 written in terms of square roots, which is what I was hoping for.

There is, of course, still the challenge of the largest (the one with the most sides) constructable *odd*-gon known today (there *might* be a bigger one—see notes 15 and 16 again): the one with $F_0F_1F_2F_3F_4 = (3)(5)(17)(257)(65,537) = 4,294,967,295$ sides. It would of course be indistinguishable from a circle! I don't think anyone will really attempt to analyze this case, but when it comes to mathematical "cranks" you can never be absolutely sure. So, that's the history of constructable *n*-gons. Now, *how* did Gauss prove that the 17-gon is constructable?

I think that question is best answered in this book by showing you how his approach works in the simpler case of $n = 5$, as the details for the regular pentagon are easy to follow and the approach extends to the $n = 17$ case. So, to start, take a look at figure 1.6.3, which shows a pentagon inscribed in a unit radius circle centered on the origin (with one vertex on the horizontal axis). The other vertices are labeled r, r^2, r^3, and r^4, where $r = e^{i\theta}$, with $\theta = 2\pi/5$ radians being the angle the radius vector to r makes with the horizontal axis (of course the vertex on the horizontal axis is $r^0 = 1$). These vertices are evenly spaced around the circle, and so, as discussed at the beginning of this section, they are solutions to $z^5 - 1 = 0$,

$$(z - 1)(z - r)(z - r^2)(z - r^3)(z - r^4) = 0, r = e^{i\,2\pi/5}.$$

And don't forget that $r^5 = (e^{i\,2\pi/5})^5 = e^{i2\pi} = 1$.

Now, following Gauss, let's form two sums, A and B, as follows:

$$A = r + r^4,$$
$$B = r^2 + r^3.$$

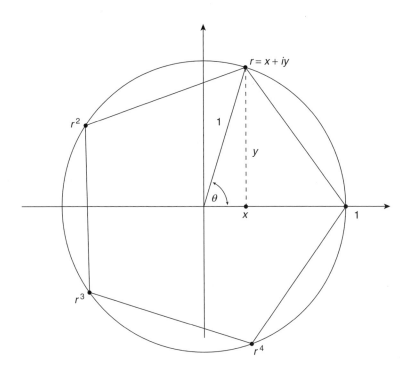

Figure 1.6.3. The regular pentagon

Then,

$$AB = (r + r^4)(r^2 + r^3) = r^3 + r^4 + r^6 + r^7.$$

But, since $r^6 = r \cdot r^5 = r \cdot 1 = r$, and $r^7 = r^2 \cdot r^5 = r^2 \cdot 1 = r^2$, we thus have

$$AB = r + r^2 + r^3 + r^4.$$

Remember next what r, r^2, r^3, r^4, and 1 *are*—they are *vectors* in the complex plane, each of unit length and originating at the same point (the origin). Can you see what their sum is? Can you see, because of their equal length and uniform spacing in direction, that their sum is *zero*? That is, $1 + r + r^2 + r^3 + r^4 = 0$, which of course means that $r + r^2 + r^3 + r^4 = -1$. Thus,

$$\boxed{AB = -1}.$$

Notice, too, that

$$A + B = r + r^2 + r^3 + r^4$$

which is the *same* as *AB*! That is,

$$\boxed{A + B = -1}.$$

It is easy to solve these two equations for *A* and *B*; all we'll actually need is *A*, which is

$$A = \frac{-1 + \sqrt{5}}{2}.$$

In the notation of figure 1.6.3, we see that the coordinates of *r* and r^4 (which are, of course, complex conjugates) are $x + iy$ and $x - iy$, respectively. From our original definition of *A* we have

$$A = r + r^4 = (x + iy) + (x - iy) = 2x,$$

and so

$$x = \frac{A}{2} = \frac{-1 + \sqrt{5}}{4}.$$

If you'll recall the earlier discussion on *constructable* (by compass-and-straightedge) quantities then you can see that *x* is such a quantity (*x* is, of course, equal to $\cos(\theta) = \cos(2\pi/5)$). So, first construct *x* along the horizontal radius, starting from the origin, then erect a perpendicular there, and its intersections with the circle will give the two vertices associated with *r* and r^4. And from that we easily have our pentagon!

This discussion of the construction of the regular 5-gon by compass-and-straightedge can be extended to the regular 17-gon. The procedure is of course a bit more complex (pun intended!), but not to the point of being terrible. The analysis follows the same line of reasoning I've used here.[17] Indeed, the analysis was extended by Gauss to any regular F_p-gon with F_p a Fermat prime, which is to say that Gauss showed it is possible to write $\cos(2\pi/F_p)$ in terms of, at worst, square roots. In this section, for example, you've already seen $\cos(2\pi/F_1) = \cos(2\pi/5)$ and $\cos(2\pi/F_2) = \cos(2\pi/17)$ written that way. (The case of $\cos(2\pi/F_0) = \cos(2\pi/3)$

doesn't even require square roots, as $\cos(2\pi/3) = -1/2$, which is obviously constructable.) What about the remaining (perhaps *last?*) two cases, those of $F_3 = 257$ and $F_4 = 65,537$? That is, what are $\cos(2\pi/257)$ and $\cos(2\pi/65,537)$ in terms of square roots?

To actually carry-out Gauss's method for those two cases has proven to be beyond human labor. With the development of electronic computers, however, a massive computational requirement is no longer a barrier. Indeed, the computer scientist Michael Trott has written a Mathematica program,[18] implementing Gauss's method, that computes symbolic expressions for $\cos(2\pi/F_p)$. He first checked the code by asking for the expressions for the first three Fermat primes; the results were in agreement with the known expressions. And the code worked for $F_3 = 257$, too. There was just one problem—the resulting expression was too immense to actually print! The program revealed to Trott that $\cos(2\pi/257)$ has over five *thousand* square roots, and that the entire expression requires over $1,300,000$ bytes of memory, (1.3 megabytes), *more than did all the text files of this book.* The corresponding numbers for $\cos(2\pi/65,537)$ would be simply astronomical—and that explains the last line in Professor Stillwell's note on the Hermes' manuscript.

1.7 Fermat's last theorem, and factoring complex numbers.
In this section I want to pursue the connection between complex numbers and prime numbers just a bit more, mostly in a historical context. I'll start with what is almost surely still the most famous problem in mathematics, the question of Fermat's last theorem, so named not because it was the last thing Fermat wrote down before dying, but rather because it was the last remaining of his *conjectures* to go unresolved (until 1995). The story of the origin of this problem is almost too well known to bear repeating, but in the interest of completeness I will *quickly* tell it here once again.

In 1670, five years after his father's death, Fermat's son Samuel arranged for a new printing of Diophantus's *Arithmetica* (circa A.D. 250). Indeed, it was just a bit more than that, as the new printing also included the marginal notes that the elder Fermat had written in his personal copy of *Arithmetica*. Fermat scribbled many assertions and conjectures in those margins, and in them he would often claim to have a proof—which he did *not* include. All these assertions and conjectures have been success-fully resolved by later mathematicians, but they usually proved to be real challenges. Here's one example, and I'll start by quoting from E. T. Bell's famous 1937 book of biographies, *Men of Mathematics*:

> Anyone playing with numbers might well pause over the curious fact that $27 = 25 + 2$. The point of interest here is that both 25 and 27 are exact powers, namely $27 = 3^3$ and $25 = 5^2$. Thus we observe that $y^3 = x^2 + 2$ has a solution in *whole numbers x, y*; the solution is $y = 3$, $x = 5$. As a sort of superintelligence test the reader may now prove that $y = 3$, $x = 5$ are the *only* whole numbers which satisfy the equation. It is not easy. In fact it requires more innate intellectual capacity to dispose of this apparently childish thing than it does to grasp the theory of relativity. (p. 20)

Well!, you might well think to yourself, that's certainly claiming a lot and, I'll admit to you, my first reaction to reading Bell's words was none too sympathetic. But I think I was wrong in that, and I think you'll soon agree. He may be guilty of just a bit of exaggeration, but actually by not very much. Continuing now with Bell,

> The equation $y^3 = x^2 + 2$, *with the restriction that the solution y, x is to be in whole numbers,* is *indeterminate* (because there are more unknowns, namely two, x and y, than there are equations, namely one, connecting them).... There is no difficulty whatever in describing an infinity of solutions *without* the restriction to whole numbers: thus we may give x *any* value we please and then deter-mine y by adding 2 to this x^2 and extracting the cube root of the result.... The solution $y = 3$, $x = 5$ is seen "by inspection": the difficulty of the problem is to prove that there are *no other* whole numbers y, x which will satisfy the equation. Fermat proved that there are none but, as usual, suppressed his proof, and it was not until many years after his death that a proof was found.

What Bell did not tell his readers is that the proof that was found is due to Euler; it appeared in his 1770 book *Algebra* and it makes use of complex numbers in factoring polynomials. I'll show it to you later in this section, and you'll see then that while it is fantastically clever there are issues with it that even Euler did not understand. Indeed, it would be nearly another hundred years before they were understood.

Another of Fermat's marginal notes in *Arithmetica* lit the famous fire that burned for over three centuries before being finally extinguished by Andrew Wiles in 1995. The note reads as follows:

> It is impossible to separate a cube into two cubes, or a biquadrate into two biquadrates, or generally any power except a square into two powers with the same exponent. I have discovered a truely marvellous proof of this, which however the margin is not large enough to contain.[19]

That is, Fermat asserted that for any integer $n > 2$ there are no integer solutions x, y, and z to the equation $x^n + y^n = z^n$. The case of $n = 2$, of course, has *infinitely many* integer solutions, à la Pythagoras!

Did Fermat actually have such a proof? Nobody really knows but, since Wiles's proof[20] requires so much high-powered mathematics invented *after* Fermat's time, most mathematicians now believe he once *thought* he had a proof but then discovered an error in it. Not realizing the furor his son would later stir up by publishing his marginal notes, Fermat neglected to go back and either add a disclaimer or simply cross it out (by then he'd probably forgotten he had even written the note). Fermat *did* have an elegant so-called "infinite descent" proof for the special case of $n = 4$, which is usually dated at 1637 (this proof he *did* write down at the end of his copy of *Arithmetica*, and you can find it in any good book on number theory). It is also believed that he had a proof for the $n = 3$ case (although that proof has never been found), because by 1638 he was posing $x^3 + y^3 = z^3$ as a challenge problem to other mathematicians. This is particularly interesting because it is yet another reason for believing that by 1638 Fermat knew his marginal note was wrong—if he had a *general* proof valid for *all* integer n, then why bother with proving special cases?

After Fermat, progress came with proofs for additional special cases. Euler published a slightly flawed (but later patched up by others) proof for the $n = 3$ case (again, using complex numbers) in 1753. He also made the prescient observation that his $n = 3$ proof had no resemblance at all to Fermat's $n = 4$ proof; the proofs for the two values of n required totally different approaches. The $n = 5$ case was established in

1825 by the joint efforts of the twenty-year-old German mathematician Lejeune Dirichlet (1805–1859)—it was his first published paper!—and the French mathematician Adrien-Marie Legendre (1752–1833). In 1832 Dirichlet also proved the truth of Fermat's assertion for $n = 14$. And in 1839 the French mathematician Gabriel Lamé (1795–1870) did the same for $n = 7$. This last result is particularly interesting because it made Dirichlet's proof for $n = 14$ "less interesting." Here's why.

Suppose there *are* integer solutions to the $n = 14$ case, that there are integers x, y, and z such that $x^{14} + y^{14} = z^{14}$, an equation that can be written as $(x^2)^7 + (y^2)^7 = (z^2)^7$. If we write $A = x^2$, $B = y^2$, and $C = z^2$, then of course A, B, and C are integers since x, y, and z are, and we would have $A^7 + B^7 = C^7$. But Lamé's proof showed that this is impossible. So our assumption that there are integer solutions to the $n = 14$ case must be false, hence, Dirichlet's proof for $n = 14$ is an immediate consequence of Lamé's $n = 7$ proof. This same argument shows the $n = 6$ case, too: *if* there were integer solutions to $x^6 + y^6 = z^6$, which can be written as $(x^2)^3 + (y^2)^3 = (z^2)^3$, *then* there would be integer solutions to $A^3 + B^3 = C^3$, which is *not* possible because of Euler's proof for the $n = 3$ case. It is curious to note that the generally authoritative *Dictionary of Scientific Biography* credits the French mathematician Joseph Liouville (1809–1882) with the 1840 (!) observation that the impossibility of integer solutions to $x^n + y^n = z^n$ immediately says the same for $x^{2n} + y^{2n} = z^{2n}$. It is difficult to believe that this (which follows from precisely the above arguments) wasn't known *long before* 1840.

Lamé's proof for the $n = 7$ case was *very* complicated, and continuing to hunt for individual proofs for specific values of n would obviously be a never-ending task; there are an infinity of integers! Of course, one doesn't really have to be concerned about *all* the integer possibilities for n. It was quickly realized that only *prime* values for n need to be considered. This is actually pretty easy to show, depending only on the fact that every integer $n \geq 3$ is either a prime, or is divisible by an odd prime, or is divisible by 4. Here's why.

It should be obvious to you that if we divide *any* integer by 4 the remainder must be either 0, 1, 2, or 3. If the remainder is 0, that's the case of the number being divisible by 4. For the other three cases, we argue as follows. With k some integer, if the remainder is

1, the number is $4k + 1$, which is odd;
2, the number is $4k + 2 = 2(2k + 1)$, which is even;
3, the number is $4k + 3$, which is odd.

In the two odd cases either the number is an odd prime or, if it is not prime, then it is a composite that can be factored into a product of odd primes. In either case the number is therefore divisible by an odd prime (perhaps itself). For the even case, the $2k + 1$ factor (which is of course odd) is either an odd prime or a composite that can be factored into a product of odd primes. In any case the $2k + 1$ factor (and so the number itself) is divisible by an odd prime. And that's it, at least for the preliminary divisibility claim.

Now, suppose m is *any* integer (equal to or greater than 3). As we've just seen, m is divisible either by 4 or by an odd prime (perhaps itself, the case if m is prime). In the first case there is an integer k such that $m = 4k$, so $x^m + y^m = z^m$ becomes $x^{4k} + y^{4k} = z^{4k}$, which is $(x^k)^4 + (y^k)^4 = (z^k)^4$. Or, using the now familar argument that $A = x^k$, $B = y^k$, and $C = z^k$ are all integers because x, y, and z are, we have A, B, and C as integer solutions to $A^4 + B^4 = C^4$. But this is impossible by Fermat's proof for the $n = 4$ case. In the second case, where m is divisible by an odd prime, we have $m = kp$ (where p is that prime) and k is some integer. Thus, $x^m + y^m = z^m$ becomes $(x^k)^p + (y^k)^p = (z^k)^p$ or, as before, A, B, and C are integer solutions to $A^p + B^p = C^p$. Whether or not this is possible is, of course, just the issue of Fermat's last theorem again, except that now we have the common power p limited to being a prime.

With Lamé's proof for $p = 7$, the next interesting case to consider would then be $p = 11$. But people were now just about out of enthusiasm for this sort of "nibbling around the edge" of the problem; a more powerful approach was needed. One early (1823) step was taken by the French mathematician Sophie Germain (1776–1831). She discovered that if p is a prime such that $2p + 1$ is also prime (as is the case for $p = 11$), then under certain conditions Fermat's conjecture is true. Such values of p are called *Germain primes*, and it is still an open question whether they are infinite in number (being only finite in number would make her result much less interesting). This was a big step forward, but still it fell far short of resolving the basic question of Fermat's last theorem.

All the specialized proofs did have one thing in common: all depended at some point on finding some factoring identity. For example, for $n = 3$ it is $x^3 + y^3 = (x + y)(x^2 - xy + y^2)$ and for $n = 7$ it is $(x + y + z)^7 - (x^7 + y^7 + z^7) = 7(x + y)(x + z)(y + z)[(x^2 + y^2 + z^2 + xy + xz + yz)^2 + xyz(x + y + z)]$. As n becomes larger such factoring identities become both more complicated and difficult to find, as the identity for one value of n tells you nothing about what to do for the next value. That was the observation, you'll recall, of Euler after finding his proof for $n = 3$ had essentially nothing in common with Fermat's proof for $n = 4$. Then, at a meeting of the Paris Academy of Sciences in March 1847, Lamé claimed to have at last solved the problem for *all* exponents. Lamé observed that there is a very direct, *general* way to factor $x^n + y^n$ into n *linear* factors, *if* one is willing to use complex numbers. As you'll recall from our discussion in the previous section on regular n-gons, the equation $X^n - 1 = 0$ has the n roots of $1, r, r^2, \ldots, r^{n-1}$, where $r = e^{i \, 2\pi/n}$, and so

$$X^n - 1 = (X - 1)(X - r)(X - r^2) \cdots (X - r^{n-1}).$$

Next, let $X = -x/y$; then

$$\left(-\frac{x}{y}\right)^n - 1 = \left(-\frac{x}{y} - 1\right)\left(-\frac{x}{y} - r\right)\left(-\frac{x}{y} - r^2\right) \cdots \left(-\frac{x}{y} - r^{n-1}\right),$$

or

$$(-1)^n \frac{x^n}{y^n} - 1 = \left[-\frac{x+y}{y}\right]\left[-\frac{x+ry}{y}\right]\left[-\frac{x+r^2y}{y}\right] \cdots \left[-\frac{x+r^{n-1}y}{y}\right],$$

which is

$$\frac{(-1)^n x^n - y^n}{y^n} = (-1)^n \frac{(x + y)(x + ry)(x + r^2 y) \cdots (x + r^{n-1} y)}{y^n}.$$

As argued before, we need only consider the case where n is an *odd* prime (*odd* is the important word here), and since $(-1)^{\text{odd}} = -1$, and canceling y^n on both sides, we arrive at

$$-x^n - y^n = -(x + y)(x + ry)(x + r^2 y) \cdots (x + r^{n-1} y).$$

At last, we have the *general* factoring identity (for *n* *odd*) of

$$x^n + y^n = (x+y)(x+ry)(x+r^2y)\cdots(x+r^{n-1}y) = z^n, r = e^{i2\pi/n}.$$

So far, all of this is okay. But then Lamé made a mistake.

Lamé asserted that, because of the above identity, since the right-hand side is a power of *n* then *each* of the factors in the middle (i.e., $(x+y)$, $(x+ry)$, etc.), if relatively prime (whatever that might mean in the realm of complex numbers), must *individually* also be powers of *n*. In the realm of the ordinary real integers, this is indeed true. That is, if we have an integer that is an *n*th power, that is, if we have N^n, where *N* is an integer, then we can write this as $N^n = N \cdot N \cdot N \cdots N$, with *N* repeated *n* times. Now, if we factor *N* into primes, the *unique factorization theorem* for the real integers tells us that this can be done only in *one* way, that is, $N = p_1 p_2 \cdots p_s$, where one or more of the p_i may in fact be equal, (i.e., one or more of the p_i may repeat). So, in $N \cdot N \cdot N \cdots N$ each prime factor will appear *n* times (once for each *N*), so N^n will be equal to $p_1^n p_2^n \cdots p_s^n$, and so each factor is indeed an *n*th power.

But this is *not* generally true with Lamé's factorization of $x^n + y^n$ into complex numbers. Indeed, when Lamé put forth his idea he was immediately rebutted by Liouville on this very point. Complex factoring was not new, and Liouville reminded Lamé of Euler's use of just such factoring decades earlier. And how did Lamé *know* his factors were "relatively prime,"? asked Liouville. Later it became known that *three years earlier*, in 1844, the German mathematician Ernst Kummer (1810–1893) had proven (in an admittedly obscure publication) that unique factorization *failed* in the complex factoring of precisely the polynomials[21] Lamé thought would solve Fermat's last theorem. The world had to wait nearly another 150 years for the resolution Lamé incorrectly thought he had achieved, and, when it did finally come, it was by entirely different means.

To end this section, let me show you how the property of unique factorization into primes can fail in complex number systems. Unique factorization doesn't *have* to fail for complex numbers—in 1832, for example, Gauss very carefully showed that the system of complex numbers with the form $a+ib$, *a* and *b* ordinary integers (the so-called *Gaussian integers*), does possess the property of unique factorization into primes—but for each new system the property does have to be validated. One way to generalize the Gaussian integers is to consider all complex number systems with the general form $a + ib\sqrt{D}$, where D is an integer with no square factors. ($D = 1$ is, of course, the case of the Gaussian integers, while $D = 2$ occurs in Euler's proof of Fermat's assertion concerning the integer solution of $y^3 = x^2 + 2$, which I'll show you at the end of this discussion.) Unique factorization into primes *fails* for $D = 6$, which I'll now demonstrate by exhibiting a specific example.[22] From now on, then, we are specifically considering numbers of the form $a + ib\sqrt{6}$, with a and b ordinary integers—this set of numbers I'll call the integers of **S**. All of the ordinary integers are obviously in **S**, with $b = 0$.

If we add any two integers in **S**, or if we multiply any two integers in **S**, we will always get a result that is also in **S**. This is probably pretty obvious,[23] but here's why anyway:

$$(a_1 + ib_1\sqrt{6}) + (a_2 + ib_2\sqrt{6}) = (a_1 + a_2) + i(b_1 + b_2)\sqrt{6},$$

$$(a_1 + ib_1\sqrt{6}) \cdot (a_2 + ib_2\sqrt{6}) = (a_1 a_2 - 6b_1 b_2) + i(a_1 b_2 + a_2 b_1)\sqrt{6}.$$

What about division? If A and B are any two of the integers of **S**, we say A *divides* B if there is an integer C in **S** such that $B = AC$. This brings us in a natural way to the question of the *primes* of **S**.

To help us understand what a prime is (in **S**), let's define what is called the *norm* of an integer A in **S**, that is, $N(A)$, where $\triangleq$ means "by definition":

$$N(A) = N(a + ib\sqrt{6}) \triangleq a^2 + 6b^2.$$

That is, the norm of A is simply the square of the absolute value of A, and it is clear that $N(A)$ is always an ordinary, nonnegative integer. The reason for introducing the norm is that it gives us a way to *order* the complex numbers in **S**. Unlike the ordinary real numbers, which are ordered from $-\infty$ to $+\infty$ (from *smaller* to *larger*), there is no intrinsic meaning to saying (for example) that $-3 + i3$ is larger (or smaller) than $2 - i7$. The ordering of the ordinary real numbers is along the real line, and is based on the *distance* of a number from the origin. The absolute value (squared) generalizes this idea from the one-dimensional real line to the two-dimensional complex plane by assigning an ordering based on the distance (squared) of the complex number from the origin.

The norm has three properties that will be of interest to us, all obvious with about five seconds of thought ($\Rightarrow$ means "implies"):

(a) $N(A) = 0 \Rightarrow A = 0$ (i.e., $a = 0$ *and* $b = 0$);

(b) $N(A) = 1 \Rightarrow A = \pm 1$ (i.e., $a = \pm 1$ *and* $b = 0$).

Finally, if B is also an integer in **S**,

(c) $N(AB) = N(A)N(B)$,

which is simply the statement that the absolute value of the product of two complex numbers is the product of their individual absolute values. Okay, let's now see what all this has to do with the primes of **S**.

Just as all the prime factors of an ordinary composite integer are *each* smaller than the integer, as well as larger than 1, we will analogously say that *each* prime factor of a nonprime integer A in **S** must have a norm both greater than 1 and less than the norm of A. That is, if $B = AC$ and if B is *not* a prime in **S**, then $1 < N(A) < N(B)$ and $1 < N(C) < N(B)$. Otherwise B *is* a prime. This definition is a generalization of how we define a prime when factoring ordinary integers. Let's do some examples.

Is 10 a prime in **S**? No, because $10 = 5 \cdot 2$ and $N(5) = 25$ and $N(2) = 4$ are both greater than 1 and less than $N(10) = 100$. That was easy! Now, what about 2; is *it* a prime in **S**? Let's suppose it isn't, and in fact has factors A and B. Then, $2 = AB$, and so $N(2) = 4 = N(AB) = N(A)N(B)$. For 2 *not* to be a prime we require both $N(A)$ and $N(B)$ to be greater than 1 and less than 4. Remembering that the norm is *always* an ordinary nonnegative integer, we see that only $N(A) = N(B) = 2$ could result in the conclusion that 2 can be factored, that is, that 2 is *not* a prime in **S**. But that says $N(A) = 2 = a^2 + 6b^2$, which has *no* integer solutions for a and b (this should be obvious). Thus, 2 *is* a prime in **S**. And what about 5; is *it* a prime in **S**? We answer this in the same way, by assuming it is *not* a prime and writing $5 = AB$. Thus, $N(5) = 25 = N(A)N(B)$ and the only possibility, for 5 to be factorable, is $N(A) = N(B) = 5$. And this says that $N(A) = a^2 + 6b^2 = 5$, which we see is impossible for integer values of a and b. Thus, 5 is a prime in **S**, too.

Now, just one more example. Is $2 + i\sqrt{6}$ a prime in **S** (you'll see why *this* particular number in just a moment)? Again, suppose it isn't prime and write $2 + i\sqrt{6} = AB$. Thus, $N(2 + i\sqrt{6}) = 10 = N(AB) = N(A)N(B)$.

The only possibilities are $N(A) = 2$ and $N(B) = 5$, or $N(A) = 5$ and $N(B) = 2$. But we've already shown that there are no integers in **S** with either of these norms. So $2 + i\sqrt{6}$ has no factors A and B and it *is* a prime in **S**.

Here's the rabbit out of the hat. Our earlier factoring of $10 = 5 \cdot 2$ expresses 10 as a product of two primes, 5 and 2. But, $10 = (2+i\sqrt{6}) \cdot (2 - i\sqrt{6})$, too, a product of two *different* primes. *Unique* prime factorization has failed in **S**.[24]

Finally, here's how Euler anticipated not just Lamé, but everybody else as well, in the use of complex factoring to solve number theory problems. You'll recall that in 1770 Euler showed that the *only* integer solutions to $y^3 = x^2 + 2$ are $y = 3$, $x = 5$. He did this as follows, using the complex number system $a + ib\sqrt{2}$ (which certainly must have seemed very mysterious to just about everybody else in the mid-eighteenth century). Factoring the right-hand side of the equation, Euler wrote

$$y^3 = (x + i\sqrt{2})(x - i\sqrt{2})$$

and argued (as loosely as would Lamé decades later) that each factor must be a cube since their product is a cube. In particular, Euler claimed that there must be integers a and b such that

$$x + i\sqrt{2} = (a + ib\sqrt{2})^3.$$

Thus,

$$x + i\sqrt{2} = (a + ib\sqrt{2})(a + ib\sqrt{2})(a + ib\sqrt{2})$$
$$= (a^2 - 2b^2 + i2ab\sqrt{2})(a + ib\sqrt{2})$$
$$= a^3 - 2ab^2 - 4ab^2 + i2a^2b\sqrt{2} + ia^2b\sqrt{2} - i2b^3\sqrt{2},$$

or,

$$\boxed{x + i\sqrt{2} = a^3 - 6ab^2 + i[3a^2b - 2b^3]\sqrt{2}}.$$

Equating imaginary parts on both sides of the equality, we have

$$1 = 3a^2b - 2b^3 = b(3a^2 - 2b^2).$$

Now b is an integer, and since a is an integer then so is $3a^2 - 2b^2$, and their product is 1. This can be true only if $b = \pm 1$ and $3a^2 - 2b^2 = \pm 1$, with each having the same sign. If $b = +1$, then we have $3a^2 - 2b^2 = 1$ or $3a^2 - 2 = 1$ or $3a^2 = 3$ or, $a = \pm 1$. And if $b = -1$ we have $-(3a^2 - 2b^2) = 1$ or $3a^2 - 2 = -1$ or $3a^2 = 1$, and this is not possible for any integer a. So, $b = +1$ and $a = \pm 1$.

Thus, equating real parts in the boxed expression, we see that $x = a^3 - 6ab^2 = a^3 - 6a$ since $b = +1$. If $a = +1$, then $x = 1 - 6 = -5$, and if $a = -1$, then $x = -1 + 6 = +5$. The fact that x is squared in the original equation makes the sign of x irrelevant; we see that $x = +5$ is the *only* positive solution (and hence $y^3 = 27$ or $y = 3$). Thus, claimed Euler, the *only* positive solution to $y^3 = x^2 + 2$ is, as originally asserted by Fermat, $(x, y) = (5, 3)$. This is a compelling analysis, but Euler (as would Lamé) glossed over the issue of the uniqueness of his factorization, as well as making the unsupported argument that the product of two complex quantities being a cube means that each factor must necessarily also be a cube.

1.8 Dirichlet's discontinuous integral.

To start the last section of this chapter, I want you to gaze upon the following mathematical expression and honestly ask yourself if you think it's possible for someone to have *just made it up*. I personally think it so spectacularly bizarre that no one could have had that much imagination! It is something that had to be discovered. Here it is, where ω is simply (for now) a dummy variable of integration:

$$\int_{-\infty}^{\infty} \frac{e^{i\omega x}}{\omega} d\omega = i\pi \, \text{sgn}(x) \quad ,$$

where $\text{sgn}(x)$ (pronounced signum x) is the discontinuous *sign function*,

$$\text{sgn}(x) = \begin{cases} +1 & \text{if } x > 0, \\ -1 & \text{if } x < 0. \end{cases}$$

The astonishing statement in the box will be very important to us later, in chapter 5, when we get to the impulse function and the Fourier transform, but for now its interest to us is simply in how we can derive it with the aid of Euler's formula.

As a last comment on sgn(x), at least for now, notice that formally we can write

$$\frac{d}{dx}|x| = \text{sgn}(x).$$

If this isn't clear, then draw a graph of the absolute function and look at its slope for the cases of $x < 0$ and $x > 0$. Thus, formally, we have the interesting expression (which you should think about, *carefully*, because you'll see it again in a *very* interesting calculation we'll be doing later in section 5.4):

$$|x| = \int_0^x \text{sgn}(s)\,ds.$$

See if you can "explain" for yourself, now, why this makes sense.

We start the calculation of $\int_{-\infty}^{\infty} e^{i\omega x}/\omega\,d\omega$ by using a "trick" that starts with an integral that defines a particular function $g(y)$:

$$g(y) = \int_0^{\infty} e^{-sy}\frac{\sin(s)}{s}\,ds, y \geq 0.$$

It probably isn't obvious where $g(y)$ comes from, and I don't think it *is* obvious—that's the trick! I don't know the history of this approach but, as you'll see, it works. *Someone* in the past was very, very clever. If we differentiate $g(y)$ using Leibniz's rule for how to differentiate an integral,[25] then

$$\frac{dg}{dy} = \int_0^{\infty} \frac{d}{dy}\left\{e^{-sy}\frac{\sin(s)}{s}\right\}ds = \int_0^{\infty} -se^{-sy}\frac{\sin(s)}{s}\,ds,$$

or

$$\frac{dg}{dy} = -\int_0^{\infty} e^{-sy}\sin(s)\,ds.$$

This last integral is not difficult to do; it yields nicely to a double application of integration by parts, with the result

$$\frac{dg}{dy} = -\frac{1}{1 + y^2}.$$

If we now integrate this last expression, then

$$g(y) = C - \tan^{-1}(y),$$

where C is the constant of indefinite integration. We could evaluate C if we just knew the value of $g(y)$ for some value of y. And, in fact, a little thought should convince you that $g(\infty) = 0$ (think about the area interpretation of the integral, about what $\sin(s)/s$ looks like, and how e^{-sy} behaves as $y \to \infty$ for *any* $s \geq 0$).[26] Thus,

$$0 = C - \tan^{-1}(\infty) = C - \frac{\pi}{2},$$

or $C = \pi/2$. Thus,

$$g(y) = \int_0^\infty e^{-sy} \frac{\sin(s)}{s} ds = \frac{\pi}{2} - \tan^{-1}(y).$$

Setting $y = 0$, we get (since $\tan^{-1}(0) = 0$) the important result (first derived—are you surprised and, if so, why?—by Euler)

$$\boxed{\int_0^\infty \frac{\sin(s)}{s} ds = \frac{\pi}{2}},$$

which appears countless times in advanced mathematics, physics, and engineering.

There is a marvelous gem tucked away in this last result that is not at all obvious. But it isn't hard to tease it out into full view, and *it* is what we are after here. If we change variables to $s = k\omega$, where k is some constant (and so $ds = k d\omega$), then if $k > 0$ we have

$$\int_0^\infty \frac{\sin(s)}{s} ds = \frac{\pi}{2} = \int_0^\infty \frac{\sin(k\omega)}{k\omega} k d\omega = \int_0^\infty \frac{\sin(k\omega)}{\omega} d\omega,$$

or

$$\int_0^\infty \frac{\sin(k\omega)}{\omega} d\omega = \frac{\pi}{2} \text{ for any } k > 0,$$

not just $k = 1$. That's pretty surprising, but we are not done yet. Suppose now that $k < 0$, that is, $k = -l$ where $l > 0$. Then,

$$\int_0^\infty \frac{\sin(k\omega)}{\omega} d\omega = \int_0^\infty \frac{\sin(-l\omega)}{\omega} d\omega = -\int_0^\infty \frac{\sin(l\omega)}{\omega} d\omega = -\frac{\pi}{2}.$$

That is, we have the result (where I've replaced l with $k > 0$, which is a trivial change in notation)

$$\int_0^\infty \frac{\sin(k\omega)}{\omega} d\omega = \begin{cases} +\frac{\pi}{2} & \text{if } k > 0, \\ -\frac{\pi}{2} & \text{if } k < 0. \end{cases}$$

or, if we replace the k with x to match our earlier notation (again, a trivial step), we have the truly wonderful

$$\int_0^\infty \frac{\sin(\omega x)}{\omega} d\omega = \begin{cases} +\frac{\pi}{2} & \text{if } x > 0 \\ -\frac{\pi}{2} & \text{if } x < 0 \end{cases},$$

a result called *Dirichlet's discontinuous integral.* It is named after the same Dirichlet in the previous section who proved the $n = 5$ and $n = 14$ cases of Fermat's last theorem. He discovered this integral result sometime before 1837. Since the integrand of Dirichlet's integral is an *even* function of ω, and using the sgn function, we can also write

$$\int_{-\infty}^\infty \frac{\sin(\omega x)}{\omega} d\omega = \pi \operatorname{sgn}(x),$$

because extending the interval of integration from 0 to ∞ to $-\infty$ to ∞ simply doubles the value of the integral.

For our final step, we again use Euler's formula. Since $\cos(\omega x)/\omega$ is an *odd* function of ω, then

$$\int_{-\infty}^{\infty} \frac{\cos(\omega x)}{\omega}\,d\omega = 0,$$

and so

$$\int_{-\infty}^{\infty} \frac{e^{i\omega x}}{\omega}\,d\omega = \int_{-\infty}^{\infty} \frac{\cos(\omega x)}{\omega}\,d\omega + i\int_{-\infty}^{\infty} \frac{\sin(\omega x)}{\omega}\,d\omega$$

$$= i\int_{-\infty}^{\infty} \frac{\sin(\omega x)}{\omega}\,d\omega = i\pi\,\mathrm{sgn}(x),$$

just as claimed at the start of this section. This derivation *has*, of course, made use of some pretty slick trickery, but our result *is* correct. It can be verified using Cauchy's theory of integration in the complex plane (that is, with a so-called *contour integral*—see the end of note 12 again), and you can find it all worked-out that way in any number of advanced books.[27] You'll see this result used to good effect in section 5.4.

Chapter 2
Vector Trips

2.1 The generalized harmonic walk.

As discussed in the opening section of chapter 1, one of the fundamental intellectual breakthroughs in the historical understanding of just what $i = \sqrt{-1}$ means, physically, came with the insight that multiplication by a complex number is associated with a *rotation* in the complex plane. That is, multiplying the vector of a complex number by the complex exponential $e^{i\theta}$ rotates that vector counterclockwise through angle θ. This is worth some additional explanation, as this elegant property of complex exponentials often pays big dividends by giving us a way to formulate, in an elementary way, seemingly very difficult problems. In *An Imaginary Tale*, for example, I showed how to use the rotation idea to solve an amusing little "treasure hunt" puzzle from George Gamow's famous 1947 book *One, Two, Three...Infinity*.[1] This chapter is devoted to much more sophisticated problems than is Gamow's, but they will yield, too, in part, because of the "multiplication is rotation" concept.

As our first warm-up exercise on thinking of vector trips in terms of complex numbers and, in particular, in terms of complex exponentials, consider the following problem. A man starts walking from the origin of a rectangular coordinate system. His first step is of unit length along the positive real axis. He then spins on his heels through the counterclockwise (positive) angle of θ and walks a second step of distance one-half unit. Then he spins on his heels through another positive angle of θ and walks a third step of distance one-third unit. He continues this process endlessly (his next step, for example, is of distance one-fourth unit after

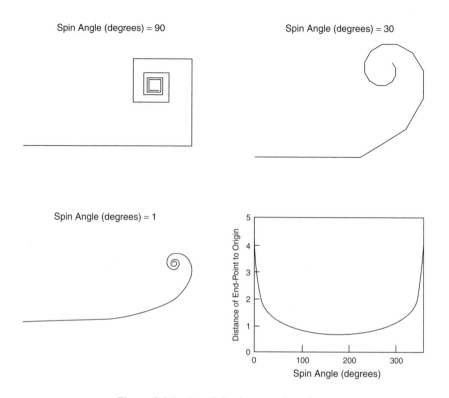

Figure 2.1.1. A walk in the complex plane

spinning on his heels through yet another positive angle of θ). If the spin angle is $\theta = 90°$, for example, his path is as shown in the upper left plot of figure 2.1.1.

The first step of the walk is a vector of length one, at an angle of zero to the real axis, that is, it is $e^{i0} = 1$. The second step is a vector of length $\frac{1}{2}$ at angle θ to the real axis, that is, it is $\frac{1}{2}e^{i\theta}$. The third step is of length $\frac{1}{3}$ at angle 2θ to the real axis, that is, it is $\frac{1}{3}e^{i2\theta}$. And so on. Thus, if $\mathbf{p}(\theta)$ is the position vector of the walker's "ultimate destination" (what I'll call the *endpoint*), then

$$\mathbf{p}(\theta) = 1 + \frac{1}{2}e^{i\theta} + \frac{1}{3}e^{i2\theta} + \frac{1}{4}e^{i3\theta} + \frac{1}{5}e^{i4\theta} + \cdots .$$

If $\theta = 90° = \pi/2$ radians, for example, then Euler's formula says

$$\mathbf{p}\left(\frac{\pi}{2}\right) = 1 + \frac{1}{2}e^{i\pi/2} + \frac{1}{3}e^{i\pi} + \frac{1}{4}e^{i3\pi/2} + \frac{1}{5}e^{i2\pi} + \cdots$$

$$= 1 + \frac{1}{2}i - \frac{1}{3} - \frac{1}{4}i + \frac{1}{5} + \frac{1}{6}i - \frac{1}{7} + \frac{1}{8}i \cdots$$

$$= \left(1 - \frac{1}{3} + \frac{1}{5} - \cdots\right) + i\left(\frac{1}{2} - \frac{1}{4} + \frac{1}{6} - \cdots\right),$$

which we could (for this particular value of θ) have written by inspection. From the Maclaurin power series expansion of $\ln(1 + x)$—look back at section 1.3—we recognize the second series as equal to $\frac{1}{2}\ln(2)$ because $\frac{1}{2} - \frac{1}{4} + \frac{1}{6} - \cdots = \frac{1}{2}(1 - \frac{1}{2} + \frac{1}{3} - \cdots)$, while the first series has been known since 1671 to be equal to $\pi/4$ (we'll derive this result in chapter 4). Thus, the final distance from the origin of our walker (for $\theta = \pi/2$) is

$$\left|\mathbf{p}\left(\frac{\pi}{2}\right)\right| = \sqrt{\left\{\frac{\pi}{4}\right\}^2 + \left\{\frac{1}{2}\ln(2)\right\}^2} = 0.8585.$$

For both $\theta = 0$ and $\theta = 2\pi$ radians ($0°$ and $360°$), it is clear that the walk is simply a straight line path out along the positive real axis, and that the endpoint is at infinity; I call these two special cases *harmonic walks.*[2] For all $\theta \neq 0, 2\pi$ it is physically clear that the *general* walk is a spiral that converges to an endpoint that is a finite distance from the origin (the top right plot of figure 2.1.1 shows the spiral for the spin angle $\theta = 30°$ and the bottom left plot shows the spiral for $\theta = 1°$). That is, $|\mathbf{p}(\theta)| < \infty$ for all $\theta \neq 0, 2\pi$. Since

$$\mathbf{p}(\theta) = \sum_{k=1}^{\infty} \frac{\cos\{(k-1)\theta\}}{k} + i\sum_{k=1}^{\infty} \frac{\sin\{(k-1)\theta\}}{k},$$

then a converging spiral for all $\theta \neq 0, 2\pi$ means that both of these sums converge for all $\theta \neq 0, 2\pi$. This is now obvious from the physical interpretation of the sums as the real and imaginary (the x-axis and the y-axis) components of a spiral walk, but it would be (I think) nontrivial tasks to prove the convergences for almost all θ by purely analytical means.

The walks in figure 2.1.1 are not drawn to the same scale. The smaller the spin angle the larger the spiral, as shown in the bottom right plot of that figure (is it a priori obvious to you that the bottom right plot should be symmetrical about the vertical line at the $180°$ spin angle?[3]). One might guess that a spin angle of $\theta = 180°$ gives the walk that *minimizes* the distance of the walk's end-point from the origin, as that walk is simply a decaying oscillation back-and-forth along the real axis. For $\theta = 180°$ we have

$$\mathbf{p} = 1 - \frac{1}{2} + \frac{1}{3} - \frac{1}{4} + \cdots = \ln(2) = 0.693,$$

and the bottom right plot of figure 2.1.1 confirms that this is, indeed, the distance of the endpoint from the origin when that distance is smallest.

2.2 Birds flying in the wind.

This section does not use complex number anywhere (although it does use vector concepts); it is a "warm-up" for the next section that does use them in similar (but more complicated) problems.[4] To begin, suppose a bird flies along a straight path from point A to point B, and then back to A. A and B are distance d apart. The bird's speed, relative to the ground, is V *when there is no wind.* The time to make the round trip, with no wind, is then obviously $T = 2d/V$. Suppose now that there *is* a wind blowing from A to B at speed $W(<V)$. When flying with the wind from A to B the bird's ground speed is $V + W$, and when flying against the wind from B back to A, the bird's ground speed is $V - W$, so the total flight time is given by

$$\frac{d}{V + W} + \frac{d}{V - W} = \frac{d(V - W) + d(V + W)}{(V + W)(V - W)}$$
$$= \frac{Vd - Wd + Vd + Wd}{V^2 - W^2} = \frac{2Vd}{V^2 - W^2}$$
$$= \frac{2Vd/V^2}{(V^2 - W^2)/V^2} = \frac{2d/V}{1 - (W/V)^2} = \frac{T}{1 - (W/V)^2} \geq T$$

with equality if $W = 0$. That is, a nonzero wind *increases* the total round trip flight time.[5]

Suppose next that the bird flies through this same wind (called a *vector field*) not along a straight, back-and-forth path, but rather along a prescribed closed-loop two-dimensional path that we'll call *C*. This means, of course, that the bird is now flying not only sometimes *with* and other times *against* the wind, but also at all possible angles to the wind, too. What can we say now about the affect of the wind on the total round trip flight time? Quite a lot, it turns out.

We imagine that the bird always attempts to fly along a straight line through its body, from tail to beak, with velocity vector **V** *relative to the ground* when there is no wind. The magnitude of **V** is a constant, but of course the direction of **V** is not. Without a wind present, that vector would indeed define the direction of the bird's motion. If there is a wind velocity vector **W** (which we will take as a constant, i.e., same speed and direction, everywhere) at some angle to **V**, however, it is the vector sum of **V** and **W** that defines the instantaneous direction of the bird's motion. That is, the bird's motion is defined by the velocity vector **v** = **V** + **W**, which is *tangent* to the path *C*. Let's suppose the instantaneous angle between **W** and **v** is θ (which will, of course, vary with the bird's instantaneous position on *C*). From the law of cosines, then, figure 2.2.1 shows that the relationship between $v = |\mathbf{v}|$, $V = |\mathbf{V}|$, and $W = |\mathbf{W}|$ is

$$V^2 = W^2 + v^2 - 2Wv\cos(\theta),$$

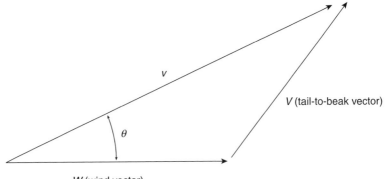

Figure 2.2.1. A bird in the wind

or

$$v^2 - [2W\cos(\theta)]v + W^2 - V^2 = 0.$$

This is easily solved for v to give

$$v = W\cos(\theta) \pm \sqrt{V^2 - W^2\sin^2(\theta)}.$$

We obviously require that $V \geq W$ to keep the square root real for all θ. Actually, we require strict *inequality*, (i.e., $V > W$) to keep $v > 0$, which means the bird is always flying *forward* along C, even when it is flying head-on into the wind (and on a closed-loop path that condition *will* happen at some point). This same requirement also tells us that we must use only the plus sign, in order to keep $v > 0$ for all V. Thus,

$$v = R + W\cos(\theta), R = \sqrt{V^2 - W^2\sin^2(\theta)}.$$

Now, the differential time required for the bird to fly along a differential segment ds of C is $dt = ds/v$. Thus, the total time to fly around C is

$$T = \oint_C \frac{ds}{v},$$

where the circle in the symbol $\oint_C$ means we add up all the differential times ds/v as the bird travels the *closed-loop* path C. Thus,

$$T = \oint_C \frac{ds}{R + W\cos(\theta)} = \oint_C \frac{R - W\cos(\theta)}{\{R + W\cos(\theta)\}\{R - W\cos(\theta)\}} ds$$

$$= \oint_C \frac{R - W\cos(\theta)}{R^2 - W^2\cos^2(\theta)} ds = \oint_C \frac{R - W\cos(\theta)}{V^2 - W^2\sin^2(\theta) - W^2\cos^2(\theta)} ds$$

$$= \oint_C \frac{R - W\cos(\theta)}{V^2 - W^2\{\sin^2(\theta) + \cos^2(\theta)\}} ds = \oint_C \frac{R - W\cos(\theta)}{V^2 - W^2} ds$$

$$= \oint_C \frac{R}{V^2 - W^2} ds - \frac{W}{V^2 - W^2} \oint_C \cos(\theta) ds.$$

The projection of the differential length ds along the (fixed) wind direction is $\cos(\theta) ds$. Since C is *closed* (the bird eventually returns to its starting point), then the total sum of all the infinitesimal projections must be *zero*. Thus, $\oint_C \cos(\theta)\,ds = 0$, and so

$$T = \oint_C \frac{R}{V^2 - W^2}\,ds = \oint_C \frac{\sqrt{V^2 - W^2 \sin^2(\theta)}}{V^2 - W^2}\,ds > \oint_C \frac{\sqrt{V^2 - W^2}}{V^2 - W^2}\,ds$$

$$= \oint_C \frac{ds}{\sqrt{V^2 - W^2}} > \oint_C \frac{ds}{V}.$$

But the last integral is the total time to fly around the closed path C with *no* wind and so, *with* a wind present, it takes the bird *longer* to fly around C. This generalizes our earlier result when the bird always flew either completely with or completely against the wind on a straight back-and-forth path.

There is an amusing (and, I think, surprising) implication of this result for the "real world" of sports. Track and field events generally have the stipulation that records don't count if the wind is too strong (or, at best, an asterisk is attached meaning "wind-aided"). The reasoning is seemingly obvious for such events as the discus, hammer, and javelin throws, as well as short sprints (if the wind is from behind the athlete). Our result here, however, shows that the "obviousness" is false for running events that involve races requiring an integral number of closed loops around a track (of any shape). Indeed, if a record is set in such an event with a constant wind blowing in *any* direction, then the record should absolutely count because the athlete would have done *even better* without the wind!

Now, with these two warm-up examples done, let me show you a vector trip problem in which complex numbers and vector rotation will be of great help in arriving at an answer even more surprising than the one above.

2.3 Parallel races.
A recent mathematical paper opens with the following passage:[6]

> It all started one day when I went running on a trail with my faithful dog Rover. Now Rover does not actually rove. In fact, Rover is so well trained that he always runs exactly one yard to my right. As long

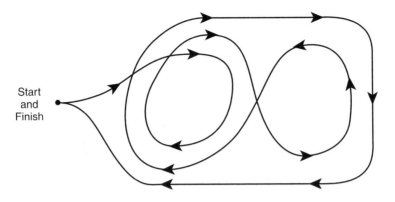

Figure 2.3.1. A complicated running trail

as I change direction smoothly, he will adjust his speed and his path perfectly so as to remain in this position. . . . On this particular day our trail was flat but curvy. We looped around several times. [see figure 2.3.1]

Since the trail in the figure is traveled in the clockwise sense, it is obvious that Rover (running on an *inside* parallel path to the *right* of his master) runs the shorter distance. The author of this paper (a professor of mathematics, of course!) then asked the following question: "How much *further* did I run than Rover?" The answer to this question is almost always a surprise to those who hear it for the first time, because it is so simple to state. The trail shown in the figure seems so convoluted and twisted that it is hard to believe the answer wouldn't also be equally complicated. We can, however, get a hint toward the answer by first considering an almost trivial special case.

Figure 2.3.2 shows the paths of two runners who start and finish a parallel run at the same time, with the inside runner following a trail consisting of the sides of a triangle. The outside runner is always a fixed distance d away from the inside runner. To satisfy this requirement, even when the inside runner is at a vertex (A, B, or C) the outside runner must swing around on the arc of a circle with radius d. Since the runners start and finish together, this requires that the outside runner perform the "swings" instantly (which is just slightly impossible!), but we'll overlook that objection (which is, of course, due to our triangle runner *not* making

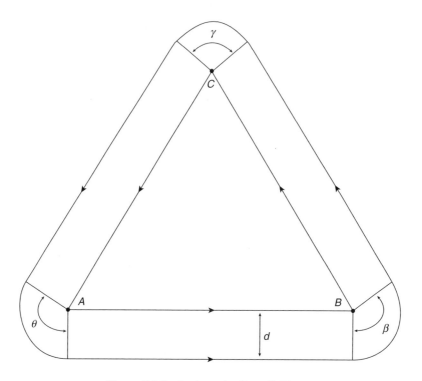

Figure 2.3.2. A triangular "parallel" run

his changes in direction "smoothly" at the vertices). All we care about, after all, is the *difference* in the total distances traveled by the two runners. Two observations can now be immediately made from the figure. First, on the straightaways the two run the same distance. Second, the outside runner has run one complete circular circumference of radius d after executing the three swings (because $\theta + \beta + \gamma = 360°$). The total of the partial circular swings is the source of the extra distance the outside runner travels. Thus, the outside runner travels precisely $2\pi d$ further than does the triangle runner, a value that is *independent* of the actual size of the triangle. That is, the only parameter that matters is d, the constant spacing between the two runners.

Wouldn't it be nice if that were the case, too, for the convoluted run of figure 2.3.1? Well, in fact it is (usually)! Under very general conditions the difference between the distances run by the dog and his master is simply equal to the product of the spacing between the two and the net angular change experienced by the running path's tangential vector. The above result for the simple triangular run of figure 2.3.2 is a special case of this, of course, with the net angular change of the path's tangential vector obviously equal to 2π (the tangential vector rotates through one complete revolution from start to finish).

How can we *prove* that the difference in the distances run is a function only of the separation distance and the angular change of the path's tangential vector? The triangular run was easy because the geometry of the run was both fixed and simple. In the general case, however, we could have much more complicated runs, with twists and loops galore, far more than the run of figure 2.3.1. How could we possibly take all that potential complexity into account? Complex numbers, and vector rotation by multiplication by $i = \sqrt{-1}$, are the key.

We'll start by taking a look at figure 2.3.3, which shows an arbitrary section of the path C traveled by our runner (we'll get to the dog in just a bit). As shown in the figure, the point O serves as our origin, with the vector $\mathbf{r}(t)$ from O to a point on C defining the runner's position vector at time t. The vector $\mathbf{r}(t)$ makes angle $\theta(t)$ with the horizontal axis

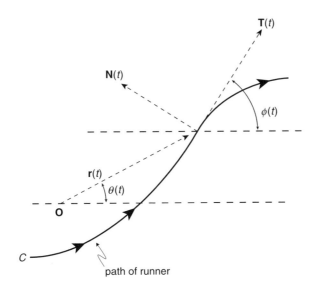

Figure 2.3.3. Defining our notation

through O. Thus, if $x(t)$ and $y(t)$ are the coordinates of the runner at time t, then we have

$$\mathbf{r}(t) = x(t) + iy(t),$$

$$|\mathbf{r}(t)| = \sqrt{x^2(t) + y^2(t)}.$$

The *velocity vector* of the runner is $\mathbf{r}'(t)$, where the prime denotes differentiation with respect to t:

$$\mathbf{r}'(t) = \frac{d}{dt}\mathbf{r}(t) = x'(t) + iy'(t).$$

The magnitude of the velocity vector is the runner's *speed*, which when integrated with respect to t gives the distance traveled along C:

$$|\mathbf{r}'(t)| = \sqrt{x'^2 + y'^2} = r'(t).$$

And finally, the *unit tangent vector* to C is $\mathbf{T}(t)$, where

$$\mathbf{T}(t) = \frac{\mathbf{r}'(t)}{|\mathbf{r}'(t)|} = \frac{\mathbf{r}'(t)}{r'(t)} = \frac{x' + iy'}{\sqrt{x'^2 + y'^2}}.$$

This is because the velocity vector $\mathbf{r}'(t)$ is tangent to C, and so dividing $\mathbf{r}'(t)$ by its own magnitude gives us a *unit* length vector in the same direction as $\mathbf{r}'(t)$. Thus, $\mathbf{r}'(t) = \mathbf{T}(t)|\mathbf{r}'(t)|$. (Notice that $\mathbf{T}(t)$ is dimensionless.)

And finally, to complete the preliminary development of our notation, imagine that we rotate $\mathbf{r}'(t)$ by $90°$, counterclockwise, to get a vector perpendicular (*normal*) to $\mathbf{T}(t)$. That is, let's multiply $\mathbf{r}'(t)$ by i, to get $i\mathbf{r}'(t) = i[x'(t) + iy'(t)] = -y'(t) + ix'(t)$. If we divide this vector by its length then we get the *unit* normal vector $\mathbf{N}(t)$ (as shown in figure 2.3.3) which points to the runner's *left*:

$$\mathbf{N}(t) = \frac{i\mathbf{r}'(t)}{|\mathbf{r}'(t)|},$$

which is, like $\mathbf{T}(t)$, dimensionless. The reason we will need $\mathbf{N}(t)$ is because it will allow us to write a mathematical expression for the *dog's* location: since the dog is at the runner's *right*, then the dog's position vector is

$$\mathbf{d}(t) = \mathbf{r}(t) - \alpha\mathbf{N}(t),$$

where α represents the constant distance the dog maintains between himself and his master (α is equal to one yard in the original problem statement).

Next, without attempting to explain immediately why we are doing the following calculation (you'll see why, soon), let's find an expression for $\theta'(t)$, the rate of change of the angle the runner's position vector makes with the horizontal axis. Since

$$\cos\{\theta(t)\} = \frac{x}{\sqrt{x^2 + y^2}},$$

then differentiation with respect to t gives (remember, θ, x, and y are all functions of t)

$$-\sin(\theta)\theta' = \frac{x'\sqrt{x^2 + y^2} - x/(2\sqrt{x^2 + y^2})(2xx' + 2yy')}{x^2 + y^2}$$

$$= \frac{x'(x^2 + y^2) - x^2 x' - xyy'}{(x^2 + y^2)^{3/2}} = \frac{x'y^2 - xyy'}{(x^2 + y^2)^{3/2}}.$$

But

$$\sin(\theta) = \frac{y}{\sqrt{x^2 + y^2}},$$

so

$$\theta' = -\frac{x'y^2 - xyy'}{\sin(\theta)(x^2 + y^2)^{3/2}} = -\frac{x'y^2 - xyy'}{(y/\sqrt{x^2 + y^2})(x^2 + y^2)^{3/2}} = -\frac{x'y - xy'}{x^2 + y^2},$$

that is,

$$\boxed{\theta' = \frac{xy' - x'y}{x^2 + y^2}}.$$

Why did we do this? That is a pertinent question to ask because we actually are *not* going to need θ' itself to solve our original problem. You'll recall that the angle we will find to be at the core of the solution is the net angular change of the path's *tangential* vector (θ is the angle of the runner's *position* vector). That is, it is $\phi(t)$, the angle $\mathbf{T}(t)$ makes with the horizontal that we are going to be interested in, not $\theta(t)$. You will be relieved to know, however, that there *is* actually a reason for what we did. The reason for calculating $\theta'(t)$ is that not only is it easy to do but knowledge of $\theta'(t)$ will guide us quickly and painlessly to $\phi'(t)$. Here's how. We start by writing our earlier expression for $\mathbf{T}(t)$ as

$$\mathbf{T}(t) = \frac{\mathbf{r}'(t)}{|\mathbf{r}'(t)|} = c(t)\mathbf{r}'(t),$$

where

$$c(t) = \frac{1}{|\mathbf{r}'(t)|} = \frac{1}{r'(t)}.$$

Thus,

$$\mathbf{T}(t) = c(t)[x'(t) + iy'(t)] = c(t)x'(t) + ic(t)y'(t).$$

Now, think about what we have here. We have a vector with given components—$c(t)x'(t)$ and $c(t)y'(t)$—that makes an angle $\phi(t)$ with the horizontal axis. But with our earlier calculation we worked out $\theta'(t)$, the *derivative* of an angle that a vector with given components—$x(t)$ and $y(t)$—makes with respect to the horizontal axis. All that is different in these two cases is the given components. So, we can simply use $c(t)x'(t)$ and $c(t)y'(t)$ for $x(t)$ and $y(t)$, respectively, in the equation for $\theta'(t)$ to get $\phi'(t)$. Thus, by inspection (!),

$$\phi'(t) = \frac{(cx')(cy'' + c'y') - (cx'' + c'x')(cy')}{c^2 x'^2 + c^2 y'^2} = \frac{c^2 x'y'' - c^2 x''y'}{c^2 x'^2 + c^2 y'^2},$$

or

$$\boxed{\phi' = \frac{x'y'' - x''y'}{x'^2 + y'^2}}.$$

Now, recall that

$$\mathbf{N}(t) = \frac{i\mathbf{r}'(t)}{|\mathbf{r}'(t)|} = \frac{-y' + ix'}{\sqrt{x'^2 + y'^2}}.$$

Therefore,

$$\mathbf{N}'(t) = \frac{\sqrt{x'^2 + y'^2}(-y'' + ix'') - (-y' + ix')(x'x'' + y'y'')/(\sqrt{x'^2 + y'^2})}{x'^2 + y'^2},$$

which reduces, after just a bit of algebra, to

$$\mathbf{N}'(t) = \frac{x' + iy'}{\sqrt{x'^2 + y'^2}} \cdot \frac{x''y' - y''x'}{x'^2 + y'^2} = \mathbf{T}(t)\{-\phi'(t)\},$$

that is, to

$$\boxed{\mathbf{N}'(t) = -\phi'(t)\mathbf{T}(t)}.$$

Recall, too, the position vector for the dog, who always stays the constant distance α to the runner's right:

$$\mathbf{d}(t) = \mathbf{r}(t) - \alpha\mathbf{N}(t).$$

Thus,

$$\mathbf{d}'(t) = \mathbf{r}'(t) - \alpha\mathbf{N}'(t) = \mathbf{r}'(t) + \alpha\phi'(t)\mathbf{T}(t).$$

You'll recall from earlier, however, that we have $\mathbf{r}'(t) = \mathbf{T}(t)|\mathbf{r}'(t)|$, and so

$$\mathbf{d}'(t) = \mathbf{T}(t)|\mathbf{r}'(t)| + \alpha\phi'(t)\mathbf{T}(t) = \{|\mathbf{r}'(t)| + \alpha\phi'(t)\}\mathbf{T}(t).$$

The dog's *speed* is simply the absolute value of $\mathbf{d}'(t)$, and so his speed is

$$|\mathbf{d}'(t)| = |\{|\mathbf{r}'(t)| + \alpha\phi'(t)\}\mathbf{T}(t)|,$$

or, since the absolute value of a product is the product of the absolute values,

$$|\mathbf{d}'(t)| = |\{|\mathbf{r}'(t)| + \alpha\phi'(t)\}||\mathbf{T}(t)|.$$

Now, recall that $\mathbf{T}(t)$ is the *unit* tangent vector to C, that is, that $|\mathbf{T}(t)| = 1$, and so, finally,

$$|\mathbf{d}'(t)| = |\{|\mathbf{r}'(t)| + \alpha\phi'(t)\}|.$$

Now we make an assumption that I will defer justifying until the end of the analysis, that $|\mathbf{r}'(t)| + \alpha\phi'(t) \geq 0$ for all t. Then the dog's speed becomes

$$|\mathbf{d}'(t)| = |\mathbf{r}'(t)| + \alpha\phi'(t)$$

where, keep in mind, the *runner's* speed is $|\mathbf{r}'(t)|$. If we integrate a speed over any time interval we'll get the distance run during that interval. So, denoting the distances run in the time interval 0 to $\widehat{T}$ by the runner and the dog as L_R and L_D, respectively, we have

$$L_R - L_D = \int_0^{\widehat{T}} |\mathbf{r}'(t)| \, dt - \int_0^{\widehat{T}} \{|\mathbf{r}'(t)| + \alpha\phi'(t)\} dt$$

$$= -\alpha \int_0^{\widehat{T}} \phi'(t) \, dt = -\alpha\{\phi(\widehat{T}) - \phi(0)\} = \alpha\{\phi(0) - \phi(\widehat{T})\}.$$

At the start of the run, $\phi(0) = 0$. If you follow Figure 2.3.1 from start to finish you should see that the tangential vector to C rotates through one-and-a-half *clockwise* revolutions by the end of the run, that is, $\phi(\widehat{T}) = -3\pi$. So, if α equals one yard, then

$$L_R = L_D + 1 \cdot \{0 - (-3\pi)\} = L_D + 3\pi.$$

That is, the runner has traveled 3π yards more than has the dog.

One final point: what about the assumption made earlier that $|\mathbf{r}'(t)| + \alpha\phi'(t) \geq 0$? We can see the physical significance of this assumption by recalling that the *radius of curvature*, at an arbitrary point on the runner's path, is $R(t)$, where

$$R(t) = \frac{\{x'^2 + y'^2\}^{3/2}}{x'y'' - x''y'}.$$

The concept of the radius of curvature (due to Newton, from 1671) is defined at every point along a curve, and refers to the radius of the circle that "most snugly fits" the curve at each point. Its reciprocal is called the *curvature*. For example, a straight line can be thought of as the circumference of a circle with infinite radius (and so the curvature of a straight line is the reciprocal of infinity, or zero). And the circle that "most snugly fits" a circular curve is the circle itself—the radius of curvature of a circle *is* the radius of the circle—and since the radius is a constant then so is the curvature. You can find the above formula for $R(t)$ derived in just about any calculus textbook.

The radius of curvature can be either positive or negative; it is negative when the path is turning clockwise, and positive when the path is turning counter-clockwise. Since, from before, we have $|\mathbf{r}'(t)| = \sqrt{x'^2 + y'^2}$ (and looking back at the boxed expression for $\phi'(t)$), you can immediately see that

$$R(t) = \frac{|\mathbf{r}'(t)|}{\phi'(t)}.$$

Thus, if $|\mathbf{r}'(t)| + \alpha\phi'(t) < 0$, then $|\mathbf{r}'(t)| < -\alpha\phi'(t)$. An absolute value is nonnegative, of course, and so the condition $|\mathbf{r}'(t)| + \alpha\phi'(t) < 0$ is equivalent to saying $0 < |\mathbf{r}'(t)| < -\alpha\phi'(t)$. (Since the physical meaning of α requires $\alpha > 0$, it is clear that $\phi'(t) < 0$.) Using the left endpoint of the double inequality on $|\mathbf{r}'(t)|$ we have $R(t) < 0$, and using the right endpoint we have $-\alpha < R(t)$. Thus,

$$-\alpha < R(t) < 0.$$

Since $R(t)$ is negative, then we see that the condition $|\mathbf{r}'(t)| + \alpha\phi'(t) < 0$ means the runner is turning clockwise (i.e., toward the dog) with a radius of curvature *less* than the separation distance (as the radius of curvature of a turn decreases the turn becomes *sharper*, i.e., more severe). When that happens, our simple result no longer applies and, as the creator of this problem wrote (see note 6 again),

> I would be turning *towards* Rover so sharply that he could not compensate by slowing down. Instead he would have to do some additional running around (perhaps on a rather small scale) in order to remain in the ideal position beside me. Our simple formula [for the difference in the distances traveled by the runner and his dog] would no longer apply. The reader might like to think about what happens if, for example, I run clockwise around a circle of radius less than one yard, while Rover remains exactly one yard to my right.

That final sentence is a good challenge problem, too. The author doesn't provide the answer but it isn't too hard to work out, and so I'll let you think about it for a while (you can find the answer in the last section of this chapter).

2.4 Cat-and-mouse pursuit.

In the previous section we used complex vector concepts to solve a racing problem. The same ideas can be used to solve a historically important chase problem, too. Suppose a mouse suddenly comes upon a snoozing cat and, startled and excited, begins running in a circle centered on the cat (mice are not very smart!). The cat awakes and, upon observing that lunch has arrived, begins its chase. The question of interest for the cat is: can I catch the mouse? The mouse, of course, has a related question: can I escape the cat?

The mathematical origin of such problems dates from 1732 and is due to the French mathematician and hydrographer Pierre Bouguer (1698–1758), who formulated his pioneering analysis in response to a most practical problem of his day—determining the path followed by a pirate ship pursuing a merchant vessel sailing on a *straight* escape path. For us, however, let's return to our hungry cat and its prey, the mouse (whose circle is a more complicated path than is the merchant vessel's). To set the cat-and-mouse problem up for mathematical analysis, you'll see that complex numbers and their associated vectors, and Euler's formula, will be most helpful. First, some assumptions.

Denoting the cat and the mouse by C and M, respectively, let's agree to start measuring time, t, at the instant the cat awakes ($t = 0$). Furthermore, let's agree to let the radius of the circle the mouse runs along be unity. This involves no loss of generality, as we can measure distance in any units we wish. Also, let's draw our coordinate system such that at $t = 0$ the mouse (point M) is at (1,0), and that the mouse runs around its circular path in the counterclockwise sense at a constant speed of unity. Again, these assumptions have no affect on the generality of our analysis—as for the speed assumption, we'll simply reference the cat's speed (also assumed to be a constant) to the mouse's unity speed: for a "slow" cat its speed is *less* than one, and for a "fast" cat its speed is *greater* than one.

Now, finally, our last major assumption: we have to decide what *strategy* the cat will employ in its chase. Assuming our cat lives its life always doing what appears to be the most direct thing to bring it instant pleasure, let's assume our kitty uses what is called a *pure pursuit* strategy. That is, at every instant of time the cat runs directly towards the instantaneous location of the mouse. So, here's what we have: C is a point pursuing M such that C's *velocity vector* always points at M's present location, and the velocity vectors of both C and M are constant. The magnitude of C's velocity vector is the cat's speed, s, while of course the magnitude of the mouse's velocity vector (its speed) is one. If we write $\mathbf{m}(t)$ as the mouse's location at time t, then the mouse's position vector is

$$\mathbf{m}(t) = \cos(t) + i\sin(t) = e^{it},$$

which agrees with our assumption that the mouse is at $(1,0)$ at $t = 0$. Notice, also, that as t increases from zero the angle of $\mathbf{m}(t)$ increases from zero, that is, $\mathbf{m}(t)$ rotates in the CCW sense. And finally, the velocity vector of the mouse is

$$\frac{d}{dt}\mathbf{m}(t) = -\sin(t) + i\cos(t),$$

so the mouse's speed is indeed the constant one because

$$|\frac{d}{dt}\mathbf{m}(t)| = |-\sin(t) + i\cos(t)| = \sqrt{\sin^2(t) + \cos^2(t)} = 1.$$

As for the cat, let's write its position vector as

$$\mathbf{c}(t) = x(t) + iy(t).$$

Then, as shown in figure 2.4.1, the geometry of the problem requires that the vector pointing from C to M be $e^{it} - \mathbf{c}(t)$. Now, since C always runs directly toward M, C's velocity vector always points toward M, and as just demonstrated, we have such a vector in $e^{it} - \mathbf{c}(t)$. To get the magnitude of the cat's velocity vector right, first notice that the *unit* vector (recall from the previous section that a unit vector has length one) pointing from C to M is simply $e^{it} - \mathbf{c}(t)$ divided by its own length,

$$\frac{e^{it} - \mathbf{c}(t)}{|e^{it} - \mathbf{c}(t)|}.$$

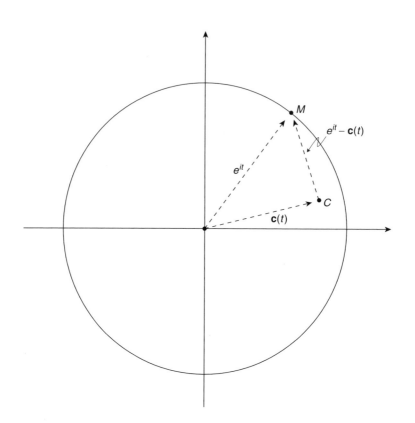

Figure 2.4.1. The geometry of cat-and-mouse pursuit

So, since we need a vector of length (magnitude) s for the cat's velocity vector, we can write that velocity vector as

$$\frac{d}{dt}\mathbf{c}(t) = \frac{dx}{dt} + i\frac{dy}{dt} = s\frac{e^{it} - \mathbf{c}(t)}{|e^{it} - \mathbf{c}(t)|}.$$

And thus, using Euler's formula to expand e^{it}, we have

$$\frac{dx}{dt} + i\frac{dy}{dt} = s\frac{\cos(t) + i\sin(t) - x - iy}{|\cos(t) + i\sin(t) - x - iy|}$$

$$= s\frac{\{\cos(t) - x\} + i\{\sin(t) - y\}}{|\{\cos(t) - x\} + i\{\sin(t) - y\}|}$$

$$= s\frac{\{\cos(t) - x\} + i\{\sin(t) - y\}}{\sqrt{\{\cos(t) - x\}^2 + \{\sin(t) - y\}^2}}.$$

If we now equate real and imaginary parts, we arrive at the following pair of differential equations:

$$\frac{dx}{dt} = s\frac{\cos(t) - x}{\sqrt{\{\cos(t) - x\}^2 + \{\sin(t) - y\}^2}} = s\frac{\cos(t) - x}{D(t, x, y)},$$

$$\frac{dy}{dt} = s\frac{\sin(t) - y}{D(t, x, y)},$$

where $D(t, x, y) = \sqrt{\{\cos(t) - x\}^2 + \{\sin(t) - y\}^2}$.

To solve these differential equations *analytically* for $x(t)$ and $y(t)$ is probably not possible, that is, *I* can't do it! But it is not at all difficult to get numerical results for any given value of s, using a computer, a good programming language (e.g., MATLAB), and the rather unsophisticated approach of approximating the derivatives with the ratios of very small (but nonzero) increments, that is, for a fixed, *small* value of Δt, we have

$$\frac{dx}{dt} \approx \frac{\Delta x}{\Delta t}, \frac{dy}{dt} \approx \frac{\Delta y}{\Delta t}.$$

Thus, for some given initial values of t, x, and y, our approach is to write the pair of differential equations as

$$\Delta x \approx \left\{ s\frac{\cos(t) - x}{D(t, x, y)} \right\} \Delta t, \Delta y \approx \left\{ s\frac{\sin(t) - y}{D(t, x, y)} \right\} \Delta t$$

and then to calculate new values of x, y, and t as

$$x_{\text{new}} = x_{\text{old}} + \Delta x, y_{\text{new}} = y_{\text{old}} + \Delta y, t_{\text{new}} = t_{\text{old}} + \Delta t.$$

If we repeat this process a large number of times, then (assuming round-off errors don't accumulate too much) we can plot these values of x and y to show the trajectory of the cat.[7] We can simultaneously plot the mouse's trajectory because we *already know* where the mouse is at each instant of time (on its circle). Figures 2.4.2 and 2.4.3 show two such plots, each using $\Delta t = 0.001$ seconds. Figure 2.4.2 is for a "fast" cat running 1.05 times as fast as the mouse, and this cat does eventually capture the mouse (the MATLAB program looped through 2,500 iterations to generate this figure, that is, the simulated duration of the pursuit is 2.5 seconds). Figure 2.4.3, however, shows that a "slow" cat running 0.9 times as fast as the mouse does *not* end in a capture—the cat's trajectory quickly settles down into a circle, itself, that remains inside and forever lagging the mouse's circular trajectory.

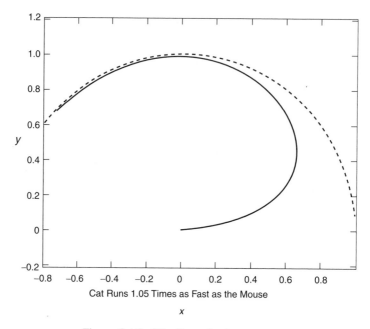

Cat Runs 1.05 Times as Fast as the Mouse

x

Figure 2.4.2. "Fast" cat chasing a mouse

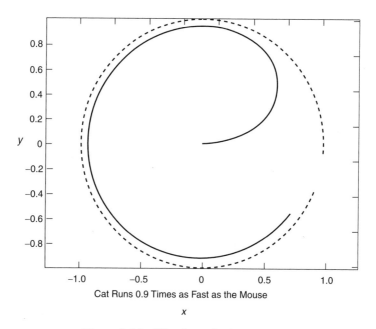

Cat Runs 0.9 Times as Fast as the Mouse

x

Figure 2.4.3. "Slow" cat chasing a mouse

2.5 Solution to the running dog problem.

Let the man run on a circular path (centered on the origin) with radius A, while his dog always stays precisely one yard to his right. We can write the parametric equations of the runner's circular path as

$$x(t) = -A\cos(t), \quad y(t) = A\sin(t).$$

These equations describe a clockwise circular run of radius A that begins (at time $t = 0$) at $x = -A$, $y = 0$ and, at some later time, returns to the starting point. Then, using the equations developed in section 2.3, the man's position vector is

$$\mathbf{r}(t) = -A\cos(t) + iA\sin(t) = A[-\cos(t) + i\sin(t)],$$

and so

$$\mathbf{r}'(t) = A[\sin(t) + i\cos(t)],$$

which says $|\mathbf{r}'(t)| = A$. This also tells us that

$$\mathbf{T}(t) = \frac{\mathbf{r}'(t)}{|\mathbf{r}'(t)|} = \sin(t) + i\cos(t),$$

and so

$$\mathbf{N}(t) = i\mathbf{T}(t) = -\cos(t) + i\sin(t) = \frac{\mathbf{r}(t)}{A}.$$

And finally, the dog's position vector is $\mathbf{d}(t) = \mathbf{r}(t) - \alpha\mathbf{N}(t)$ (remember, $\alpha = 1$ yard) and so

$$\mathbf{d}(t) = \mathbf{r}(t) - \frac{\mathbf{r}(t)}{A} = \left(1 - \frac{1}{A}\right)\mathbf{r}(t).$$

By studying this last result, for various values of A, we discover a very interesting behavior. First, if $A > 1$ (the man runs around in a "big" circle) then the factor $(1 - 1/A)$ is such that $0 < (1 - 1/A) < 1$ and the dog simply runs around (always one yard to his master's right) on a "smaller" circle inside the man's circle. If we now imagine A shrinking, then when $A = 1$ we have $\mathbf{d}(t) = 0$: the dog doesn't really "run" at all, but merely spins around on its feet while remaining on the origin! If A continues to shrink, then for $A < 1$ we see that the factor $(1 - 1/A)$ is such that $-\infty < (1 - 1/A) < 0$. The fact that the factor is negative means $\mathbf{d}(t)$ is now pointing in the direction *opposite* to $\mathbf{r}(t)$: the nature of the problem has suddenly and dramatically changed because, for $A < 1$, the man and the dog are not really running *together* anymore; the man and the dog are now on *opposite* sides of the origin! In the special case of $A = 1/2$, for example, where $\mathbf{d}(t) = -\mathbf{r}(t)$, we actually have a *chase* rather than a parallel run: the man and the dog are running on the *same* circle, half a circumference apart (it isn't at all clear who is chasing who, however!) and yet, the dog *is* always exactly one yard to the man's right.

The case where A shrinks even more $\left(A < \frac{1}{2} \right)$, is shown in figure 2.5.1, where the radius of the runner's circle is A and the radius of the dog's circle is $|A(1 - 1/A)|$. We need the absolute value signs since the radius of a circle is always nonnegative and, for this situation, $1 - 1/A$ is negative. Notice that, for $A < 1$,

$$\left| A \left(1 - \frac{1}{A} \right) \right| = |A - 1| = 1 - A,$$

and so the distance between the runner and his dog is, indeed, still one (yard) even when $A < 1$ because $(A) + (1 - A) = 1$. Most surprising of all, perhaps, is that it is now clear that it is the *dog*, not the man, that runs on the larger circle! And finally, as A shrinks even more, until it finally reaches zero, we see that it is the *man* who has now stopped running—it is he who is now spinning around on his feet while remaining on the origin (as did the dog when $A = 1$). When $A = 0$ the dog runs along a circular path (of radius one yard) around the stationary (spinning) man.

We see from all this that our original result, that the man runs a distance greater than the dog's distance by an amount equal to 2π times the net number of clockwise rotations of $\mathbf{T}(t)$, fails for $A < 1$. For a circular path the net number of rotations of $\mathbf{T}(t)$ is obviously one, for any A; our formula gives the man's total run distance as 2π yards greater

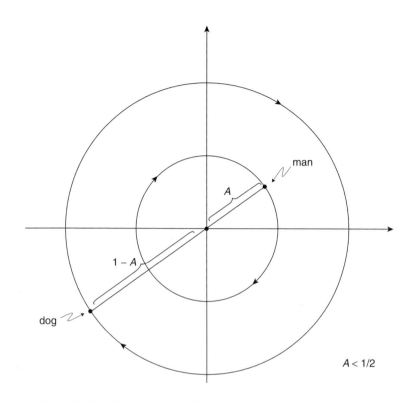

Figure 2.5.1. The geometry of the man running on a "small" circle
$$(A < \tfrac{1}{2})$$

than the dog's for any A. But for $A < 1$ we see that actually the man
runs a total distance of $L_R = 2\pi A$ and the dog runs a total distance of
$L_D = 2\pi(1 - A)$. Thus,

$$L_R - L_D = 2\pi A - 2\pi(1 - A) = 4\pi A - 2\pi = 2\pi(2A - 1),$$

which is *less* than 2π since $A < 1$. Indeed, for $A < 1/2$, $L_R - L_D < 0$; it is
the *dog* that runs further (as clearly shown in figure 2.5.1), not the man.

Chapter 3
The Irrationality of π^2

3.1 The irrationality of π.

The search for ever more digits of π is many centuries old, but the question of its irrationality seems to date only from the time of Euler. It wasn't until the 1761 proof by the Swiss mathematician Johann Lambert (1728–1777) that π was finally shown to be, in fact, irrational. Lambert's proof is based on the fact that $\tan(x)$ is irrational if $x \neq 0$ is rational. Since $\tan(\pi/4) = 1$ is *not* irrational, then $\pi/4$ cannot be rational, i.e., $\pi/4$ is irrational, and so then π too must be irrational. Lambert, who for a while was a colleague of Euler's at Frederick the Great's Berlin Academy of Sciences, started his proof by first deriving[1] a continued fraction expansion for $\tan(x)$,

$$\tan(x) = \cfrac{1}{\cfrac{1}{x} - \cfrac{1}{\cfrac{3}{x} - \cfrac{1}{\cfrac{5}{x} - \cfrac{1}{\cfrac{7}{x} - \cdots}}}}$$

Lambert's derivation was done in a less than iron-clad way, however, and a really solid mathematical demonstration wasn't published until the French mathematician Adrien-Marie Legendre (1752–1833) brought out the seventh edition of his book *Eléments de géométrie* (1808). In fact, Legendre did more than show π is irrational; he showed that π^2 is irrational. If π^2 is irrational then so must be π. (The converse does *not* follow, e.g., $\sqrt{2}$ is irrational but its square is obviously rational.) After all, if π were rational, that is, if there exist integers p and q such that $\pi = p/q$, then $\pi^2 = p^2/q^2$, which is also a ratio of integers, and so π^2 would be rational as well. If one has, however, previously shown that π^2 is *not* rational then we have a contradiction with the assumption that $\pi = p/q$, which shows that π itself can *not* be rational.

Legendre suspected that π's irrationality is fundamentally different from $\sqrt{2}$'s, writing in his book that "It is probable that the number π is not even included in the algebraic irrationals [as is $\sqrt{2}$], but it appears to be very difficult to demonstrate this proposition rigorously."[2] Indeed, it wasn't until 1882 that the German mathematician Ferdinand Lindemann (1852–1939) showed that π is transcendental (with the aid of Euler's formula). The transcendental problem is an amusing one because *most* real numbers *are* transcendental[3] and yet, paradoxically, demonstrating that any *particular* real number is transcendental can be very difficult (as Legendre noted in the case of π). This was recognized by the German mathematician David Hilbert (1862–1943), who, in 1900, presented the mathematical world with a list of twenty-three unsolved problems. Number seven on that famous list[4] was a challenge to study the transcendence problem, and in particular to determine the nature of $2^{\sqrt{2}}$.

In 1920 the Russian mathematician Alexander Gelfond (1906–1968) made some progress by showing that a^b, where $a \neq 0$ or 1 but algebraic and $b = i\sqrt{c}$, where c is a positive, nonsquare integer, is transcendental. In 1930 his fellow Russian Rodion Kuzmin (1891–1949) extended this to include the case where $b = \sqrt{c}$ is real. Kuzmin's result thus answered Hilbert's specific question in the affirmative; $2^{\sqrt{2}}$ *is* transcendental. And finally, in 1936 the German mathematician Carl Siegel (1896–1981) showed that a^b ($a \neq 0$ or 1 but algebraic and b not real *and* rational) is *always* transcendental.[5] An important special case is $a = i$ (which is algebraic—see note 2 again) and $b = -2i$ (which is certainly not real). Thus, i^{-2i} must be transcendental. But, using Euler's formula,

$$i^{-2i} = (e^{i\pi/2})^{-2i} = e^{(i\pi/2)(-2i)} = e^{\pi}.$$

Thus, e^{π} is transcendental (the natures of "similar" numbers such as π^e, π^{π}, e^e, $e + \pi$, and πe, however, are still unknown).

Now, before starting the math of this chapter, first a little philosophy. Does it *really matter* that π is irrational? Do physicists or engineers *really need* to know the value of π beyond, say, 3.14159265 (or even the crude approximation that schoolchildren are often taught, 22/7)? Probably not. As the Plumber says in *Kandelman's Krim,*

> It is now an accepted fact among plumbers, a fact impressed on all
> apprentices, that the circumference of a pipe is found by multiply-
> ing the diameter by π ... I am of course perfectly well aware of the
> irrationality of π, but, on the job, π is $3\frac{1}{7}$, or 3 if I am in a hurry.[6]

The question of π's irrationality isn't a question of accuracy, however,
but rather it is a *spiritual* issue. As one mathematician wrote in a book
for a lay audience,[7]

> What does it matter whether π is rational or irrational? A mathe-
> matician faced with [this] question is in much the same position as
> a composer of music being questioned by someone with no ear for
> music. Why do you select some sets of notes and have them repeated
> by musicians, and reject others as worthless? It is difficult to answer
> except to say that there are harmonies in these things which we
> find that we can enjoy. It is true of course that some mathematics
> is useful. [The applications of logarithms, differential equations,
> and linear operators are then mentioned.] But the so-called pure
> mathematicians do not do mathematics for such [practical appli-
> cations]. *It can be of no practical use*[8] *to know that π is irrational, but if
> we can know it would surely be intolerable not to know.*

A demonstration of π's irrationality is not often presented in undergraduate texts, but it can in fact be done with freshman calculus concepts
and Euler's formula. In the next several sections I'll follow in Legendre's
footsteps and show you the stronger result that π^2 is irrational, although
the approach I'll use is a modern one, not Legendre's original proof.
(An immediate consequence of this is that Euler's sum of the reciprocals
squared—equal to $\pi^2/6$—is irrational.) The proof (one by contradiction)
that I'll show you is essentially the one given in Carl Siegel's beautiful
little 1946 book *Transcendental Numbers*. That book is a reprint of a series
of typed lectures Siegel gave at Princeton in the spring of 1946, and
those notes are (very!) terse. Siegel was speaking to graduate students
and professors of mathematics, and often leaped over many intervening
steps from one equation to the next, steps which he clearly thought to
be self-evident to his audience.[9] Some of the leaps were of Olympic size.
I have filled in those missing steps. So, we begin.

3.2 The $R(x) = B(x)e^x + A(x)$ equation, D-operators, inverse operators, and operator commutativity.

We know that the power series expansion of e^x is one of "infinite degree" that is, there is no largest n for the general term of x^n in the expansion. Suppose, however, that we attempt to approximate e^x in the neighborhood of $x = 0$ by the *ratio* of two finite polynomials, each of degree n. That is, if $A(x)$ and $B(x)$ are each of degree n, then we wish to find $A(x)$ and $B(x)$ such that, for $x \approx 0$,

$$e^x \approx -\frac{A(x)}{B(x)}.$$

With this goal in mind, we'll start by defining the function $R(x)$ to be

$$\boxed{R(x) = B(x)e^x + A(x)} = B(x)\left[e^x + \frac{A(x)}{B(x)}\right],$$

which means $R(x)$ will be "small" for $x \approx 0$ *if* we have selected $A(x)$ and $B(x)$ properly. In the next section I'll use the results of this section to solve the $R(x)$ equation for $A(x)$ and $B(x)$.

Since $A(x)$ and $B(x)$ are each of degree n, we can write

$$A(x) = a_0 + a_1 x + a_2 x^2 + \cdots + a_n x^n,$$
$$B(x) = b_0 + b_1 x + b_2 x^2 + \cdots + b_n x^n,$$

and so it is clear, by direct multiplication and the power series of e^x, that

$$R(x) = (a_0 + b_0) + (a_1 + b_0 + b_1)x$$
$$+ \left(a_2 + b_1 + \frac{1}{2!}b_0 + b_2\right)x^2 + \cdots.$$

Let's next suppose that we wish $R(x)$ to *start* with the x^{2n+1} term, that is, let's demand that the coefficients of the first $2n + 1$ terms of $R(x)$ vanish. This means that, as $x \to 0$, $R(x)$ will vanish as fast as x^{2n+1} vanishes (which is, of course, faster than the vanishing of x itself). This requirement serves as a measure of how well we have determined $A(x)$ and $B(x)$. The faster $R(x)$ vanishes as $x \to 0$ (the larger is n) the better our approximation of e^x. So, setting the first $2n+1$ coefficients of $R(x)$ to zero, we have

$$a_0 + b_0 = 0,$$

$$a_1 + b_0 + b_1 = 0,$$

$$a_2 + b_1 + \frac{1}{2!}b_0 + b_2 = 0,$$

and so on. Notice, carefully, that this gives us $2n+1$ equations for the $2n+2$ coefficients of $A(x)$ and $B(x)$, that is, we have more unknowns than equations, so the coefficients of $A(x)$ and $B(x)$ are *under*determined by our requirement that the first $2n+1$ terms of $R(x)$ vanish. It is clear, therefore, that there *must* exist a nontrivial solution for the $n+1$ *a*s and the $n+1$ *b*s.

We can find general formulas for $A(x)$ and $B(x)$ (in the next section) as soon as we develop some preliminary results concerning what is called the *differentiation operator*. We start by defining this operator, denoted by **D**, as follows: if n is a positive integer then $\mathbf{D}^n\phi(x)$ will mean the nth derivative of the function $\phi(x)$. Thus, we actually have an *infinite* set of operators defined as

$$\boxed{\mathbf{D}^n\phi(x) = \frac{d^n}{dx^n}\phi(x), \quad n = 1, 2, 3, \cdots}.$$

Since differentiation is a linear operation, it is clear that if c_1 and c_2 are constants then

$$\mathbf{D}^n\{c_1\phi_1(x) + c_2\phi_2(x)\} = c_1\mathbf{D}^n\phi_1(x) + c_2\mathbf{D}^n\phi_2(x).$$

We can extend the meaning of $\mathbf{D}^n\phi(x)$ to the case where n is a nonpositive integer ($n \leq 0$) as follows, with two observation. First, if $n = 0$, we make the plausible argument that since $n = 0$ means *zero* differentiations (which leaves $\phi(x)$ unaffected), it "must" be so that

$$\mathbf{D}^0\phi(x) = \phi(x),$$

which means, *formally*, that we can think of the operator $\mathbf{D}^0$ as equivalent to simply multiplying by one. Indeed, we will think of an *operator function* of $\mathbf{D}$, for example, $g(\mathbf{D}) = 1+\mathbf{D}$, as equivalent to unity when applied zero times, i.e., $g(\mathbf{D})^0 = (1 + \mathbf{D})^0 = 1$. Second, since $(d^k/dx^k)\{d^j\phi/dx^j\} = d^{k+j}\phi/dx^{k+j}$, for k and j both positive integers, then formally we have $\mathbf{D}^k\mathbf{D}^j\phi(x) = \mathbf{D}^{k+j}\phi(x)$; we'll extend this to include *all* integer values of k and j, even negative ones.

Now, what could $\mathbf{D}^{-1}$ possibly mean? It can't mean doing *less than zero* differentiations, right? Well, formally, $\mathbf{D}^{-1}\mathbf{D}^1 = \mathbf{D}^0 = 1$, so $\mathbf{D}^{-1}$ should be the mathematical operation that "undoes" what $\mathbf{D}^1$ does. That is, $\mathbf{D}^{-1}$ is the *inverse* operator of $\mathbf{D}^1$; $\mathbf{D}^{-1}$ should be an *integration*. So, let's *define* $\mathbf{D}^{-1}$ as

$$\mathbf{D}^{-1}\phi(x) = \int_0^x \phi(t)\,dt \ .$$

The t is, of course, simply a dummy variable of integration.

Two operators are said to *commute* when the order of their application doesn't matter. So, if the operators $\mathbf{D}^1$ and $\mathbf{D}^{-1}$ commute, then it would be true that $\mathbf{D}^1\mathbf{D}^{-1}\phi(x) = \mathbf{D}^{-1}\mathbf{D}^1\phi(x)$. It is easy to show, however, that this is generally *not* the case for these two operators. That is, from Leibniz's rule for differentiating an integral,[10] we have

$$\mathbf{D}^1\mathbf{D}^{-1}\phi(x) = \frac{d}{dx}\left\{\int_0^x \phi(t)\,dt\right\} = \phi(x),$$

while

$$\mathbf{D}^{-1}\mathbf{D}^1\phi(x) = \mathbf{D}^{-1}\frac{d}{dx}\phi(x)$$
$$= \int_0^x \frac{d}{dt}\phi(t)\,dt = \int_0^x \frac{d\phi}{dt}\,dt = \int_0^x d\phi = \phi(x) - \phi(0).$$

So, $\mathbf{D}^1\mathbf{D}^{-1} \neq \mathbf{D}^{-1}\mathbf{D}^1$ unless it just happens that $\phi(0) = 0$. You'll see, as we proceed into the development of our proof for the irrationality of π^2, that matters will be cleverly arranged so that this condition for the commutativity of $\mathbf{D}^1$ and $\mathbf{D}^{-1}$ is always satisfied, even though it is not true in general. We will just be very careful to work only with ϕ-functions such that $\phi(0) = 0$.

Now, what if $n = -2$—what could $\mathbf{D}^{-2}$ mean? We might think that since $\mathbf{D}^{-1}$ is an integration then $\mathbf{D}^{-2}$ should mean *two* integrations, and indeed, that is so. But there is a clever trick that will allow us to reduce the two integrations back down to one. This is one of the points I mentioned earlier that Seigel passes over without comment in his book, so here is how it is done. We start by writing

$$\mathbf{D}^{-2}\phi(x) = \mathbf{D}^{-1}\mathbf{D}^{-1}\phi(x) = \mathbf{D}^{-1}\int_0^x \phi(t)\,dt.$$

Next, if we write

$$f(x) = \int_0^x \phi(t)\,dt = \int_0^x \phi(s)\,ds$$

(the two integrals are equal because all that has "changed" is the dummy variable of integration, from t to s, which is really no change at all), then

$$\mathbf{D}^{-2}\phi(x) = \mathbf{D}^{-1}f(x) = \int_0^x f(t)\,dt = \int_0^x \left\{ \int_0^t \phi(s)\,ds \right\} dt.$$

The double integral has a very nice geometric interpretation, as shown in figure 3.2.1. It represents the integration of ϕ (in vertical strips) over the two-dimensional triangular region shown in the figure. To see this, think of what the double integral notation is "saying": pick a value for t from the interval 0 to x (the outer integral) and then, for that value of t, integrate $\phi(s)$ from $s = 0$ to $s = t$ (the inner integral), which is a vertical integration path (a strip of width dt). Picking a new value for t simply selects a new vertical integration path (strip) and, as t varies from 0 to x, the total of all these strips covers the triangular region.

From the same figure you can see, however, that we can integrate ϕ over the same region in *horizontal* strips (of width ds) just as well, by writing the double integral as $\int_0^x \left\{ \int_s^x \phi(s)\,dt \right\} ds$. Since we can pull $\phi(s)$

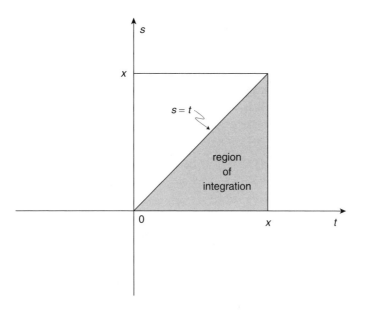

Figure 3.2.1. Region of integration for $\mathbf{D}^{-2}$

out of the inner integral (the inner integration variable is t, not s), we therefore have

$$\mathbf{D}^{-2}\phi(x) = \int_0^x \left\{ \int_s^x \phi(s)\,dt \right\} ds = \int_0^x \phi(s) \left\{ \int_s^x dt \right\} ds = \int_0^x \phi(s)(x-s)\,ds,$$

and so, as claimed, $\mathbf{D}^{-2}$ is a one-dimensional integral, just as is $\mathbf{D}^{-1}$. Well, then what about $\mathbf{D}^{-3}$? Can we repeat the above arguments to see what $\mathbf{D}^{-3}$ is? Yes!

Since $\mathbf{D}^{-3}\phi(x) = \mathbf{D}^{-1}\mathbf{D}^{-2}\phi(x)$, we have

$$\mathbf{D}^{-3}\phi(x) = \mathbf{D}^{-1} \int_0^x \phi(s)(x-s)\,ds = \mathbf{D}^{-1}f(x) = \int_0^x f(t)\,dt,$$

where

$$f(x) = \int_0^x \phi(s)(x-s)\,ds.$$

So

$$\mathbf{D}^{-3}\phi(x) = \int_0^x \left\{ \int_0^t \phi(s)(t-s)\,ds \right\} dt,$$

and this double integral is, using the same argument as before,

$$\mathbf{D}^{-3}\phi(x) = \int_0^x \left\{ \int_s^x \phi(s)(t-s)\,dt \right\} ds = \int_0^x \phi(s) \left\{ \int_s^x (t-s)\,dt \right\} ds.$$

The inner integral is easy to do (change variable to $u = t - s$ and you should see that the inner integral equals $\frac{1}{2}(x-s)^2$). Thus,

$$\mathbf{D}^{-3}\phi(x) = \int_0^x \phi(s)\frac{(x-s)^2}{2}\,ds.$$

If you continue in this manner (or use mathematical induction) you should be able to convince yourself that the general result is

$$\mathbf{D}^{-n-1}\phi(x) = \int_0^x \phi(s)\frac{(x-s)^n}{n!}\,ds, \quad n = 0, 1, 2, 3, \cdots.$$

So far I've not said anything about the nature of $\phi(x)$. To be quite specific on this point, we'll limit ourselves from now on to functions that have the form $\phi(x) = e^{\lambda x}P(x)$, where λ is a constant and $P(x)$ is a polynomial such that $P(0) = 0$ (which means $\phi(0) = 0$, and so $\mathbf{D}^1$ and $\mathbf{D}^{-1}$ *will* commute). This *is* pretty specific, but it will prove to be all we'll need to establish the irrationality of π^2. Then,

$$\mathbf{D}^1\phi(x) = \mathbf{D}^1\left\{e^{\lambda x}P(x)\right\}$$

$$= \frac{d}{dx}\left\{e^{\lambda x}P(x)\right\}$$

$$= e^{\lambda x}\frac{dP}{dx} + \lambda e^{\lambda x}P(x)$$

$$= e^{\lambda x}\left\{\lambda P(x) + \frac{dP}{dx}\right\}$$

$$= e^{\lambda x}\left\{\lambda P(x) + \mathbf{D}^1 P(x)\right\},$$

or, finally, if we adopt the simpler notation of just writing $\mathbf{D}$ for $\mathbf{D}^1$,

$$\mathbf{D}\phi(x) = e^{\lambda x}(\lambda + \mathbf{D})P(x).$$

Continuing in the same fashion, we can write

$$\mathbf{D}^2\phi(x) = \mathbf{D}\left\{\mathbf{D}\phi(x)\right\}$$

$$= \frac{d}{dx}\left\{e^{\lambda x}\left[\lambda P(x) + \frac{dP}{dx}\right]\right\}$$

$$= e^{\lambda x}\frac{d}{dx}\left[\lambda P(x) + \frac{dP}{dx}\right] + \lambda e^{\lambda x}\left[\lambda P(x) + \frac{dP}{dx}\right]$$

$$= e^{\lambda x}\left[\lambda\frac{dP}{dx} + \frac{d^2P}{dx^2}\right] + \lambda^2 e^{\lambda x}P(x) + \lambda e^{\lambda x}\frac{dP}{dx}$$

$$= e^{\lambda x}\left[\lambda\mathbf{D}P(x) + \mathbf{D}^2 P(x)\right] + \lambda^2 e^{\lambda x}P(x) + \lambda e^{\lambda x}\mathbf{D}P(x)$$

$$= e^{\lambda x}\left[\mathbf{D}^2 + 2\lambda\mathbf{D} + \lambda^2\right]P(x) = e^{\lambda x}(\mathbf{D} + \lambda)^2 P(x),$$

or, finally,

$$\mathbf{D}^2\phi(x) = e^{\lambda x}(\lambda + \mathbf{D})^2 P(x).$$

If you repeat this process over and over (or use induction) it is easy to see that, in general,

$$\mathbf{D}^n\phi(x) = \mathbf{D}^n e^{\lambda x}P(x) = e^{\lambda x}(\lambda + \mathbf{D})^n P(x), \quad n = 0, 1, 2, \cdots$$

.

We have now developed all the mathematical machinery we need to solve the equation $R(x) = B(x)e^x + A(x)$ for $A(x)$ and $B(x)$, the topic of the next section.

3.3 Solving for $A(x)$ and $B(x)$.

Differentiation of $R(x) = B(x)e^x + A(x)$ a total of $n + 1$ times gives

$$\mathbf{D}^{n+1} R(x) = \mathbf{D}^{n+1} \{B(x)e^x + A(x)\} = \mathbf{D}^{n+1} \{B(x)e^x\} + \mathbf{D}^{n+1} A(x).$$

Since $A(x)$ is (by our assumption at the start of the last section) of degree n, then $\mathbf{D}^{n+1} A(x) = 0$, and so, from the final, boxed result of the last section (with $\lambda = 1$), we have

$$\mathbf{D}^{n+1} \{B(x)e^x\} = \boxed{e^x(1 + \mathbf{D})^{n+1} B(x) = \mathbf{D}^{n+1} R(x)}.$$

From this we can immediately write that

$$(1 + \mathbf{D})^{n+1} B(x) = e^{-x} \mathbf{D}^{n+1} R(x).$$

Since we earlier assumed that $R(x)$ *starts* with the x^{2n+1} term, we can write $R(x) = r_1 x^{2n+1} + r_2 x^{2n+2} + \cdots$. Thus,

$$\mathbf{D}^{n+1} R(x) = (2n + 1)(2n)(2n - 1) \cdots (n + 1) r_1 x^n$$

plus terms of degree $n + 1$ and higher, and so, with $r_0 = (2n + 1)(2n) \cdots (n + 1) r_1$, we have

$$(1 + \mathbf{D})^{n+1} B(x) = e^{-x} \left[r_0 x^n + \cdots \right] = (1 - x + \frac{x^2}{2!} - \cdots)(r_0 x^n + \cdots),$$

which is simply $r_0 x^n$ plus terms which are all of degree $n+1$ and higher.

Now, since we are assuming that $B(x)$ is a polynomial of degree n, $(1 + \mathbf{D})^{n+1} B(x)$ is a polynomial of degree n, too. This is so because the operator $(1 + \mathbf{D})^{n+1}$ is, by a formal application of the binomial theorem, equivalent to a sum of various $\mathbf{D}$ operators, i.e.,

$$(1 + \mathbf{D})^{n+1} = \sum_{j=0}^{n+1} \binom{n+1}{j}(1)^{n+1-j} \mathbf{D}^j,$$

which is simply the sum of the operator $(1)^{n+1} = 1$ (for $j = 0$) and the operators $\binom{n+1}{j}\mathbf{D}^j$ (for $1 \leq j \leq n+1$). Each of the $\mathbf{D}^j$ operators, when applied to $B(x)$, reduces the degree of $B(x)$ by j (i.e., by at least one), while the 1 operator simply reproduces $B(x)$. So, since $B(x)$ is of degree n, $(1 + \mathbf{D})^{n+1}B(x)$ is also of degree n. Therefore, we keep just the leading term of $e^{-x}\mathbf{D}^{n+1}R(x)$, as all the other terms are of *higher* degree (those other terms must, of course, all have zero coefficients as there *are* no terms of degree higher than n in $(1 + \mathbf{D})^{n+1}B(x)$), that is, $(1 + \mathbf{D})^{n+1}B(x) = r_0 x^n$. Now, solving for $B(x)$ by applying the operator $(1+\mathbf{D})^{-n-1}$ to both sides (and using the fact that $(1+\mathbf{D})^{-n-1}(1+\mathbf{D})^{n+1} = (1 + \mathbf{D})^0 = 1$), we have

$$\boxed{B(x) = r_0(1 + \mathbf{D})^{-n-1}x^n}.$$

We can solve for $A(x)$ in almost the same way. Starting now with the alternative form of the $R(x)$ equation gotten by multiplying through by e^{-x}, that is, starting with $R(x)e^{-x} = B(x) + A(x)e^{-x}$, differentiating a total of $n + 1$ times gives

$$\mathbf{D}^{n+1}\left\{R(x)e^{-x}\right\} = \mathbf{D}^{n+1}B(x) + \mathbf{D}^{n+1}\left\{A(x)e^{-x}\right\},$$

or, as $\mathbf{D}^{n+1}B(x) = 0$ (because, remember, by assumption $B(x)$ is of degree n),

$$\mathbf{D}^{n+1}\left\{A(x)e^{-x}\right\} = \mathbf{D}^{n+1}\left\{R(x)e^{-x}\right\}.$$

From the final, boxed result of the last section, where we now use $\lambda = -1$, we see that

$$e^{-x}(-1 + \mathbf{D})^{n+1}A(x) = e^{-x}(-1 + \mathbf{D})^{n+1}R(x),$$

or

$$(-1 + \mathbf{D})^{n+1}A(x) = (-1 + \mathbf{D})^{n+1}R(x).$$

Now, as before, $(-1 + \mathbf{D})^{n+1}A(x)$ is a polynomial of degree n because, by assumption, $A(x)$ is a polynomial of degree n (simply repeat the argument I used in the discussion about $B(x)$). Therefore we are interested only in the terms of degree n (or less) generated by $(-1 + \mathbf{D})^{n+1}R(x)$.

Since $R(x)$ is, by assumption, a polynomial that *starts* with the x^{2n+1} term, then only the $\mathbf{D}^{n+1}$ operator in the binomial expansion of $(-1 + \mathbf{D})^{n+1}$ will generate an x^n term, that is, as before $\mathbf{D}^{n+1}R(x)$ is $r_0 x^n$ plus terms of degree $n+1$ and higher (which must of course have zero coefficients). So, we keep only the $r_0 x^n$ term and write $(-1 + \mathbf{D})^{n+1}A(x) = r_0 x^n$, or, finally,

$$A(x) = r_0(-1 + \mathbf{D})^{-n-1} x^n \quad .$$

We have thus found $A(x)$ and $B(x)$, each to within the common scale factor of r_0. The specific value of r_0 is actually unimportant because $A(x)$ and $B(x)$ appear as a *ratio* in our approximation of e^x, that is, recall from the start of the previous section that, as $x \to 0$,

$$e^x \approx -\frac{A(x)}{B(x)}.$$

Thus, the r_0 scale factor cancels out; the unimportance of its specific value is a result of $A(x)$ and $B(x)$ being underdetermined, as I mentioned earlier. So, without any loss of generality (and a nice gain in simplicity), let's simply take $r_0 = 1$ and arrive at

$$\begin{aligned} A(x) &= (-1 + \mathbf{D})^{-n-1} x^n \\ B(x) &= (1 + \mathbf{D})^{-n-1} x^n \end{aligned} \quad .$$

At this point I can almost surely predict your reaction to the above boxed expressions: "Wow, what a wild ride! I followed all the individual steps okay, and I 'see' where those expressions for $A(x)$ and $B(x)$ came from. Still, I wonder what they *mean!*" That's a perfectly respectable question to ask, too, and so I'm going to take a break from pushing forward our proof of the irrationality of π^2 (we are at about the halfway point). What I want to do before going any further is *show* you what those formulas mean. To say that the above $A(x)$ and $B(x)$ equations are "solutions" is to say that, for any particular given value of n (a positive integer), we can calculate the specific polynomials that are $A(x)$ and $B(x)$. Our proof of the irrationality of π^2 actually does *not* require that we do this, but I think it valuable to do it anyway for two important reasons. First, once you see that the $A(x)$ and $B(x)$ polynomials we arrive at *actually do* achieve the original goal (i.e., $-A(x)/B(x) \approx e^x$ for $x \approx 0$), then I think you'll gain confidence in the boxed expressions, as well as confidence that what admittedly have been some "wild" operator manipulations really do make sense. And second, in calculating some specific polynomials you'll see that they have one very special property, a property that will prove to be absolutely essential in completing the proof of the irrationality of π^2.

To start, let's back up one step, to the expressions we derived just before the boxed formulas. That is, to (with $r_0 = 1$)

$$(-1 + \mathbf{D})^{n+1} A(x) = x^n,$$
$$(1 + \mathbf{D})^{n+1} B(x) = x^n.$$

Now, suppose that $n = 1$. Then, $A(x) = a_0 + a_1 x$ and $B(x) = b_0 + b_1 x$ and so

$$(-1 + \mathbf{D})^2 (a_0 + a_1 x) = x,$$
$$(1 + \mathbf{D})^2 (b_0 + b_1 x) = x.$$

Thus,

$$(1 - 2\mathbf{D} + \mathbf{D}^2)(a_0 + a_1 x) = x,$$
$$(1 + 2\mathbf{D} + \mathbf{D}^2)(b_0 + b_1 x) = x,$$

and so, after doing the indicated operations, we arrive at

$$(a_0 + a_1 x) + (-2a_1) = x = (a_0 - 2a_1) + a_1 x,$$
$$(b_0 + b_1 x) + (2b_1) = x = (b_0 + 2b_1) + b_1 x.$$

Equating coefficients of equal powers on both sides of these expressions, we see that $a_1 = 1$ and $a_0 - 2a_1 = 0$. Thus, $a_0 = 2a_1 = 2$. Also, $b_1 = 1$ and $b_0 + 2b_1 = 0$. Thus, $b_0 = -2b_1 = -2$. So,

$$A(x) = 2 + x,$$
$$B(x) = -2 + x,$$

and our approximation is

$$e^x \approx -\frac{A(x)}{B(x)} = -\frac{2+x}{-2+x}.$$

Let's see how "good" an approximation this is (it's obviously exact *at* $x = 0$). For $x = 0.1$, which is "small" but not really *very* small, we have

$$e^x = e^{0.1} = 1.105170918,$$

while

$$-\frac{2+0.1}{-2+0.1} = -\frac{2.1}{-1.9} = \frac{2.1}{1.9} = 1.105263158.$$

We have perfect agreement with the first three decimal places of $e^{0.1}$, and that's not too bad. The approximation should be even better, however, for $n = 2$. I'll let you repeat the above process and verify that then

$$A(x) = -12 - 6x - x^2,$$
$$B(x) = 12 - 6x + x^2,$$

and so the approximation is, at $x = 0.1$,

$$-\frac{A(x)}{B(x)} = -\frac{-12 - 6(0.1) - (0.1)^2}{12 - 6(0.1) + (0.1)^2} = \frac{12.61}{11.41} = 1.105170903.$$

This agrees exactly with the first *seven* decimal places of $e^{0.1}$, which is *very* impressive!

Finally, you may have noticed for the $n = 1$ and $n = 2$ cases that the polynomial coefficients *were all integers*. Upon a little reflection of the details for the calculation of $A(x)$ and $B(x)$, you should be able to convince yourself that this will *always* be the case, for *any* value of n. We'll use this property, in particular for the $A(x)$ polynomial, at the end of the proof.

3.4 The value of $R(\pi i)$.

Our next step is to determine the value of $R(x)$ for $x = \pi i$. This isn't an obvious step by any means, but you'll soon see how it comes into play. From the first boxed equation at the start of the last section, recall that

$$\mathbf{D}^{n+1} R(x) = e^x (1 + \mathbf{D})^{n+1} B(x).$$

Thus, using our formal solution for $B(x)$ in the last box of the previous section,

$$\mathbf{D}^{n+1} R(x) = e^x (1 + \mathbf{D})^{n+1} \left[(1 + \mathbf{D})^{-n-1} x^n \right] = e^x x^n,$$

and so

$$R(x) = \mathbf{D}^{-n-1} \left\{ e^x x^n \right\}.$$

In the discussion in section 3.2 we found that

$$\mathbf{D}^{-n-1} \phi(x) = \int_0^x \phi(s) \frac{(x - s)^n}{n!} \, ds,$$

and so, with $\phi(x) = e^x x^n$, we have

Box #1 $\qquad\boxed{R(x) = \frac{1}{n!} \int_0^x (x - s)^n e^s s^n \, ds}\qquad.$

We can put this integral into more convenient form by changing variable to $u = s/x$. Then $du = (1/x) ds$, and so $ds = x(du)$. Thus,

$$R(x) = \frac{1}{n!} \int_0^1 (x - ux)^n e^{ux} (ux)^n x(du) = \frac{1}{n!} \int_0^1 x^n (1 - u)^n e^{ux} u^n x^n x(du),$$

or

Box #2 $\qquad\boxed{R(x) = \frac{x^{2n+1}}{n!} \int_0^1 (1 - u)^n u^n e^{ux} \, du}\qquad.$

We can further transform this result with another change of variable, to $t = 1 - u$ (and so $dt = -du$). Then,

$$R(x) = \frac{x^{2n+1}}{n!} \int_1^0 t^n (1 - t)^n e^{(1-t)x} (-dt),$$

or

$$R(x) = \frac{x^{2n+1}}{n!} \int_0^1 t^n (1-t)^n e^{(1-t)x} \, dt.$$

Notice that the expression for $R(x)$ in Box #2 can be written (if we simply replace the dummy variable of integration u with t) as

$$R(x) = \frac{x^{2n+1}}{n!} \int_0^1 t^n (1-t)^n e^{tx} \, dt.$$

Adding these last two expressions for $R(x)$ together (and dividing by 2) gives

$$R(x) = \frac{x^{2n+1}}{n!} \int_0^1 t^n (1-t)^n \frac{e^{tx} + e^{(1-t)x}}{2} \, dt.$$

Now

$$\frac{e^{tx} + e^{(1-t)x}}{2} = \frac{e^{tx} + e^{x-tx}}{2} = \frac{e^{x/2} e^{(tx-x/2)} + e^{x/2} e^{(x/2-tx)}}{2}$$

$$= e^{x/2} \frac{e^{(t-1/2)x} + e^{(1/2-t)x}}{2} = e^{x/2} \frac{e^{(t-1/2)x} + e^{-(t-1/2)x}}{2}$$

and so

$$R(x) = \frac{x^{2n+1}}{n!} e^{x/2} \int_0^1 t^n (1-t)^n \frac{e^{(t-1/2)x} + e^{-(t-1/2)x}}{2} \, dt.$$

If we now set $x = \pi i$ then the factor outside the integral (ignoring the factorial) becomes

$$(\pi i)^{2n+1} e^{\pi i/2} = \pi^{2n+1} (i^{2n+1}) i = \pi^{2n+1} i^{2n} i^2$$

$$= \pi^{2n+1} (i^2)^n (-1)$$

$$= \pi^{2n+1} (-1)^{n+1}.$$

Also, inside the integral we have (using Euler's formula several times)

$$\frac{e^{(t-1/2)\pi i} + e^{-(t-1/2)\pi i}}{2} = \frac{e^{-\pi i/2}e^{i\pi t} + e^{\pi i/2}e^{-i\pi t}}{2}$$

$$= \frac{-ie^{i\pi t} + ie^{-i\pi t}}{2} = -i\frac{2i\sin(\pi t)}{2} = \sin(\pi t).$$

Thus, *at last*, we have

$$R(\pi i) = (-1)^{n+1}\frac{\pi^{2n+1}}{n!}\int_0^1 t^n(1-t)^n \sin(\pi t)\,dt.$$

The reason I say *at last* is that we are *not* going to evaluate the integral. You probably have two reactions to this—first, relief (it *is* a pretty scary-looking thing) and, second, shock (why did we go through all the work needed to derive it?) In fact, all we will need for our proof that π^2 is irrational are the following two observations about $R(\pi i)$. First, and most obviously, it is *real*. And second, $R(\pi i) \neq 0$. This is because, over the entire interval of integration, the integrand (for any integer n) is positive (except at the two endpoints where the integrand is zero). Thus, the integral is certainly nonzero. $R(\pi i)$ itself can be either negative or positive, depending on whether n is even or odd, respectively, but the *sign* of $R(\pi i)$ won't matter to our proof. All that *will* matter is that $R(\pi i) \neq 0$. You'll see why soon, but first we need to establish one more result.

Recall the expression that started our analysis, the one in the box at the beginning of section 3.2. That is,

Box #3 $\boxed{\; B(x)e^x + A(x) = R(x) \;}$.

From this it immediately follows that

$$B(-x)e^{-x} + A(-x) = R(-x)$$

and so, multiplying through by e^x, we have

Box #4 $\boxed{\; A(-x)e^x + B(-x) = e^x R(-x) \;}$.

Next, recall from the expression for $R(x)$ in Box #1 that we have

Box #5 $\qquad \boxed{R(x) = \frac{1}{n!} \int_0^x (x-s)^n e^s s^n \, ds}$.

Thus,

$$e^x R(-x) = \frac{e^x}{n!} \int_0^{-x} (-x-s)^n e^s s^n \, ds.$$

Changing variable to $t = -s$ (and so $ds = -dt$), we have

$$e^x R(-x) = \frac{e^x}{n!} \int_0^x (-x+t)^n e^{-t} (-t)^n (-dt)$$

$$= \frac{e^x}{n!} \int_0^x [-(x-t)]^n \, e^{-t} (-1)^n t^n (-dt)$$

$$= -\frac{e^x}{n!} \int_0^x (-1)^n (x-t)^n e^{-t} (-1)^n t^n \, dt,$$

or, as $(-1)^n (-1)^n = (-1)^{2n} = 1$, and taking the e^x inside the integral, we have

$$e^x R(-x) = -\frac{1}{n!} \int_0^x (x-t)^n e^{(x-t)} t^n \, dt.$$

Changing variables once more, to $u = x - t$ (and so $du = -dt$), we have

$$e^x R(-x) = \frac{1}{n!} \int_x^0 u^n e^u (x-u)^n \, du = -\frac{1}{n!} \int_0^x (x-u)^n u^n e^u \, du,$$

or, if we make the trivial change in dummy variable from u to s,

$$e^x R(-x) = -\frac{1}{n!} \int_0^x (x-s)^n s^n e^s \, ds.$$

Comparing this result with the expression for $R(x)$ in Box #5, we immediately see that

$$e^x R(-x) = -R(x).$$

Inserting this result into the expression in Box #4, we have

$$A(-x)e^x + B(-x) = -R(x).$$

That is, recalling Box #3, we now have the pair of statements

$$B(x)e^x + A(x) = R(x),$$
$$-A(-x)e^x - B(-x) = R(x).$$

Subtracting the second from the first, we have

$$e^x \left[B(x) + A(-x) \right] + \left[B(-x) + A(x) \right] = 0,$$

and this can be true *for all* x only if $B(x) + A(-x)$ and $B(-x) + A(x)$ *each* identically vanish. That is, we have the conditions $B(x) = -A(-x)$ and $B(-x) = -A(x)$, and it is now obvious that these are equivalent statements and that what we thought to be *two* conditions are actually just alternative forms of a single condition.

Now, setting $x = \pi i$ in $B(x)e^x + A(x) = R(x)$, we have

$$B(\pi i)e^{\pi i} + A(\pi i) = R(\pi i),$$

or, as Euler's formula tells us that $e^{\pi i} = -1$,

$$-B(\pi i) + A(\pi i) = R(\pi i).$$

But $B(\pi i) = -A(-\pi i)$, and so $A(-\pi i) + A(\pi i) = R(\pi i)$. But, since $R(\pi i) \neq 0$, then

$$\boxed{A(-\pi i) + A(\pi i) \neq 0}.$$

3.5 The last step (at last!).

The expression $A(x) + A(-x)$ is equal to the sum of the following two polynomials (each with *integer coefficients*, as discussed earlier at the end of section 3.3):

$$a_0 + a_1 x + a_2 x^2 + a_3 x^3 + \cdots + a_n x^n,$$
$$a_0 - a_1 x + a_2 x^2 - a_3 x^3 + \cdots \pm a_n x^n,$$

where the sign of the $\pm a_n x^n$ term in the second polynomial depends on whether n is even (positive) or odd (negative). Thus,

$$A(x) + A(-x) = 2a_0 + 2a_2 x^2 + 2a_4 x^4 + \cdots$$

out to a final term that is $2a_{n-1} x^{n-1}$ if n is odd or $2a_n x^n$ if n is even. This means $A(x) + A(-x)$ can be thought of as a polynomial in the variable $u \, (= x^2)$ of degree $(n-1)/2$ if n is odd, or of degree $n/2$ if n is even. In the usual mathematical notation, $[m]$ means the *integer part* of the real number m (e.g., $[7.3] = 7$), and so the degree of $A(x) + A(-x)$ is, in the variable u, $[n/2]$, in general. That is,

$$A(x) + A(-x) = 2a_0 + 2a_2 u + 2a_4 u^2 + \cdots + 2a_{2[n/2]} u^{[n/2]}$$

where $u = x^2$ and all of the a_i are integers.

Now, suppose that π^2 is rational, that is, suppose there are two integers p and q such that $\pi^2 = p/q$. Then, with $x = \pi i \, (u = -\pi^2)$, it follows that

$$A(\pi i) + A(-\pi i) = 2a_0 - 2a_2 \frac{p}{q} + 2a_4 \frac{p^2}{q^2} - \cdots \pm 2a_{2[n/2]} \frac{p^{[n/2]}}{q^{[n/2]}}.$$

Multiplying through both sides by the *integer* $q^{[n/2]}$, it is clear that $q^{[n/2]} \{A(\pi i) + A(-\pi i)\}$ must be an integer because every term on the right-handside is an integer. That is, there must be some *nonzero* integer (because, as we just showed in the last section, $A(\pi i) + A(-\pi i) = R(\pi i) \neq 0$), which I'll write as j (which may be either positive or negative) such that

$$q^{[n/2]} R(\pi i) = j.$$

Therefore, taking absolute values of both sides of this, we have

$$|q^{[n/2]}R(\pi i)| = q^{[n/2]}|R(\pi i)| = |j| > 0.$$

One last step and we are finished with our proof. We have

$$|R(\pi i)| = \left|(-1)^{n+1}\frac{\pi^{2n+1}}{n!}\int_0^1 t^n(1-t)^n \sin(\pi t)\,dt\right|$$

$$= \frac{\pi^{2n+1}}{n!}\int_0^1 t^n(1-t)^n \sin(\pi t)\,dt.$$

As n becomes arbitrarily large we have, by inspection, both

$$\lim_{n\to\infty}\frac{\pi^{2n+1}}{n!} = 0$$

and

$$\lim_{n\to\infty}\int_0^1 t^n(1-t)^n \sin(\pi t)\,dt = 0.$$

It is therefore clear that, whatever the integer q may be, we can always pick n large enough to make $q^{[n/2]}\,|R(\pi i)| < 1$. Thus, for j some integer, we reach the conclusion that $1 > q^{[n/2]}|R(\pi i)| = |j| > 0$. That is, $1 > |j| > 0$. But there is no integer *between* 0 and 1 and we clearly have an absurdity. The only conclusion, then, is that our starting assumption that $\pi^2 = p/q$ must be wrong, that is, there are no integers p and q, and so π^2 *is irrational* and we are done.

Chapter 4
Fourier Series

4.1 Functions, vibrating strings, and the wave equation.
This entire chapter is devoted to those trigonometric series satisfying certain conditions that go under the general title of *Fourier series*—named after the French mathematician Joseph Fourier (1768–1830)—but the story of these series begins well before Fourier's birth.[1] And, as you must suspect by now, Euler's formula plays a prominent role in that story. The prelude to the tale begins with a fundamental question: what is a *function?*

Modern analysts answer that question by saying that a function $f(t)$ is simply a *rule* that assigns a value to f that is determined by the value of t, that is, a function *maps t into f*. The mapping rule (the function) might be an analytical formula, but it doesn't have to be. It could, for example, take the form simply of a table or a list (perhaps infinitely large) of numbers; you look up the value of t on the left, and to the right is the value of f for that t. There were those in the eighteenth century, however, who did not accept such a broad interpretation. In particular, the French mathematician Jean Le Rond D'Alembert (1717–1783) held to the strict interpretation that a function *must* be expressible via the ordinary processes of algebra and calculus. Not all mathematicians of that time were so rigid, however, and in particular Euler was much more open-minded. Originally of the same view as D'Alembert, Euler eventually came to the conclusion that a function was defined if you could simply *draw* the curve of $f(t)$ versus t. This difference of interpretation led to a famous controversy in mathematical physics, and that in turn led to the formal development of Fourier series.

Before getting into that controversy, let me first tell you that, while Euler's liberal view of a function sounds plausible, there are astounding surprises associated with it. For example, "drawing a curve" implies that, at *nearly* every instant of time, there is a *direction* to the tip of the pencil or pen doing the drawing. That is, the curve has a *tangent* at nearly every point along the curve, which means the derivative exists at nearly every point. (If a curve has a *finite* number of points without a derivative, e.g., $f(t) = |t|$, which has no derivative at $t = 0$, we could of course still draw it.) In 1872, however, the German mathematician Karl Weirstrass (1815–1897) showed the possibility of a function that is *everywhere* continuous but which is *nowhere* differentiable, that is, it does not have a tangent *anywhere*, and so one could *not* draw it. This function is a trigonometric series (but it isn't, as you'll see later, a *Fourier* series): it is given by $\sum_{n=1}^{\infty} b^n \cos(a^n \pi x)$, where b is a fixed positive number less than one, and a is any fixed odd integer such that $ab > 1 + (3/2)\pi$.

Weirstrass's infinite sum has a lot of what engineers call "high-frequency content," which is the physical origin of its failure to have a derivative. (A "high frequency" function is, by definition, changing its value rapidly as the independent variable changes.) That is, since $1 + 3/2\pi \approx 5.7$, then the conditions $ab > 5.7$ and $b < 1$ mean that the *smallest a* can be is 7 (the first *odd* integer greater than 5.7). The frequencies of the terms of the sum increase as a^n, so for values of b just slightly less than one (which means the amplitude factor b^n doesn't get small very fast as n increases), we have the frequencies of the terms increasing as 7^n, which gets big, fast. Alternatively, if we select a small b to make the amplitudes decay quickly as n increases, then the smallest possible a is even larger than 7 and so the frequencies increase *even faster* than does 7^n. So, big or small for b, we cannot avoid a significant high-frequency presence in Weirstrass's function. It is that high frequency content, in the limit as $n \to \infty$, that results in a sum that has an infinite number of local extrema ("peaks and valleys") in any finite interval of x, no matter how small that interval may be, and it is *that* which causes the failure of the derivative to exist anywhere.

The inspiration for Weirstrass's very unusual function was the earlier function cooked-up in 1861 by the German mathematical genius G.F.B. Riemann (1826–1866), $\sum_{n=1}^{\infty} (\sin(n^2 x)/n^2)$, a continuous function that Riemann *speculated* would fail (but offered no proof) to have

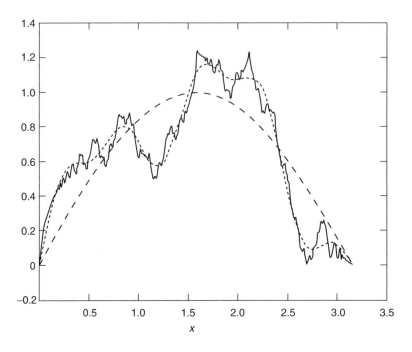

Figure 4.1.1. Riemann's function

a derivative anywhere.[2] This is a far simpler function than Weirstrass's: figure 4.1.1 shows that "simpler" is a relative term, however, because Riemann's function is pretty wild too—the figure shows three partial sums over the interval $0 < x < \pi$, using just the first term (dashed curve), the first three terms (dotted curve), and the first eighteen terms (solid curve).

With this preliminary discussion done, we are now ready to discuss the origin of the analyses that led to Fourier series; it was a problem in *physics*, motivated by musical instruments that have vibrating components, such as a piano wire or a violin string. That is, our question here is: how does a stretched, perfectly elastic string (with its two ends fixed), of uniform mass density along its length, move when set in motion? Certain specialized analyses had been carried out by the Swiss mathematician Johann Bernoulli (1667–1748) as early as 1728, who studied the motion of a *massless* string that supports a finite number of evenly spaced *point* masses (the so-called *loaded, ideal string*). It was, however, D'Alembert (in 1747), Euler (in 1748), and Johann's own son Daniel (in 1753) who were the major players in this story. And, as you'll see, while Euler and D'Alembert were "mostly right," they were also confused by just what is a function, and it was Daniel (1700–1782) who was the "most right" of the three. But first, to understand the mathematical positions of those three men, we need to discuss one of the most famous equations in physics, the so-called *one-dimensional wave equation.*

Imagine a perfectly elastic, stretched string initially lying motionless along the x-axis, with its two ends fixed at $x = 0$ and $x = l$. The *elastic* qualification means that the string is not able to support shear or bending forces (forces perpendicular to the string); it can support only forces along its own length (tension). The *stretched* qualification means that the motionless string has a built-in nonzero tension, which we'll denote as T. Let's further suppose that the mass per unit length of the stretched string is a constant, denoted by ρ. Now, imagine that at time $t = 0$ the string is given a *very slight* deflection so that it no longer lies entirely along the x-axis, except that the fixed ends are just that—*fixed.* What a *very slight* deflection means is that the tension at every point along the string remains *unchanged* by the resulting very small change in the length of the deflected string. The deflected string is illustrated in figure 4.1.2, which also shows a very short element of the string, between x and $x + \Delta x$. The curve describing the shape of

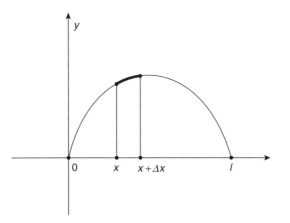

Figure 4.1.2. A vibrating string

the string is $y = y(x, t)$, that is, y is the deflection of the string from the horizontal axis for arbitrary values of x and t (notice, carefully, that we have *two* independent variables), and so the initial deflection is the given $y(x, 0) = f(x)$. This *temporal* requirement is called an *initial* condition. Another initial condition I'll impose here is that we'll release the deflected string *from rest*, that is, $\partial y / \partial t|_{t=0} = 0$ for all x. That is, we are modeling, here, what is called a *plucked* string. The derivative is a *partial* derivative because $y(x, t)$ is a function of more than one independent variable. The condition of *fixed ends* requires, in addition, that $y(0, t) = y(l, t) = 0$, for all $t \geq 0$, which are *spatial* requirements called *boundary* conditions.

What the early analysts wanted to calculate was $y(x, t)$ for all x in the interval 0 to l, for all $t > 0$. To do that they needed an equation for $y(x, t)$ that they could then solve, subject to the given initial and boundary conditions. That equation is a *partial* (because we have more than one independent variable) differential equation, called the *one-dimensional* (we have only one spatial variable, x) *wave* equation (you'll see why soon) that is actually pretty easy to derive. Here's how.

If we look at just the short string element of figure 4.1.2 in detail, then figure 4.1.3 shows that there are only *two* forces acting on it (there would be *three* if we consider the gravitational force on the string element, that is, the *weight* of the string element, but I'm going to ignore that

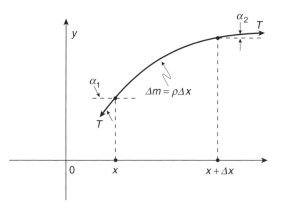

Figure 4.1.3. Free-body diagram of an isolated string element

complication). That is, figure 4.1.3 shows the string element isolated in what engineers and mathematicians call a *free-body diagram,* which completely describes the "world as seen by the string element." The angle α is the angle the tangent line to the curve of the string makes with the horizontal (obviously, $\alpha = \alpha(x, t)$), with α_1 and α_2 being the values of α at the left and right ends of the string element, respectively. The two forces acting on the string element are the tensions at each end, equal in magnitude (T) but at the angles α_1 and α_2, which are generally different because the string is, generally, curved.

Because of our assumption of small amplitude vibrations, we can take the length of the string element as very nearly always equal to Δx, and so the mass of the string element is simply $\Delta m = \rho \Delta x$. The vertical motion of the string, that is, its *vibration*, is described by Newton's second law of motion, his famous "force equals mass times acceleration,"[3] where the "force" is the net vertical force acting on the string element. This net vertical force is, from figure 4.1.3, given by $T\sin(\alpha_2) - T\sin(\alpha_1) = T[\sin(\alpha_2) - \sin(\alpha_1)]$. Because of our small amplitude assumption we can replace the sine with the tangent (you'll see why this is desirable in just a moment) and so, as the vertical acceleration is $\partial^2 y/\partial t^2$, we can write

$$\rho \Delta x \frac{\partial^2 y}{\partial t^2} = T[\tan(\alpha_2) - \tan(\alpha_1)],$$

or

$$\frac{\partial^2 y}{\partial t^2} = \frac{T}{\rho} \cdot \frac{\tan[\alpha(x + \Delta x)] - \tan[\alpha(x)]}{\Delta x}.$$

If we now imagine $\Delta x \to 0$, then the factor on the far right is the very definition of the (partial) derivative of $\tan(\alpha)$ with respect to x, that is,

$$\frac{\partial^2 y}{\partial t^2} = \frac{T}{\rho} \cdot \left\{ \frac{\partial}{\partial x} \tan(\alpha) \right\}.$$

To complete our derivation, all we need to do is to notice that the *geometry* of figure 4.1.3 says

$$\tan(\alpha) = \frac{\partial y}{\partial x}.$$

Now you can see why it was advantageous to replace the sine function with the tangent, because, at last, we have as our result that the equation of a vibrating string is the second order partial differential equation

$$\frac{\partial^2 y}{\partial t^2} = \frac{T}{\rho} \cdot \frac{\partial^2 y}{\partial x^2}.$$

That T/ρ factor is particularly interesting because of its *units*. A quick dimensional analysis shows us, in fact, that T/ρ has the units of

$$\frac{\text{force}}{\text{mass/length}} \implies \frac{\text{mass} \cdot \text{acceleration} \cdot \text{length}}{\text{mass}}$$

$$\implies \frac{\text{length} \cdot \text{length}}{(\text{time})^2}$$

$$\implies \left(\frac{\text{length}}{\text{time}}\right)^2 \implies (\text{speed})^2,$$

that is, T/ρ is a *speed* (of *something*) squared. *What* "something"? you might ask. We'll get to that soon. For now, let's just write $\sqrt{T/\rho} = c$ (with the units of speed), and so our vibrating string equation reduces to the form as you'll generally see it written in textbooks:

$$\boxed{\frac{\partial^2 y}{\partial x^2} = \frac{1}{c^2} \cdot \frac{\partial^2 y}{\partial t^2}}.$$

This is the equation that both D'Alembert (who first derived it in 1747) and Euler solved, in general, arriving at the same answer for $y(x, t)$ but with very different interpretations of what that answer *means*.

It is actually not difficult to understand the D'Alembert-Euler solution. Our basic assumption is that $y(x, t)$ is twice differentiable with respect to both x and t. With that assumption, we begin by immediately making a change of variables to

$$u = ct - x,$$
$$v = ct + x.$$

That is,

$$x = \frac{v - u}{2},$$

$$t = \frac{v + u}{2}.$$

Notice, carefully, that u and v are *independent* variables, just as x and t are. This is because knowledge of the value of u tells you nothing about the values of x and t, themselves, and so nothing of the value of v, and vice-versa.

Then, by the chain rule from calculus, we can write

$$\frac{\partial y}{\partial v} = \frac{\partial y}{\partial x} \cdot \frac{\partial x}{\partial v} + \frac{\partial y}{\partial t} \cdot \frac{\partial t}{\partial v} = \frac{\partial y}{\partial x} \cdot \frac{1}{2} + \frac{\partial y}{\partial t} \cdot \frac{1}{2c}.$$

And so, differentiating one more time and again using the chain rule,

$$\frac{\partial}{\partial u} \left(\frac{\partial y}{\partial v} \right) = \frac{\partial^2 y}{\partial u \partial v} = \frac{\partial}{\partial x} \left(\frac{\partial y}{\partial v} \right) \cdot \frac{\partial x}{\partial u} + \frac{\partial}{\partial t} \left(\frac{\partial y}{\partial v} \right) \cdot \frac{\partial t}{\partial u},$$

or, using our first result for $\partial y / \partial v$,

$$\frac{\partial^2 y}{\partial u \partial v} = \left[\frac{\partial^2 y}{\partial x^2} \cdot \frac{1}{2} + \frac{\partial^2 y}{\partial x \partial t} \cdot \frac{1}{2c} \right] \cdot \left(-\frac{1}{2} \right) + \left[\frac{\partial^2 y}{\partial t \partial x} \cdot \frac{1}{2} + \frac{\partial^2 y}{\partial t^2} \cdot \frac{1}{2c} \right]$$

$$= -\frac{1}{4} \cdot \frac{\partial^2 y}{\partial x^2} - \frac{1}{4c} \cdot \frac{\partial^2 y}{\partial x \partial t} + \frac{1}{4c} \cdot \frac{\partial^2 y}{\partial t \partial x} + \frac{1}{4c^2} \cdot \frac{\partial^2 y}{\partial t^2}$$

$$= -\frac{1}{4} \left\{ \frac{\partial^2 y}{\partial x^2} - \frac{1}{c^2} \cdot \frac{\partial^2 y}{\partial t^2} \right\}.$$

(Notice that I am assuming $\partial^2 y / \partial t \partial x = \partial^2 y / \partial x \partial t$, i.e., that the order of the two partial differentiations doesn't matter. That is not always true, but this assumption will not get us into trouble here.) The expression in the final set of braces is, of course, zero, since y *by definition* satisfies the differential equation of the vibrating string, and so

$$\frac{\partial^2 y}{\partial u \partial v} = 0.$$

This equation is now immediately integrable, *by inspection* (which is the reason for the change of variables from x and t to u and v!), to give

$$y(u, v) = \phi(u) + \psi(v),$$

where ϕ and ψ are each twice-differentiable (but otherwise arbitrary) functions of u and v, respectively. That is, the general solution to the wave equation is

$$y(x, t) = \phi(ct - x) + \psi(ct + x)$$

You can now see where the adjective *wave* comes from, as well as what the physical significance is of c. If one has a function $\psi(x)$, and if $x_0 > 0$, then $\psi(x_0 + x)$ is simply $\psi(x)$ shifted to the left by x_0 (if $x_0 < 0$, then the shift is actually to the right). Thus, $\psi(ct+x)$ is $\psi(x)$—which is $\psi(ct+x)$ at time $t = 0$—shifted to the left by the amount ct. That is, the *shape* $\psi(x)$ has traveled through distance ct in time t (that is, at a *speed* of c). A traveling shape is more commonly called a *wave*, and so *there's* the explanation for both the wave equation's name and of the physical meaning of c. The same argument applies to $\phi(ct - x)$, except of course that represents a wave traveling to the *right* at speed c. Now, at last, we are ready to do some mathematics.

The detailed plucked string problem facing D'Alembert was that of solving the equation

$$\frac{\partial^2 y}{\partial x^2} = \frac{1}{c^2} \cdot \frac{\partial^2 y}{\partial t^2},$$

subject to the boundary conditions

(1) $y(0, t) = 0$
(2) $y(l, t) = 0$

as well as the initial conditions

(3) $\partial y / \partial t|_{t=0} = 0$
(4) $y(x, 0) = f(x)$.

Applying (1) to the general solution in the above box gives

$$\phi(ct) + \psi(ct) = 0,$$

which says that $\phi = -\psi$, a conclusion that reduces the general solution to

$$y(x, t) = \phi(ct - x) - \phi(ct + x).$$

If we now apply (2) to this we arrive at

$$\phi(ct - l) = \phi(ct + l),$$

which is, since it is true for *any t*, an *identity* in *t*. There is a very interesting implication in this result, namely, that ϕ must be a *periodic function* of time (a function that endlessly repeats its values from some finite interval of time of duration *T*). I'll discuss periodic functions in more detail in the next section but, quickly, a function $s(t)$ is said to be periodic with period *T* if there is a $T > 0$ such that $s(t) = s(t+T)$ *for all t* (which is just the situation we have for ϕ). The important observation for us is simply that the difference between the two arguments of $s(t)$ and $s(t+T)$ is the *period*. So, since $\phi(ct - l)$ and $\phi(ct + l)$ have arguments differing by $2l$, $2l$ is the value of cT (where *T* is the period of ϕ), that is, the period of ϕ is $2l/c$, the time it takes a wave to travel from one end of the string to the other end, and back.

We could continue on in this fashion, applying next the boundary conditions (3) and (4) to the general solution, but I'm not going to other. D'Alembert had at this point clearly shown the periodic nature of the solutions (which, of course, is what one actually observes when looking at a vibrating string), and so he believed he had found *the* solution; a solution nicely expressed by a well-behaved, twice-differentiable function. Euler took exception with that, however, observing that it is perfectly obvious that a string can be set into motion from an initial state that is *not* twice differentiable, for instance, a plucked string as shown in figure 4.1.4. Therefore, concluded Euler, while D'Alembert's solution is *a* solution, it could not be the *general* solution. To further complicate matters, just a few years later Daniel Bernoulli solved the wave equation in an entirely different way and arrived at a *completely different looking* solution. That was, as you can appreciate, a very puzzling development, and it had far-reaching implications in mathematics, ones that continue to this day.

Bernoulli's solution to the wave equation can be established by *separating the variables*, that is, by assuming that $y(x, t) = X(x)T(t)$. This is, in fact, the standard approach used in virtually every textbook on partial differential equations published during the last one hundred and fifty years. $X(x)$ and $T(t)$ are functions *only* of *x* and *t*, respectively (e.g., *xt* is obviously separable, while x^t is not). Substituting this $y(x, t)$ into the

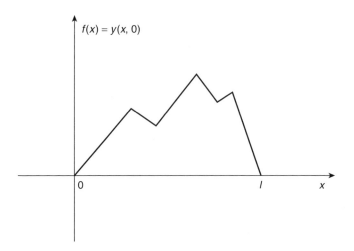

Figure 4.1.4. An initial string deflection which is not twice
differentiable at all points

wave equation gives us

$$T\frac{d^2X}{dx^2} = \frac{1}{c^2} \cdot X\frac{d^2T}{dt^2},$$

or

$$\frac{1}{X}\frac{d^2X}{dx^2} = \frac{1}{c^2} \cdot \frac{1}{T}\frac{d^2T}{dt^2}.$$

Since the left-hand side is a function only of x, and the right-hand side
is a function only of t, then the only way both sides can be equal for *all*
x and *all* t is if they are each equal to the same *constant*; let's call that
constant k. Thus,

$$\frac{d^2X}{dx^2} - kX = 0,$$

$$\frac{d^2T}{dt^2} - kc^2T = 0.$$

Whenever a mathematician sees a function that is proportional to its
own derivative (or second derivative, in this case), the idea of *exponential*
pops into her mind. So, to solve a generic equation that models both of

our equations, let's imagine that we have

$$\frac{d^2 Z}{dz^2} - aZ = 0,$$

and assume that $Z(z) = Ce^{\alpha z}$, where C and α are constants. Then,

$$\alpha^2 C e^{\alpha z} - aCe^{\alpha z} = 0,$$

$$\alpha^2 - a = 0,$$

$$\alpha = \pm\sqrt{a},$$

and so the general solution is

$$Z(z) = Ae^{z\sqrt{a}} + Be^{-z\sqrt{a}},$$

where A and B are constants. So, for our functions $X(x)$ and $T(t)$, we have (with $a = k$ for X and $a = kc^2$ for T)

$$X(x) = A_1 e^{x\sqrt{k}} + B_1 e^{-x\sqrt{k}},$$

$$T(t) = A_2 e^{ct\sqrt{k}} + B_2 e^{-ct\sqrt{k}}$$

and so

$$y(x, t) = [A_1 e^{x\sqrt{k}} + B_1 e^{-x\sqrt{k}}][A_2 e^{ct\sqrt{k}} + B_2 e^{-ct\sqrt{k}}].$$

Using boundary condition (1), $y(0, t) = 0$, we thus have, *for all $t \geq 0$,*

$$[A_1 + B_1][A_2 e^{ct\sqrt{k}} + B_2 e^{-ct\sqrt{k}}] = 0,$$

meaning either that $A_2 = B_2 = 0$ (which we reject since that results in the trivial solution of $y(x, t) = 0$), or that $A_1 + B_1 = 0$ (which we accept because it leads to a *non*-trivial solution). That is, $B_1 = -A_1$ or, dropping subscripts,

$$y(x, t) = A[e^{x\sqrt{k}} - e^{-x\sqrt{k}}][A_2 e^{ct\sqrt{k}} + B_2 e^{-ct\sqrt{k}}].$$

Turning next to boundary condition (2), $y(l, t) = 0$, we have

$$A[e^{l\sqrt{k}} - e^{-l\sqrt{k}}][A_2 e^{ct\sqrt{k}} + B_2 e^{-ct\sqrt{k}}] = 0,$$

again for all $t \geq 0$. This is cause for some alarm because it asks for something that is simply impossible(!), except for the trivial cases of $A = 0$ and/or $A_2 = B_2 = 0$ (*trivial* because these lead to $y(x, t) = 0$, which certainly does satisfy initial condition (3) but does *not* satisfy initial condition (4) unless we have the physically uninteresting case of $f(x) = 0$, a *non*-vibrating string). Our quandary is resolved once we realize we have been tacitly assuming the arbitrary constant k is *positive*. But nothing requires that; suppose instead that we assume $k < 0$, and so replace $\sqrt{k}$ with $i\sqrt{k}$ where now $k > 0$. Then,

$$y(x, t) = A[e^{ix\sqrt{k}} - e^{-ix\sqrt{k}}][A_2 e^{ict\sqrt{k}} + B_2 e^{-ict\sqrt{k}}],$$

and now $y(l, t) = 0$ leads to a nontrivial conclusion. That is, using Euler's formula on the first expression in brackets gives

$$y(x, t) = 2iA \sin(x\sqrt{k})[A_2 e^{ict\sqrt{k}} + B_2 e^{-ict\sqrt{k}}],$$

and so, for *all* t,

$$2iA \sin(l\sqrt{k})[A_2 e^{ict\sqrt{k}} + B_2 e^{-ict\sqrt{k}}] = 0,$$

which says $l\sqrt{k} = n\pi$, where n is *any* nonzero integer. That is, our "arbitrary" constant k isn't all that arbitrary at all; rather, it is

$$k = \frac{n^2 \pi^2}{l^2}, n = \cdots, -2, -1, 1, 2, \cdots.$$

And so, absorbing the $2i$ into the constant A, we have

$$y(x, t) = A \sin\left(\frac{n\pi}{l} x\right) \left[A_2 e^{ict(n\pi/l)} + B_2 e^{-ict(n\pi/l)}\right].$$

Thus,

$$\frac{\partial y}{\partial t} = A \sin\left(\frac{n\pi}{l} x\right) \left[A_2 \frac{icn\pi}{l} e^{ict(n\pi/l)} - B_2 \frac{icn\pi}{l} e^{-ict(n\pi/l)}\right],$$

and then applying initial condition (3) gives us, for all x,

$$A \sin\left(\frac{n\pi}{l} x\right) \frac{icn\pi}{l} [A_2 - B_2] = 0.$$

This is true if $A = 0$ (which we reject because it leads to the trivial $y(x, t) = 0$) or if $A_2 = B_2$. So, again dropping subscripts, we have

$$y(x, t) = A \sin\left(\frac{n\pi}{l}x\right) B\frac{icn\pi}{l}\left[e^{ict(n\pi/l)} + e^{-ict(n\pi/l)}\right]$$

$$= AB \sin\left(\frac{n\pi}{l}x\right) \frac{icn\pi}{l} 2\cos\left(\frac{nc\pi}{l}t\right)$$

or, absorbing all the constants into just one constant, which may be different for each value of n and so we'll call them c_n, we have

$$y_n(x, t) = c_n \sin\left(\frac{n\pi}{l}x\right)\cos\left(\frac{n\pi c}{l}t\right).$$

Our most general solution is the sum of all these particular solutions indexed on n. That is,

$$y(x, t) = \sum_{n=-\infty}^{\infty} y_n(x, t) = \sum_{n=-\infty}^{\infty} c_n \sin\left(\frac{n\pi}{l}x\right)\cos\left(\frac{n\pi c}{l}t\right).$$

We can simplify this a bit by noticing that, for any particular pair of values for $n = \pm k$, we can write the sum of those two terms as

$$c_{-k} \sin\left(\frac{-k\pi}{l}x\right)\cos\left(\frac{-k\pi c}{l}t\right) + c_k \sin\left(\frac{k\pi}{l}x\right)\cos\left(\frac{k\pi c}{l}t\right)$$

$$= (c_k - c_{-k}) \sin\left(\frac{k\pi}{l}x\right)\cos\left(\frac{k\pi c}{l}t\right).$$

Thus, simply redefining our arbitrary constants and observing that the $n = 0$ term in the summation is zero, we can start the sum at $n = 1$ and write, as did Bernoulli,

$$y(x, t) = \sum_{n=1}^{\infty} c_n \sin\left(\frac{n\pi}{l}x\right)\cos\left(\frac{n\pi c}{l}t\right).$$

This would be our solution *if* we knew what that infinity(!) of arbitrary c_n coefficients are. We can start to answer that by applying the one remaining initial condition (4), $y(x, 0) = f(x)$, to get

$$f(x) = \sum_{n=1}^{\infty} c_n \sin\left(\frac{n\pi}{l} x\right),$$

an astounding result that claims we should (if we just knew the values of those c_n) be able to add up an infinity of various amplitude sinusoidal functions to get the given *arbitrary* $f(x)$—which is, remember, the shape of the plucked string just before we release it.

Briefly putting aside the question of the c_n, let me first tell you that Euler didn't like Bernoulli's solution at all, arguing that it was clearly absurd to imagine adding up an infinity of *odd, periodic* functions (the sines) and to believe you could arrive at the arbitrary initial string deflection $f(x)$, since $f(x)$ is generally neither odd nor periodic. Bernoulli's trigonometric expansion, said Euler, couldn't be true *in general* (but might be okay for *particular* $f(x)$). To his credit, Bernoulli stood his ground and said that, no matter what recognized masters like Euler and D'Alembert (who agreed with Euler's rejection of Bernoulli's solution) might say, his solution *was* right. I'll discuss in section 4.3 how Euler's objections lose all their force when the concept of a function is properly understood.

Now, as my final comment in this section, it is amusing to note that it was the nonbeliever Euler who took the last step and showed how to calculate the c_n (for the particular trigonometric expansions of $f(x)$ that Euler thought did make sense). In a paper written in 1777 (but not published until 1793) Euler evaluated the c_n coefficients using what is now the standard textbook approach. I'll defer the details until later, however, to when we get to the work of Fourier; for now, what we need to do next is to take a closer look at periodic functions, in general.

4.2 Periodic functions and Euler's sum.

As you saw in the previous section, expressing functions as trigonometric series has a long, pre-Fourier history. The French mathematician Charles Bossut (1730–1814), for example, wrote a number of such series with a finite number (n) of terms, and, in 1733 Daniel Bernoulli took Bossut's formulas and simply let $n \to \infty$. Bernoulli was pretty casual with his manipulations (a common feature of eighteenth-century mathematics, and of Euler's in particular), and his results make sense only by making the meaningless argument that $\sin(\infty)$ and $\cos(\infty)$ are zero.[4] Even before Bernoulli, however, we can find the fingerprints of Euler on Fourier series.

For example, in a 1744 letter to a friend Euler wrote the following remarkable claim:

$$x(t) = \frac{\pi - t}{2} = \sum_{n=1}^{\infty} \frac{\sin(nt)}{n} = \sin(t) + \frac{\sin(2t)}{2} + \frac{\sin(3t)}{3} + \cdots.$$

This was probably (almost certainly) the first "Fourier series," although of course Euler didn't call it that since Fourier wouldn't be born until twenty-four years later. In addition to this glaring apparent anachronism (at the end of this section I'll show you how Euler derived this Fourier series decades before Fourier), there appears to be a more immediate technical question—is it even *correct*? It looks, for example, like there is a dramatic problem at $t = 0$, as Euler's formula seems to be a claim that $\pi/2 = 0$ (which I think we'd all agree isn't right). The key to understanding just what is going on here is the concept of a periodic function, and that's where I'll start, with a little review.

If $x(t)$ is a *periodic* function with fundamental period T (which we'll take to be a real number), then the function satisfies the condition

$$x(t) = x(t + T), -\infty < t < \infty,$$

where T is the *smallest* possible *positive* T, i.e., $T > 0$. (Demanding $T > 0$ eliminates the specific case of $x(t) = $ constant being called a periodic function; there is no *smallest* $T > 0$ such that $x(t) = x(t+T)$, because for every $T > 0$ there is yet a smaller $T > 0$, for instance, the T that is half of the previous T.) In simple language, a periodic function is one that endlessly "repeats itself" in both directions as the independent variable (in our case here, t) goes off to $\pm\infty$.

Probably the most obvious such functions are sine and cosine; for example, $x(t) = \sin(t)$ is periodic with period $T = 2\pi$. Notice that $\sin(t) = \sin(t + k2\pi)$ for all t for *any* positive integer k, but we take the *smallest* such k ($k = 1$) as defining the *fundamental* period T. The concept of periodicity is so simple that it is easy to fall into the trap of thinking it is nearly trivial. Believe me, it isn't! For example, what's your answer to this: is the sum of two periodic functions periodic? Most people blurt out a *yes, of course* with hardly a missed heartbeat, but the correct answer is *it depends.* That is, sometimes *yes* and other times *no.* For example, both $\cos(t)$ and $\cos(t\sqrt{2})$ are periodic; $\cos(t)$ has period 2π and $\cos(t\sqrt{2})$ has period $2\pi/\sqrt{2}$. But their sum is *not* periodic. Here's why. Let's suppose there *is* a T such that

$$x(t) = \cos(t) + \cos(t\sqrt{2}) = x(t + T) = \cos(t + T) + \cos\{(t + T)\sqrt{2}\}.$$

This is to be true for *all* t, and in particular for $t = 0$, which says that

$$x(0) = \cos(0) + \cos(0) = 2 = \cos(T) + \cos(T\sqrt{2}).$$

Since the maximum value of the *cosine* function is one, then it follows that $\cos(T) = \cos(T\sqrt{2}) = 1$. That is, there must be two (obviously different) integers m and n such that $T = 2\pi n$ and $T\sqrt{2} = 2\pi m$. But *that* means

$$\frac{T\sqrt{2}}{T} = \frac{2\pi m}{2\pi n} = \sqrt{2} = \frac{m}{n},$$

which says $\sqrt{2}$ is rational. But it *isn't*! (See the discussion on this very point in the Introduction.) We thus have a contradiction, the origin of which is our initial assumption (false, as we now know) that there *is* a period (T) to our $x(t)$. Figure 4.2.1 shows plots of $\cos(t)$, $\cos(t\sqrt{2})$, and their sum, and the sum plot "looks" anything *but* periodic. The general rule (of which you should be able to convince yourself with just a little thought) is that a sum of periodic terms will itself be periodic *if* all possible ratios of the individual periods, taken two at a time, are rational.

Here's another "simple" question about periodicity. Can two periodic functions $x_1(t)$ and $x_2(t)$ add to give a sum that is periodic with a period *less* than either of the original periods, that is, if T_1 and T_2 are the periods of $x_1(t)$ and $x_2(t)$, respectively, and if $x_1(t) + x_2(t)$ is periodic with period T, is it possible to have $T < \min(T_1, T_2)$? The answer is *yes* (which often surprises too-quick-to-answer students), as shown by the example of figure 4.2.2.

Sometimes students think that last example to be just a bit of a "cheat" since it depends on the two pulse-like functions to be *precisely* aligned,

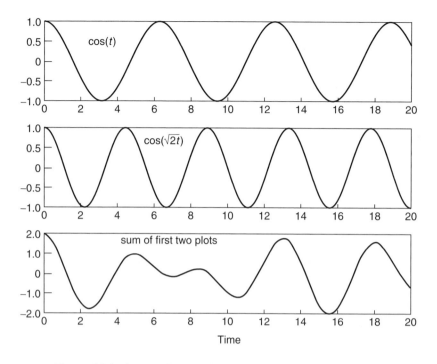

Figure 4.2.1. A sum of two periodic functions that is *not* periodic

with the pulses of one function occurring *exactly* midway between the pulses of the other function. If such a perfect alignment is absent, then the sum is still periodic but with a period that is the same as that of the original functions. So, here's an even more astounding example, of two functions of equal, *fixed* fundamental periods that have a sum with a fundamental period that can be *as small as we wish*. Suppose that N is any positive integer, and we define $x_1(t)$ and $x_2(t)$ as follows:

$$
x_1(t) = \begin{cases} \sin(2N\pi t), & 0 \le t \le \dfrac{1}{N}, \\ 0, & \dfrac{1}{N} < t < 1, \\ x_1(t+1), & -\infty < t < \infty \end{cases}
$$

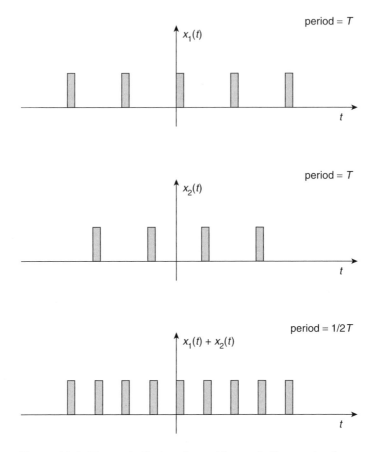

Figure 4.2.2. Two periodic functions with a periodic sum that has a *smaller* period

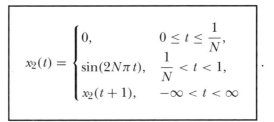

$$x_2(t) = \begin{cases} 0, & 0 \le t \le \dfrac{1}{N}, \\ \sin(2N\pi t), & \dfrac{1}{N} < t < 1, \\ x_2(t+1), & -\infty < t < \infty \end{cases}.$$

Clearly, $x_1(t)$ and $x_2(t)$ are each periodic with a fundamental period of one, independent of the value of N. But, consider their sum,

$$x_1(t) + x_2(t) = \begin{cases} \sin(2N\pi t), & 0 \le t < 1, \\ x_1(t+1) + x_2(t+1), & -\infty < t < \infty \end{cases}.$$

This sum obviously has fundamental period T, where $2N\pi T = 2\pi$, that is, $T = 1/N$. Thus, we can make T *as small as desired* by making N arbitrarily large.

Now, just one more question: can a *time-varying* $x(t)$ satisfy the condition $x(t) = x(t + T)$ for all t and yet not have a fundamental period (remember, the condition $T > 0$ eliminates *constant* functions from being called periodic)? This is a bit harder to see, but again the answer is *yes*. Here's an example. Define $x_1(t)$ and $x_2(t)$ as follows:

$$x_1(t) = \begin{cases} 1 & \text{if } t \text{ is an integer,} \\ 0 & \text{if } t \text{ is not an integer,} \end{cases},$$

$$x_2(t) = \begin{cases} 1 & \text{if } t \text{ is rational but } not \text{ an integer,} \\ 0 & \text{if } t \text{ is irrational } or \text{ is an integer} \end{cases}.$$

It is obvious that $x_1(t)$ is a perfectly "respectable" function, that is, we can actually sketch it on a piece of paper. On the other hand, $x_2(t)$ is pretty wild (try sketching it!) because between any two rationals there are an infinite number of irrationals (and vice versa). Nevertheless, $x_1(t)$ and $x_2(t)$ are each periodic with equal fundamental periods of $T = 1$ (just apply the very definition of periodicity). Now, consider their sum, which we can write as

$$s(t) = x_1(t) + x_2(t) = \begin{cases} 1 & \text{if } t \text{ is rational} \\ 0 & \text{if } t \text{ is irrational} \end{cases}.$$

This sum satisfies the periodicity condition $s(t) = s(t+T)$ for T *any* rational number, but there is no fundamental period for the sum because there is no *smallest positive* rational. The examples given here are, admittedly, carefully crafted to have particularly weird behaviors; engineers and mathematicians alike call them *pathological,* a term motivated by the medical meaning of pathology, the science of disease.

Now, to end this section, let me show you how Euler arrived at the trigonometric series that opened this discussion. When we are done you'll see the explanation for that puzzling special case at $t = 0$, which seems to claim that $\pi/2 = 0$ "!" In addition, this calculation is a good example of Euler's spectacularly inventive genius that, while often successful, would earn him some raised eyebrows from modern mathematicians. Euler started with the geometric series

$$S(t) = e^{it} + e^{i2t} + e^{i3t} + \cdots$$

and then, oblivious to any questions of convergence (e.g., $S(0) = 1+1+1+1+\cdots = \infty$), he "summed" it in the usual way. That is, multiplying through by e^{it} he arrived at

$$e^{it} S(t) = e^{i2t} + e^{i3t} + \cdots,$$

and so

$$S(t) - e^{it} S(t) = e^{it}.$$

Thus,

$$S(t) = \frac{e^{it}}{1 - e^{it}} = \frac{e^{it}(1 - e^{-it})}{(1 - e^{it})(1 - e^{-it})} = \frac{e^{it} - 1}{1 - e^{it} - e^{-it} + 1};$$

then, using his own formula that names this book to unwrap the complex exponentials,

$$S(t) = \frac{\cos(t) + i\sin(t) - 1}{2 - (e^{it} + e^{-it})}$$

$$= \frac{\cos(t) - 1 + i\sin(t)}{2 - 2\cos(t)} = \frac{-[1 - \cos(t)] + i\sin(t)}{2[1 - \cos(t)]}$$

and so, finally,

$$S(t) = -\frac{1}{2} + i\frac{1}{2} \cdot \frac{\sin(t)}{1 - \cos(t)}.$$

Returning now to Euler's original geometric series for $S(t)$, he again used his formula to write

$$S(t) = \cos(t) + \cos(2t) + \cdots + i\{\sin(t) + \sin(2t) + \cdots\},$$

and then equated the real part of this with the real part of the expression for $S(t)$ in the box. That is, he wrote

$$\cos(t) + \cos(2t) + \cos(3t) + \cdots = -\frac{1}{2}.$$

Then (being a genius!) he indefinitely integrated term by term to arrive at

$$\sin(t) + \frac{\sin(2t)}{2} + \frac{\sin(3t)}{3} + \cdots = -\frac{1}{2}t + C,$$

where C is the arbitrary constant of indefinite integration. To find C, Euler observed that if one substitutes $t = \pi/2$ this last expression becomes

$$1 - \frac{1}{3} + \frac{1}{5} - \frac{1}{7} + \cdots = -\frac{\pi}{4} + C.$$

The expression on the left is equal to $\pi/4$ (we used this fact in section 2.1, and we'll *derive* it later in this chapter), which means that C must equal $\pi/2$. Thus, as Euler wrote his friend,

$$\sum_{n=1}^{\infty} \frac{\sin(nt)}{n} = \sin(t) + \frac{\sin(2t)}{2} + \frac{\sin(3t)}{3} + \cdots$$

$$= -\frac{1}{2}t + \frac{\pi}{2} = \frac{\pi - t}{2} = x(t).$$

One can only look at all this, mouth agape, with a combination of awe and horror (in about equal parts). It is something that only a genius or a failing calculus student could have written! In the next section I'll show you the modern derivation of this formula, but we can do something right now with this astounding formula that Euler could only have dreamed about doing—we can perform several million arithmetic operations in mere seconds on a computer, *plot* the left-hand side (the actual summation of the sines) as a function of t, and then simply see if it *looks* like the plot for $(\pi - t)/2$. Figures 4.2.3 through 4.2.6 do just

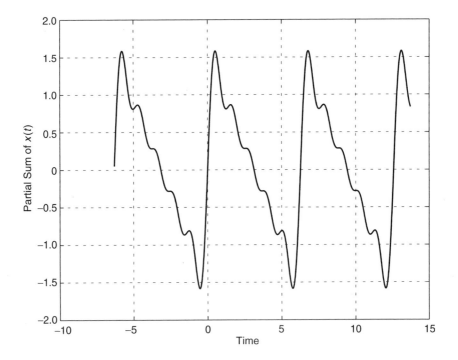

Figure 4.2.3. Sum of first five terms of Euler's series

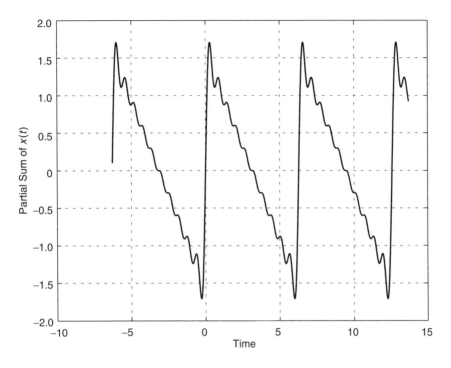

Figure 4.2.4. Sum of first ten terms of Euler's series

that, using progressively more terms to form partial sums. Even just a glance at these plots prompts several interesting observations:

(1) The Euler series is a periodic function, with period 2π.
(2) The series is an odd function.
(3) In the interval $0 < t < 2\pi$ the series *does* approximate $(\pi - t)/2$.
(4) The series makes very sudden jumps between its most negative and most positive values at the beginning (end) of each new period.
(5) Looking at the period $0 < t < 2\pi$, where the series does approximate $(\pi - t)/2$, the approximation does *not* improve *everywhere* in that interval with an increasing number of terms in the partial sum. (What *are* those wiggles around $t = 0$ and $t = 2\pi$, and why don't they go away as we add more terms?)

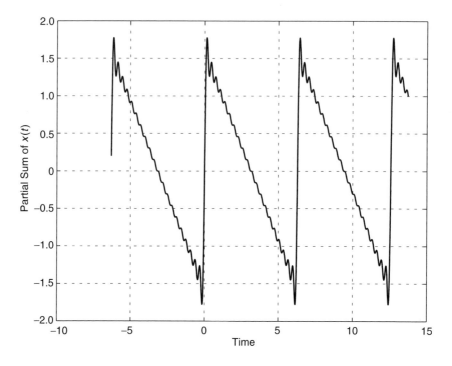

Figure 4.2.5. Sum of first twenty terms of Euler's series

Observations (1) and (2) should really be no surprise—after all, the terms in the series are all sine functions (which are, individually, odd), with periods whose ratios taken two at a time are all rational. Observation (3) is the one of real interest here, of course, as it seems to verify Euler's (outrageous) calculations, *provided* we add the qualification that the series approximates $(\pi - t)/2$ if $0 < t < 2\pi$ (but the approximation fails if t is outside of that interval). And finally, observations (4) and (5) are most curious, indeed, representing behavior that Euler completely missed, as did all those who came after him for many decades, including Fourier himself. Observations (1), (2), and (3) will be addressed in more detail in the next section. Observations (4) and (5) are given a section of their own (section 4.4) because not only is there some quite interesting mathematics involved, there is also a fascinating *human* story of a talented mathematician that has been almost lost to history (until now).

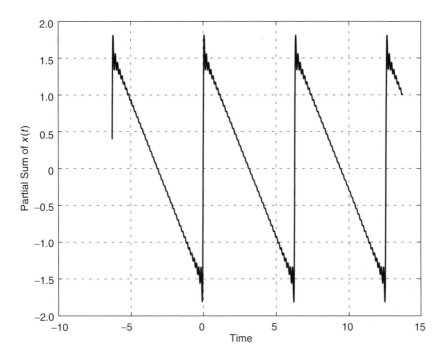

Figure 4.2.6. Sum of first forty terms of Euler's series

4.3 Fourier's theorem for periodic functions and Parseval's theorem.

In December 1807, twenty-five years after Daniel Bernoulli's death (during which matters concerning trigonometric series had remained mostly a subject of confusion), Joseph Fourier presented an astonishing paper to the Academy of Science in Paris. In it he asserted that any arbitrary function could be written in the form of such a series. Indeed, he went even further than that (which had already been the much earlier position of Bernoulli, of course). Fourier specifically stated that, contrary to Euler's view, an infinity of individually odd sine functions could represent non-odd functions, and the same could be said for cosine trigonometric series. Indeed (according to Fourier), one could even express the sine *itself* as the sum of an infinity of *cosines*. I'll show you how in the next section.

These statements drove Fourier's fellow countryman, the aging grand master of French mathematics Joseph-Louis Lagrange (1736–1813) to assert—in very strong terms—that Fourier's claims were simply impossible. It is actually no surprise that Lagrange should have felt that way—early in his career, a half-century before, he had sided with D'Alembert and Euler in opposing Bernoulli's trigonometric series solution to the wave equation. For Lagrange (with a nod to Yogi Berra), Fourier's claims for trigonometric series must have seemed like déjà vu all over again.

Fourier's claims were based on his work with a different partial differential equation from physics, the so-called *heat equation*, which he studied in its one-, two-, and three-dimensional forms. Like the wave equation, it yields to separation of variables. In the one-dimensional form, it is

$$\frac{\partial^2 u}{\partial x^2} = \frac{1}{k} \cdot \frac{\partial u}{\partial t},$$

where $u(x, t)$ is the temperature along a heat conductive (thin) rod, at position x and time t.[5] There are, of course, initial and boundary conditions that go along with this fundamental equation for any particular problem, just as with the wave equation. The physical properties of the rod material (heat capacity, cross-sectional area, and thermal conductivity) are contained in the constant k. This equation is also called the *diffusion equation*, as it describes how many quantities (in addition to heat) spread or *diffuse* with time through space. In its three-dimensional form, for example, it describes how an ink drop spreads through water, or how a drop of cream or a dissolving sugar cube spreads through a cup of coffee. The diffusion equation was brilliantly used in the 1850s by the Scottish mathematical physicist William Thomson (1804–1907), later Lord Kelvin, to describe how electricity "spreads" through a very long telegraph cable (in particular, the Atlantic cable that electrically connected England and the North American continent[6]). Thomson was an early admirer of Fourier's work—his first published paper, written when he was just *fifteen*, was on Fourier series. Later in his career Thomson used the heat equation in an attempt to calculate the age of the Earth by rational science rather than through biblical study. In that paper he called Fourier's theory a "mathematical poem."[7]

As mentioned not all were convinced by Fourier's 1807 paper;[8] still, his critics did make the question of the theory of heat propagation the Academy's subject for its mathematics prize competition of 1812. That was probably done as an attempt to encourage Fourier to clarify and expand on his ideas, and if so it worked—Fourier submitted just such an entry and won the prize. Alas, his critics were still not convinced and his new paper was not published. Then, in 1822, Fourier published his work in book form, *Théorie analytique de la chaleur*, and it was no longer possible to push aside what he had done. It was that book, in fact, that fell into the hands of the young William Thomson and so inspired him. Fourier's book is enormous, in both size and scope, and it is impossible to do it justice in my very limited space here, and so I'll limit myself to some of the more dramatic mathematical aspects of it.[9] In the rest of this section we'll suppose that we have a function $f(t)$ whose values are defined over some interval of finite length. It is *only* in this interval that we really care about what $f(t)$ is "doing."

First, with no loss of generality, we can imagine that we are working with a function $f(t)$ defined on the interval $0 < t < L$. Starting the interval at zero can be justified by simply *defining* the instant $t = 0$ to be the start of the interval, and ending the interval at L can be justified by defining our time scale (the "unit of time") properly. Or, if you prefer, we can map any finite interval $a < t < b$ into $0 < t < L$ with the simple linear rule

$$\frac{L}{b-a}(t-a) \longrightarrow t,$$

as this maps $t = a$ into $t = 0$ and $t = b$ into $t = L$.

Second—this appears to be the seminal insight due to Fourier alone, an insight missed by Euler, D'Alembert, Bernoulli, Lagrange, and everyone else until Fourier—we can extend the definition of $f(t)$ from the interval $0 < t < L$ to the interval $-L < t < L$ by defining $f(t)$ in the extension $-L < t < 0$ *any way we want*. This is because, since we are actually only interested in $f(t)$ in the interval $0 < t < L$, *who cares* how the extended $f(t)$ behaves in the interval $-L < t < 0$? The answer is, *nobody cares*. The reason for making such an extension is that if we make it cleverly then we can reap interesting, indeed astonishing, mathematical simplicity. And valuable results, too. (I'll show you some examples of this soon.) Two special ways of extending $f(t)$ into $-L < t < 0$ are to make an *even* extension (as shown in figure 4.3.1) or an *odd* extension (as shown in figure 4.3.2).

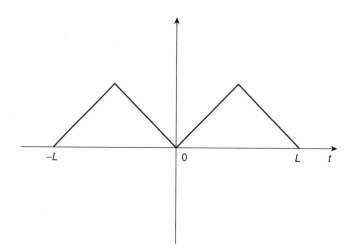

Figure 4.3.1. The even extension of $f(t)$

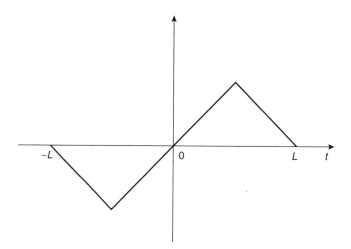

Figure 4.3.2. The odd extension of $f(t)$

Now, one last step. Let's consider the behavior of $f(t)$ in the interval $-L < t < L$ as defining a single period of the *periodic extension* of our extended $f(t)$, that is, let's extend the behavior of $f(t)$ over the interval $-L < t < L$ to $-\infty < t < \infty$. We thus have a periodic function with period $T = 2L$ and so, according to Fourier, the periodic extension of $f(t)$ (which from now on will be what I mean when writing $f(t)$) can be expressed as the trigonometric series

$$f(t) = a_0 + \sum_{k=1}^{\infty} \{a_k \cos(k\omega_0 t) + b_k \sin(k\omega_0 t)\},$$

where the infinity of *a*s and *b*s are all constants and $\omega_0 = 2\pi/T$ (which is called the *fundamental frequency*). It is very convenient, at this point, to write the sum in terms of complex exponentials because they are generally easier to work with than are sines and cosines. So, using Euler's formula,

$$f(t) = a_0 + \sum_{k=1}^{\infty} \left\{ a_k \frac{e^{ik\omega_0 t} + e^{-ik\omega_0 t}}{2} + b_k \frac{e^{ik\omega_0 t} - e^{-ik\omega_0 t}}{2i} \right\}$$

$$= a_0 + \sum_{k=1}^{\infty} \left\{ \frac{a_k}{2} + \frac{b_k}{2i} \right\} e^{ik\omega_0 t} + \left\{ \frac{a_k}{2} - \frac{b_k}{2i} \right\} e^{-ik\omega_0 t}.$$

Then, letting our summation index run from $k = -\infty$ to $k = \infty$, we can write $f(t)$ as

$$\boxed{f(t) = \sum_{k=-\infty}^{\infty} c_k e^{ik\omega_0 t}, \quad \omega_0 T = 2\pi}\quad,$$

where the c_k are constants (generally, *complex* constants). Indeed, if $f(t)$ is a real-valued function of *t*, as it will be for us, *usually* (but *not always*—read on for a beautiful example of the "real value" of a complex-valued function), and since the conjugate of a real quantity is equal to itself (the *conjugate* of $x + iy$, written as $(x + iy)^*$, is $x - iy$), we see that, whatever the coefficients c_k are, it must be true that

$$f(t) = \sum_{k=-\infty}^{\infty} c_k e^{ik\omega_0 t} = f^*(t) = \sum_{k=-\infty}^{\infty} c_k^* e^{-ik\omega_0 t}.$$

This, in turn, says $c_k^* = c_{-k}$ (just set the coefficients of matching exponential terms on each side of the equality equal to each other). And notice, too, that for the special case of $k = 0$ we have $c_0^* = c_0$, which says that for *any* real-valued $f(t)$ it will always be true that c_0 is real.

With all of this preliminary stuff behind us we can no longer avoid the question of just *what are* the Fourier coefficients—what are the c_k? They are actually easy to determine. For k any particular integer from $-\infty$ to ∞ (let's say, $k = n$), multiply both sides of the boxed expression for $f(t)$ by $e^{-n\omega_0 t}$ and integrate over a period, that is, over *any* interval of length T. Then,

$$\int_{t'}^{t'+T} f(t)e^{-in\omega_0 t}\, dt = \int_{t'}^{t'+T} \left\{ \sum_{k=-\infty}^{\infty} c_k e^{ik\omega_0 t} \right\} e^{-in\omega_0 t}\, dt$$

$$= \sum_{k=-\infty}^{\infty} c_k \int_{t'}^{t'+T} e^{i(k-n)\omega_0 t}\, dt.$$

The last integral is easy to do; we'll consider the two cases of $n \neq k$ and $n = k$ separately. First, if $n \neq k$ then

$$\int_{t'}^{t'+T} e^{i(k-n)\omega_0 t}\, dt = \left(\frac{e^{i(k-n)\omega_0 t}}{i(k-n)\omega_0} \right) \Bigg|_{t'}^{t'+T}$$

$$= \frac{e^{i(k-n)\omega_0(t'+T)} - e^{i(k-n)\omega_0 t'}}{i(k-n)\omega_0}$$

$$= \frac{e^{i(k-n)\omega_0 t'}[e^{i(k-n)\omega_0 T} - 1]}{i(k-n)\omega_0}.$$

But, since $\omega_0 T = 2\pi$, and since $k - n$ is an integer, Euler's formula says we have $e^{i(k-n)\omega_0 T} = 1$, and so the integral is zero. On the other hand, if $n = k$ then the integral becomes

$$\int_{t'}^{t'+T} e^0\, dt = (t|_{t'}^{t'+T} = T.$$

In summary, then,

$$\int_{\text{period}} e^{i(k-n)\omega_0 t}\, dt = \begin{cases} T, & k \neq n, \\ 0, & k = n. \end{cases}$$

Thus,

$$\int_{\text{period}} f(t)e^{-in\omega_0 t}\, dt = c_n T,$$

so, at last, we have the elegant result that the Fourier coefficients are given by

$$\boxed{\; c_n = \tfrac{1}{T} \cdot \int_{\text{period}} f(t)e^{-in\omega_0 t}\, dt,\; \omega_0 T = 2\pi \;}.$$

As I mentioned at the end of section 4.1, this approach to finding the coefficients in the trigonometric series expansion for a periodic $f(t)$ is due to Euler.[10]

Now we can see how a modern analyst would derive Euler's 1744 series discussed in the previous section. As the plots there show (see figures 4.2.3 through 4.2.6 again), Euler was actually working with a *periodic* function, with period $T = 2\pi$ (one such period is $0 < t < 2\pi$). We therefore have $\omega_0 = 1$ (from $\omega_0 T = 2\pi$) and so, in the interval $0 < t < 2\pi$,

$$f(t) = \frac{\pi - t}{2} = \sum_{n=-\infty}^{\infty} c_n e^{int},$$

where

$$c_n = \frac{1}{2\pi}\int_0^{2\pi} \frac{\pi - t}{2} e^{-int}\, dt = \frac{1}{4}\int_0^{2\pi} e^{-int}\, dt - \frac{1}{4\pi}\int_0^{2\pi} t e^{-int}\, dt.$$

The $n = 0$ case is easy to do:

$$c_0 = \frac{1}{4}\int_0^{2\pi} dt - \frac{1}{4\pi}\int_0^{2\pi} t\, dt = \frac{1}{4}\left(t \Big|_0^{2\pi} \right) - \frac{1}{4\pi}\left(\frac{1}{2}t^2 \Big|_0^{2\pi} \right)$$

$$= \frac{2\pi}{4} - \frac{4\pi^2}{8\pi} = 0.$$

For the case of $n \neq 0$, the first integral for c_n is

$$\frac{1}{4} \int_0^{2\pi} e^{-int} \, dt = \frac{1}{4} \left(\frac{e^{-int}}{-in} \Big|_0^{2\pi} \right) = \frac{e^{-i2\pi n} - 1}{-i4n} = 0,$$

since $e^{-i2\pi n} = 1$ for *any* integer n. Thus, our formula for c_n reduces to

$$c_n = -\frac{1}{4\pi} \int_0^{2\pi} t e^{-int} \, dt.$$

Using integration by parts (or simply a good integral table) we find that, for any constant a,

$$\int t e^{at} \, dt = \frac{e^{at}}{a} \left(t - \frac{1}{a} \right).$$

So, setting $a = -in$, we have

$$c_n = -\frac{1}{4\pi} \left\{ \frac{e^{-int}}{-in} \left(t - \frac{1}{-in} \right) \right\} \Big|_0^{2\pi}$$

$$= -\frac{i}{4\pi n} \left\{ e^{-int} \left(t - \frac{i}{n} \right) \right\} \Big|_0^{2\pi}$$

$$= -\frac{i}{4\pi n} \left[e^{-i2\pi n} \left(2\pi - \frac{i}{n} \right) + \frac{i}{n} \right],$$

or, as $e^{-i2\pi n} = 1$,

$$c_n = -\frac{i}{4\pi n} \left(2\pi - \frac{i}{n} + \frac{i}{n} \right) = -\frac{i}{2n}.$$

Thus, in the interval $0 < t < 2\pi$ we have

$$f(t) = \frac{\pi - t}{2} = \sum_{n=-\infty, \neq 0}^{\infty} -\frac{i}{2n} e^{int},$$

or, writing the summation out in *pairs* of terms ($n = \pm 1, \pm 2, \pm 3$, etc.) we have

$$
\begin{aligned}
\frac{\pi - t}{2} &= -\frac{i}{2}\left[\left(\frac{e^{it} - e^{-it}}{1}\right) + \left(\frac{e^{i2t} - e^{-i2t}}{2}\right) + \left(\frac{e^{i3t} - e^{-i3t}}{3}\right)\right. \\
&\qquad \left. + \left(\frac{e^{i4t} - e^{-i4t}}{4}\right) + \cdots\right] \\
&= -\frac{i}{2}\left[2i\sin(t) + \frac{2i\sin(2t)}{2} + \frac{2i\sin(3t)}{3} + \frac{2i\sin(4t)}{4} + \cdots\right] \\
&= \sin(t) + \frac{\sin(2t)}{2} + \frac{\sin(3t)}{3} + \frac{\sin(4t)}{4} + \cdots,
\end{aligned}
$$

which is the expression Euler wrote in his 1744 letter. Our derivation here, of course, has avoided the brilliant (but outrageous) excesses of Euler's derivation given in the previous section. And notice one little tidbit we get "for free," so to speak: if we set $t = \pi/2$, then

$$
\frac{\pi}{4} = \sin\left(\frac{\pi}{2}\right) + \frac{\sin(\pi)}{2} + \frac{\sin(3\pi/2)}{3} + \frac{\sin(2\pi)}{4} + \frac{\sin(5\pi/2)}{5} + \cdots,
$$

or

$$
\frac{\pi}{4} = 1 - \frac{1}{3} + \frac{1}{5} - \frac{1}{7} + \cdots,
$$

which is usually called Leibniz's series, after the German mathematician Gottfried Leibniz (1646–1716)—you'll recognize it as having been used before in this book without proof.

We can do a couple of very interesting things with Euler's series; first, integrating both sides from 0 to x we get

$$
\int_0^x \frac{\pi - t}{2}\,dt = \int_0^x \sum_{n=1}^{\infty} \frac{\sin(nt)}{n}\,dt
$$

$$
= \frac{\pi}{2}x - \frac{x^2}{4} = \sum_{n=1}^{\infty}\frac{1}{n}\int_0^x \sin(nt)\,dt
$$

$$= \sum_{n=1}^{\infty} \frac{1}{n} \left(-\frac{\cos(nt)}{n} \right) \Big|_0^x = \sum_{n=1}^{\infty} \frac{1 - \cos(nx)}{n^2}$$

$$= \sum_{n=1}^{\infty} \frac{1}{n^2} - \sum_{n=1}^{\infty} \frac{\cos(nx)}{n^2}.$$

As Euler showed in 1734 (and we'll derive it, too, later in this section),

$$\sum_{n=1}^{\infty} \frac{1}{n^2} = \frac{\pi^2}{6},$$

and so

$$\boxed{\sum_{n=1}^{\infty} \frac{\cos(nx)}{n^2} = \frac{\pi^2}{6} - \frac{\pi}{2} x + \frac{x^2}{4} = \frac{3x^2 - 6\pi x + 2\pi^2}{12}}.$$

Setting $x = \pi/2$, we get

$$\sum_{n=1}^{\infty} \frac{\cos(n\pi/2)}{n^2} = -\frac{1}{2^2} + \frac{1}{4^2} - \frac{1}{6^2} + \frac{1}{8^2} - \cdots = \frac{3\pi^2/4 - 6\pi^2/2 + 2\pi^2}{12},$$

or

$$-\frac{1}{(1 \cdot 2)^2} + \frac{1}{(2 \cdot 2)^2} - \frac{1}{(3 \cdot 2)^2} + \frac{1}{(4 \cdot 2)^2} - \cdots = \frac{3\pi^2/4 - \pi^2}{12} = -\frac{\pi^2}{48},$$

or

$$-\left[\frac{1}{4 \cdot 1^2} - \frac{1}{4 \cdot 2^2} + \frac{1}{4 \cdot 3^2} - \frac{1}{4 \cdot 4^2} + \cdots \right] = -\frac{\pi^2}{48},$$

or, at last, the beautiful formula

$$\boxed{\frac{1}{1^2} - \frac{1}{2^2} + \frac{1}{3^2} - \frac{1}{4^2} + \cdots = \frac{\pi^2}{12}}.$$

This is easily and quickly "checked" on a computer ($\pi^2/12 = 0.82246703342411$ while the first ten thousand terms of the alternating series itself directly sum to 0.822467028442461).

Now, go back to the boxed expression for $\sum_{n=1}^{\infty} \cos(nx)/n^2$ and integrate *it* from 0 to u:

$$\int_0^u \sum_{n=1}^{\infty} \frac{\cos(nx)}{n^2} \, dx = \int_0^u \frac{3x^2 - 6\pi x + 2\pi^2}{12} \, dx$$

$$= \sum_{n=1}^{\infty} \frac{1}{n^2} \int_0^u \cos(nx) \, dx$$

$$= \left(\frac{x^3}{12} - \frac{\pi x^2}{4} + \frac{\pi^2 x}{6} \right) \Big|_0^u,$$

or

$$\sum_{n=1}^{\infty} \frac{1}{n^2} \left(\frac{\sin(nx)}{n} \right) \Big|_0^u = \sum_{n=1}^{\infty} \frac{\sin(nu)}{n^3} = \frac{u^3}{12} - \frac{\pi u^2}{4} + \frac{\pi^2 u}{6}.$$

And so, setting $u = \pi/2$,

$$\sum_{n=1}^{\infty} \frac{\sin(n\pi/2)}{n^3} = \frac{1}{1^3} - \frac{1}{3^3} + \frac{1}{5^3} - \frac{1}{7^3} + \cdots = \frac{\pi^3}{12 \cdot 8} - \frac{\pi^3}{4 \cdot 4} + \frac{\pi^3}{12}$$

$$= \frac{\pi^3}{32}.$$

Again, this is easily checked; $\pi^3/32 = 0.968946146259937$ while the first one thousand terms of the alternating series itself directly sum to 0.96894614619687. Our formula in this case is particularly interesting because the sum of the reciprocals of *all* the integers cubed is still an open question,

$$\sum_{n=1}^{\infty} \frac{1}{n^3} = \frac{1}{1^3} + \frac{1}{2^3} + \frac{1}{3^3} + \frac{1}{4^3} + \cdots = ?$$

and our calculation, of the reciprocals of the *odd* integers cubed with alternating signs would seem to be closely related. But, as Groucho Marx used to say, "close, but no cigar." One could continue on in this fashion literally forever—integrating and inserting particular values for the independent variable—thereby deriving an infinity of ever more exotic

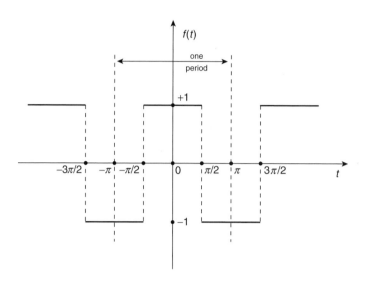

Figure 4.3.3. A "square-wave"

formulas involving the sums of the reciprocals of ever higher powers of the integers, but I think I'll stop here with that sort of thing and move on to something else.

An historically important example (that we'll return to in the next section) is the periodic function with a period defined as

$$f(t) = \begin{cases} +1, & |t| < \dfrac{\pi}{2}, \\ -1, & \dfrac{1}{2}\pi < |t| < \pi, \end{cases}$$

and illustrated in figure 4.3.3 (the illustration makes it obvious, I think, why engineers, physicists, and mathematicians call this periodic function a *square wave*). Therefore, following Fourier (with $T = 2\pi$ and $\omega_0 = 1$),

$$f(t) = \sum_{n=-\infty}^{\infty} c_n e^{int},$$

$$c_n = \frac{1}{2\pi} \int_{-\pi}^{\pi} f(t) e^{-int} \, dt.$$

The coefficients are particularly easy to calculate for this particular problem, as we have for $n = 0$

$$c_0 = \frac{1}{2\pi} \int_{-\pi}^{\pi} f(t)\,dt = 0,$$

and for $n \neq 0$

$$
\begin{aligned}
c_n &= \frac{1}{2\pi} \left[-\int_{-\pi}^{-\pi/2} e^{-int}\,dt + \int_{-\pi/2}^{\pi/2} e^{-int}\,dt - \int_{\pi/2}^{\pi} e^{-int}\,dt \right] \\
&= \frac{1}{2\pi} \left[\left(\frac{e^{-int}}{in} \Big|_{-\pi}^{-\pi/2} \right) - \left(\frac{e^{-int}}{in} \Big|_{-\pi/2}^{\pi/2} \right) + \left(\frac{e^{-int}}{in} \Big|_{\pi/2}^{\pi} \right) \right] \\
&= \frac{1}{2\pi in} [e^{in\pi/2} - e^{in\pi} - e^{-in\pi/2} + e^{in\pi/2} + e^{-in\pi} - e^{-in\pi/2}] \\
&= \frac{1}{2\pi in} [2(e^{in\pi/2} - e^{-in\pi/2}) - (e^{in\pi} - e^{-in\pi})] \\
&= \frac{1}{2\pi in} \left[4i \sin\left(n\frac{\pi}{2}\right) - 2i\sin(n\pi) \right] = \frac{2}{n\pi} \sin\left(n\frac{\pi}{2}\right).
\end{aligned}
$$

Therefore,

$$
\begin{aligned}
f(t) &= \frac{2}{\pi} \sum_{n=-\infty,\, n\neq 0}^{\infty} \frac{\sin(n\pi/2)}{n} e^{-int} \\
&= \frac{2}{\pi} \left[(e^{it} + e^{-it}) + \left(\frac{-e^{i3t}}{3} + \frac{e^{-i3t}}{-3} \right) + \left(\frac{e^{i5t}}{5} + \frac{-e^{-i5t}}{-5} \right) + \cdots \right] \\
&= \frac{2}{\pi} \left[2\cos(t) - 2\frac{\cos(3t)}{3} + 2\frac{\cos(5t)}{5} - \cdots \right] \\
&= \frac{4}{\pi} \left[\cos(t) - \frac{\cos(3t)}{3} + \frac{\cos(5t)}{5} - \cdots \right].
\end{aligned}
$$

Notice that substituting $t = 0$ (which means $f = +1$) gives us Leibniz's series again. Figure 4.3.4 shows a partial sum (with ten terms) of this Fourier series and, as with Euler's series, we again see those curious wiggles in the neighborhoods of sudden transitions of the function.

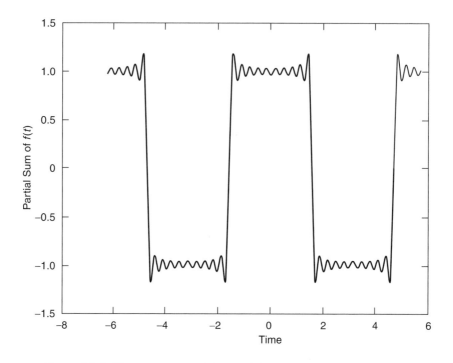

Figure 4.3.4. The ten-term partial sum of the square-wave Fourier series

A famous result lies hidden in this Fourier series. Since $f^2(t) = 1$ for all t,

$$\frac{16}{\pi^2}\left[\cos(t) - \frac{1}{3}\cos(3t) + \frac{1}{5}\cos(5t) - \cdots\right]^2 = 1.$$

Now, suppose we integrate term by term on both sides of this equality, from 0 to 2π (when we square the left-hand side we get the squares of all the individual terms *plus* all possible cross products). Using the easy-to establish result

$$\int_0^{2\pi} \cos(mt)\cos(nt)\,dt = \begin{cases} 0, & m \neq n, \\ \pi, & m = n \end{cases}$$

(just replace the cosines with their complex exponential equivalents and integrate), we arrive at

$$\frac{16}{\pi^2}\left[\pi + \pi\frac{1}{3^2} + \pi\frac{1}{5^2} + \cdots\right] = 2\pi,$$

or

$$\frac{1}{1^2} + \frac{1}{3^2} + \frac{1}{5^2} + \cdots = \frac{\pi^2}{8}.$$

That is, we have the value of the sum of the reciprocals of the odd integers squared, which we write as

$$\sum_{n=1}^{\infty}\frac{1}{(2n-1)^2} = \frac{\pi^2}{8}.$$

Now, since all the integers can be separated into two sets, the evens and the odds, we can write the sum of the reciprocals of *all* the integers squared as (the first sum on the right of the first equality is the evens, and the second sum is the odds)

$$\sum_{n=1}^{\infty}\frac{1}{n^2} = \sum_{n=1}^{\infty}\frac{1}{(2n)^2} + \sum_{n=1}^{\infty}\frac{1}{(2n-1)^2} = \frac{1}{4}\sum_{n=1}^{\infty}\frac{1}{n^2} + \frac{\pi^2}{8},$$

or

$$\frac{3}{4}\sum_{n=1}^{\infty}\frac{1}{n^2} = \frac{\pi^2}{8},$$

which immediately gives us Euler's result, which made him world-famous (his derivation was, of course, based on a *non*-Fourier method):

$$\sum_{n=1}^{\infty}\frac{1}{n^2} = \frac{\pi^2}{6}.$$

I used this result, you'll recall, earlier in this section, and we can easily "check" it:

$$\frac{\pi^2}{6} = 1.64493406684823$$

while the first ten thousand terms of the series directly sum to 1.64483407184807.

Finally, here's a really spectacular use of Fourier series. Let $f(t) = \cos(\alpha t)$, $-\pi < t < \pi$, where α is any real, *non*-integer (you'll see why soon) number. If we imagine that we periodically extend this $f(t)$ onto $-\infty < t < \infty$, with period 2π, then $\omega_0 = 1$ and

$$f(t) = \cos(\alpha t) = \sum_{n=-\infty}^{\infty} c_n e^{int},$$

where

$$
\begin{aligned}
c_n &= \frac{1}{2\pi} \int_{-\pi}^{\pi} \cos(\alpha t) e^{-int}\, dt \\[2mm]
&= \frac{1}{2\pi} \int_{-\pi}^{\pi} \frac{e^{i\alpha t} + e^{-i\alpha t}}{2} e^{-int}\, dt \\[2mm]
&= \frac{1}{4\pi} \left[\int_{-\pi}^{\pi} e^{i(\alpha-n)t}\, dt + \int_{-\pi}^{\pi} e^{-i(\alpha+n)t}\, dt \right] \\[2mm]
&= \frac{1}{4\pi} \left[\frac{e^{i(\alpha-n)t}}{i(\alpha-n)} + \frac{e^{-i(\alpha+n)t}}{-i(\alpha+n)} \right]_{-\pi}^{\pi} \\[2mm]
&= \frac{1}{4\pi i} \left[\frac{e^{i(\alpha-n)\pi} - e^{-i(\alpha-n)\pi}}{\alpha-n} - \frac{e^{-i(\alpha+n)\pi} - e^{i(\alpha+n)\pi}}{\alpha+n} \right] \\[2mm]
&= \frac{1}{4\pi i} \cdot \frac{
\begin{array}{c}
\alpha e^{i\alpha\pi} 2\cos(n\pi) - \alpha e^{-i\alpha\pi} 2\cos(n\pi) \\
- n e^{i\alpha\pi} 2i\sin(n\pi) - n e^{-i\alpha\pi} 2i\sin(n\pi)
\end{array}
}{\alpha^2 - n^2} \\[2mm]
&= \frac{1}{4\pi i} \cdot \frac{2\alpha \cos(n\pi)(e^{i\alpha\pi} - e^{-i\alpha\pi})}{\alpha^2 - n^2} \\[2mm]
&= \frac{2\alpha \cos(n\pi) 2i \sin(\alpha\pi)}{4\pi i(\alpha^2 - n^2)} \\[2mm]
&= \frac{\alpha \cos(n\pi)\sin(\alpha\pi)}{\pi(\alpha^2 - n^2)} = \frac{\alpha(-1)^n \sin(\alpha\pi)}{\pi(\alpha^2 - n^2)}.
\end{aligned}
$$

So,

$$\cos(\alpha t) = \sum_{n=-\infty}^{\infty} \frac{\alpha(-1)^n \sin(\alpha\pi)}{\pi(\alpha^2 - n^2)} e^{int},$$

or

$$\cos(\alpha t) = \frac{\alpha \sin(\alpha\pi)}{\pi\alpha^2} + \frac{1}{\pi} \sum_{n=1}^{\infty} \frac{\alpha(-1)^n \sin(\alpha\pi)}{\pi(\alpha^2 - n^2)} [e^{int} + e^{-int}],$$

or

$$\cos(\alpha t) = \frac{\sin(\alpha\pi)}{\pi\alpha} + \frac{\alpha \sin(\alpha\pi)}{\pi} \sum_{n=1}^{\infty} \frac{(-1)^n}{\alpha^2 - n^2} 2\cos(nt),$$

or,

$$\boxed{\cos(\alpha t) = \frac{\sin(\alpha\pi)}{\pi} \left[\frac{1}{\alpha} + 2\alpha \sum_{n=1}^{\infty} \frac{(-1)^n}{\alpha^2 - n^2} \cos(nt) \right]}.$$

This last expression is stuffed with marvelous special cases. For example, if $t = 0$ then

$$\frac{\pi}{\sin(\alpha\pi)} = \frac{1}{\alpha} + 2\alpha \sum_{n=1}^{\infty} \frac{(-1)^n}{\alpha^2 - n^2},$$

which, for $\alpha = 1/2$, reduces to

$$\pi = 2 + \sum_{n=1}^{\infty} \frac{(-1)^n}{(1/2)^2 - n^2} = 2 + 4 \sum_{n=1}^{\infty} \frac{(-1)^n}{1 - 4n^2}.$$

This is easily "checked" numerically; using the first ten thousand terms of the series gives, for the right-hand side, a value of 3.14159264859029, which *is* pretty nearly equal to pi. Or, if we set $t = \pi$ in the boxed expression, then as $\cos(n\pi) = (-1)^n$ (and as $(-1)^n(-1)^n = (-1)^{2n} = 1$) we have

$$\cos(\alpha\pi) = \frac{\sin(\alpha\pi)}{\pi} \left[\frac{1}{\alpha} + 2\alpha \sum_{n=1}^{\infty} \frac{1}{\alpha^2 - n^2} \right],$$

or, alternatively, we have the dazzling result

$$\frac{\pi}{\tan(\alpha\pi)} = \frac{1}{\alpha} + 2\alpha \sum_{n=1}^{\infty} \frac{1}{\alpha^2 - n^2}.$$

Before doing any further particular Fourier series expansions, let me show you an extremely useful formula that holds for *all* Fourier series. To begin, I'll define the so-called *energy* of the function $f(t)$, *in a period*, as

$$W = \int_{period} f^2(t)\, dt.$$

The origin of this definition is from physics, but we can treat it here as simply a definition. (You'll soon see that it is an extraordinarily good definition because we will be able to use it to do some wonderful, *easy* calculations.) Next, let's insert the Fourier series expansion for $f(t)$ into the energy integral. Writing $f^2(t) = f(t) \cdot f(t)$ and using a different index of summation for each $f(t)$, we have

$$W = \int_{period} \left\{ \sum_{m=-\infty}^{\infty} c_m e^{im\omega_0 t} \right\} \left\{ \sum_{n=-\infty}^{\infty} c_n e^{in\omega_0 t} \right\} dt$$

$$= \sum_{m=-\infty}^{\infty} \sum_{n=-\infty}^{\infty} c_m c_n \int_{period} e^{i(m+n)\omega_0 t}\, dt.$$

As we've already seen, when deriving the general formula for the Fourier coefficients, this last integral is zero for m and n such that $m + n \neq 0$, and is equal to T (the period) if $m + n = 0$, that is, if $m = -n$. Our result thus reduces to what is called *Parseval's formula*, named after the French mathematician Antoine Parseval des Chenes (1755–1836),

$$W = \sum_{k=-\infty}^{\infty} c_k c_{-k} T = T \sum_{k=-\infty}^{\infty} c_k c_k^* = T \sum_{k=-\infty}^{\infty} |c_k|^2,$$

where I've used the fact that $c_{-k} = c_k^*$ if $f(t)$ is a real-valued function. This result is often written in the alternative form, once it is noticed that $|c_{-k}| = |c_k|$, as

$$\frac{W}{T} = c_0^2 + 2 \sum_{k=1}^{\infty} |c_k|^2 = \frac{1}{T} \int_{\text{period}} f^2(t)\, dt,$$

and W/T is called the *power* of the function $f(t)$, because in physics power is energy per unit time.

Here's a really spectacular application of Parseval's formula. Let $f(t) = e^{-t}$ over the interval $0 < t < 2\pi$, and then imagine that the definition is extended over the real line, $-\infty < t < \infty$, with $0 < t < 2\pi$ representing one period. We can express the resulting periodic function as the Fourier series (remember, $\omega_0 T = 2\pi$ and so, in this problem, we have $\omega_0 = 1$)

$$f(t) = \sum_{n=-\infty}^{\infty} c_n e^{int},$$

where

$$c_n = \frac{1}{2\pi} \int_0^{2\pi} e^{-t} e^{-int}\, dt.$$

The integral is easy to do:

$$c_n = \frac{1}{2\pi} \int_0^{2\pi} e^{-(1+in)t}\, dt = \frac{1}{2\pi} \left\{ \frac{e^{-(1+in)t}}{-(1+in)} \Bigg|_0^{2\pi} \right.$$

$$= \frac{1}{2\pi} \cdot \frac{1 - e^{-(1+in)2\pi}}{1+in}$$

$$= \frac{1}{2\pi} \cdot \frac{1 - e^{-2\pi} e^{-in2\pi}}{1+in} = \frac{1 - e^{-2\pi}}{2\pi(1+in)}.$$

Thus,

$$|c_n|^2 = \frac{(1 - e^{-2\pi})^2}{4\pi^2(1 + n^2)},$$

and so

$$\sum_{n=-\infty}^{\infty} |c_n|^2 = \frac{(1 - e^{-2\pi})^2}{4\pi^2} \sum_{n=-\infty}^{\infty} \frac{1}{1+n^2}.$$

By Parseval's formula this is equal to

$$\frac{1}{T} \int_{\text{period}} f^2(t)\, dt = \frac{1}{2\pi} \int_0^{2\pi} e^{-2t}\, dt$$

$$= \frac{1}{2\pi} \left(-\frac{e^{-2t}}{2} \Big|_0^{2\pi} \right)$$

$$= \frac{1}{4\pi}(1 - e^{-4\pi}) = \frac{1}{4\pi}(1 - e^{-2\pi})(1 + e^{-2\pi}).$$

That is,

$$\frac{1}{4\pi}(1 - e^{-2\pi})(1 + e^{-2\pi}) = \frac{(1 - e^{-2\pi})^2}{4\pi^2} \cdot \sum_{n=-\infty}^{\infty} \frac{1}{1+n^2}$$

and, suddenly, we have the pretty result that

$$\sum_{n=-\infty}^{\infty} \frac{1}{1+n^2} = \pi \frac{1 + e^{-2\pi}}{1 - e^{-2\pi}}.$$

I'll show you an even *more* beautiful generalization of this in section 5.5. Of course, one might wonder if this result is "right," and two such easy-to-calculate expressions simply cry out for a numerical test:

$$\pi \frac{1 + e^{-2\pi}}{1 - e^{-2\pi}} = 3.15334809493716,$$

while letting n run from $-100,000$ to $100,000$ (there is nothing special about $100,000$—I simply picked a number "big enough" to get a reasonably good approximation to the actual value of the sum) gives

$$\sum_{n=-100,000}^{100,000} \frac{1}{1+n^2} = 1 + 2 \sum_{n=1}^{100,000} \frac{1}{1+n^2} = 3.15332809503716.$$

As wonderful as this last example is, let me now show you another use of Parseval's formula which I think tops it. First, let's extend our

definition of the energy per period of a periodic function to include
complex-valued functions. One way to do this, while keeping the energy
itself real (thereby keeping at one toe in the "real world"), is to write

$$W = \int_{\text{period}} f(t)f^*(t)\, dt = \int_{\text{period}} |f(t)|^2\, dt.$$

This reduces, if $f(t)$ is real, to our earlier definition for the energy of a
real-valued $f(t)$. With our extended definition of energy we have

$$W = \int_{\text{period}} \left\{ \sum_{m=-\infty}^{\infty} c_m e^{im\omega_0 t} \right\} \left\{ \sum_{n=-\infty}^{\infty} c_n e^{in\omega_0 t} \right\}^*\, dt,$$

and, since the conjugate of a sum is the sum of the conjugates and the
conjugate of a product is the product of the conjugates,

$$W = \int_{\text{period}} \left\{ \sum_{m=-\infty}^{\infty} c_m e^{im\omega_0 t} \right\} \left\{ \sum_{n=-\infty}^{\infty} c_n^* e^{-in\omega_0 t} \right\}\, dt$$

$$= \sum_{m=-\infty}^{\infty} \sum_{n=-\infty}^{\infty} c_m c_n^* \int_{\text{period}} e^{i(m-n)\omega_0 t}\, dt.$$

Again, since the integral is zero for m and n such that $m - n \neq 0$, and is
equal to T if $m - n = 0$ (that is, if $m = n$), we have

$$W = \sum_{k=-\infty}^{\infty} c_k c_{-k}^* T = T \sum_{k=-\infty}^{\infty} c_k c_{-k}^* = T \sum_{k=-\infty}^{\infty} |c_k|^2,$$

which is, again, Parseval's formula.

Notice, *very carefully*, that in this derivation we did *not* use $c_{-k} = c_k^*$,
which is good since for complex-valued functions it isn't true! This is
quite interesting, in fact, because since we have now shown that Parseval's
formula is true for complex-valued functions (which include real-valued
functions as a special case), our earlier use of $c_{-k} = c_k^*$ for real-valued
functions was actually demanding *more* of the Fourier coefficients than
necessary. That is, analyzing the problem for the *general* case requires
fewer conditions than does the analysis for the *special* case! Yes, the
mathematics gods do often work in mysterious ways.

Now, suppose $f(t) = e^{i\alpha t}$ over the interval $-\pi < t < \pi$, where α is any real constant that is not an integer (you'll see why we need this condition soon), and then imagine that $f(t)$ is periodically extended over the entire real line $-\infty < t < \infty$. We thus have a *complex*-valued periodic function with period 2π. Since $\omega_0 = 1$, Fourier's theorem tells us that

$$f(t) = \sum_{k=-\infty}^{\infty} c_k e^{ikt},$$

where

$$c_k = \frac{1}{2\pi} \int_{-\pi}^{\pi} e^{i\alpha t} e^{-ikt} \, dt = \frac{1}{2\pi} \int_{-\pi}^{\pi} e^{i(\alpha-k)t} \, dt = \frac{1}{2\pi} \left\{ \frac{e^{i(\alpha-k)t}}{i(\alpha - k)} \right\} \Bigg|_{-\pi}^{\pi}$$

$$= \frac{1}{2\pi} \cdot \frac{e^{i(\alpha-k)\pi} - e^{-i(\alpha-k)\pi}}{i(\alpha - k)}$$

$$= \frac{1}{2\pi} \cdot \frac{i2 \sin\{(\alpha - k)\pi\}}{i(\alpha - k)} = \frac{\sin\{\pi(\alpha - k)\}}{\pi(\alpha - k)}.$$

If we assume that α is any real number *not* equal to an integer, then we see that this result for c_k is *always* defined (that is, there is no integer k for which the denominator is zero), and so our expression for c_k holds for *all* integer k.

The energy of our complex-valued function is

$$W = \int_{\text{period}} f(t) f^*(t) \, dt = \int_{-\pi}^{\pi} e^{i\alpha t} e^{-i\alpha t} \, dt = \int_{-\pi}^{\pi} dt = 2\pi.$$

Parseval's formula then tells us that

$$\frac{W}{T} = \sum_{k=-\infty}^{\infty} |c_k|^2 = \frac{2\pi}{2\pi} = 1 = \sum_{k=-\infty}^{\infty} \frac{\sin^2\{\pi(\alpha - k)\}}{\pi^2(\alpha - k)^2}.$$

Or, if we make the change of variable $\alpha = u/\pi$, then

$$1 = \sum_{k=-\infty}^{\infty} \frac{\sin^2\{\pi((u/\pi) - k)\}}{\pi^2((u/\pi) - k)^2} = \sum_{k=-\infty}^{\infty} \frac{\sin^2\{u - k\pi\}}{(u - k\pi)^2}.$$

Since $\sin^2(u - k\pi) = \sin^2(u)$ for all integer k,

$$1 = \sum_{k=-\infty}^{\infty} \frac{\sin^2(u)}{(u - k\pi)^2} = \sin^2(u) \sum_{k=-\infty}^{\infty} \frac{1}{(u + k\pi)^2},$$

where the second equality follows since the summation index k runs through *all* the integers, positive *and* negative. And so, finally, writing $u = \pi\alpha$, we have

$$\frac{1}{\sin^2(\pi\alpha)} = \sum_{k=-\infty}^{\infty} \frac{1}{(\pi\alpha + k\pi)^2},$$

which becomes the beautiful identity (known by other means to Euler as early as 1740)[11]

$$\sum_{k=-\infty}^{\infty} \frac{1}{(k + \alpha)^2} = \frac{\pi^2}{\sin^2(\pi\alpha)}.$$

Inserting particular values for α reduces this result to specific, testable cases. For example, if $\alpha = 1/2$, then

$$\sum_{k=-\infty}^{\infty} \frac{1}{(k + 1/2)^2} = \pi^2.$$

I personally don't find this very "obvious" and, you too might well ask, *is it true?* Repeating our earlier approach of doing a numerical check, we find that

$$\pi^2 = 9.86960440108936,$$

while

$$\sum_{k=-10,000}^{10,000} \frac{1}{(k + 1/2)^2} = 9.86940441108855.$$

Well, I'm convinced! I hope you are, too. But maybe not.

So, let me end this section with the admission that I have willingly (perhaps a little too willingly) avoided saying anything about the *convergence* of the Fourier series I have been manipulating with abandon.

Certain *sufficiency* conditions *have* been established that ensure convergence; these are the so-called *strong Dirichlet conditions,* formulated in
1829. If $f(t)$ is a periodic function, then the partial sums of its Fourier
series do indeed converge to $f(t)$ for all t (other than at discontinuities)
if, besides being absolutely integrable,

   (1) $f(t)$ has a finite number of finite discontinuities in a period

and

   (2) $f(t)$ has a finite number of extrema in a period.

These are called strong conditions because they are known *not* to be
necessary conditions, that is, there are $f(t)$ that do not satisfy all the
strong conditions but nonetheless do have convergent Fourier series
for all t (other than at discontinuities). The strong conditions are sufficient conditions that are mild enough so that all functions an engineer
would encounter in "real life" do satisfy them. (Functions that don't
satisfy all of the strong Dirichlet conditions are sufficiently weird that
they are called pathological. For example, figure 4.3.5 shows one period
of $f(t) = t * \sin(1/t)$, a function with an *infinite* number of extrema in
the interval $0 < t < 1$; indeed, there are an infinity of extrema in *any*

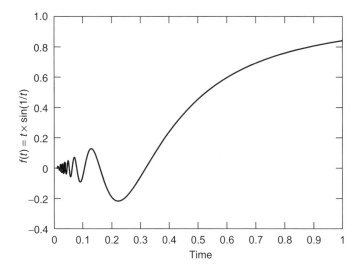

Figure 4.3.5. One period of a function that violates a Dirichlet
strong condition

finite interval $0 < t < \epsilon$, no matter how small is ϵ!) In 1904 the Hungarian mathematician Lipót Fejér (1880–1959) discovered convergence conditions weaker than the Dirichlet strong conditions, but even they are stronger than *necessary and sufficient* conditions; such conditions are still unknown.

4.4 Discontinuous functions, the Gibbs phenomenon, and Henry Wilbraham.

We are now at the point where we have to say something about those funny wiggles that won't go away in Euler's series for $(\pi - t)/2$. As Figures 4.2.3 through 4.2.6 indicate, adding in more and more terms in that Fourier series does indeed result in the partial sums becoming better and better approximations to $(\pi - t)/2$ over *most* of the period $0 < t < 2\pi$, but that is not so as we approach either $t = 0$ from the right or $t = 2\pi$ from the left. This sort of "wiggly" behavior (electrical engineers call it "ringing") is, in fact, characteristic of the Fourier series for *any* discontinuous periodic function near a discontinuity (we saw it, too, in the square-wave Fourier series partial sum of figure 4.3.4) but, astonishingly, the wiggles escaped the notice of mathematicians for many years after Fourier.[12]

Except for one.

But that's getting ahead of my story here. The story of that single exception is, I think, the most interesting part of the whole "tale of the wiggles," and I'll save it for last (there is historical information at the end of this section that is published for the first time).

The traditional tale of how those wiggles finally came to be noticed begins with an 1898 letter[13] to the British science journal *Nature*, written by the American Albert Michelson (1852–1931), head of the University of Chicago's physics department. In his letter Michelson disputed the possibility that a Fourier series could represent a discontinuous function. As he wrote in his opening sentences,

In all expositions of Fourier's series which have come to my notice, it is expressly stated that the series can represent a discontinuous function. The idea that a real discontinuity can replace a sum of continuous curves [the individual terms of the series] is so utterly at variance with the physicists' notions of quantity, that it seems

to me to be worth while giving a very elementary statement of the
problem in such simple form that the mathematicians can at once
point to the inconsistency if any there be.

In particular, Michelson did not believe that the Fourier series for the
periodically extended function $f(t) = t$, $-\pi < t < \pi$, could converge
to zero at the instants of the discontinuities (odd multiples of π). These
zero values would represent "isolated points" and Michelson offered an
argument that he thought showed this to be nonsense.

A reply[14] was published just a week later, from A.E.H. Love (1863–
1940), then a professor of mathematics at Cambridge University. Love
correctly pointed out Michelson's mathematical errors in reasoning,
beginning several of his sentences with "It is not legitimate to . . . " and
"The processes employed are invalid . . . ," and ending with the observation that Michelson had *started* his analysis with a series that doesn't even
converge! Michelson replied several weeks later with a short note, stating
that he was unconvinced; immediately following his letter, in the same
issue of *Nature*, was one from a new correspondent, the American mathematical physicist J. W. Gibbs (1839–1903) at Yale, as well as a new letter
from Love.[15] Love attempted once again to identify Michelson's mis-
steps, and Gibbs, too, made criticism of Michelson's reasoning. There
was no reply from Michelson, and then several months later Gibbs wrote
a fateful follow-up letter,[16] one that is often cited in modern textbooks.
The other *Nature* letters have been forgotten by all but historians.

It is a short letter, written only, said Gibbs, to "correct a careless error"
and "unfortunate blunder" that he had made in his first letter, but it
contained, almost as a throwaway line, the seminal statement that, as the
number of terms in the partial sum of a Fourier series for a discontinuous
periodic function is increased, the amplitude of the wiggles will indeed
decrease everywhere *except* in the neighborhood of the discontinuity.
There, the partial sums will *overshoot and undershoot* the original function,
with a maximum overshoot amplitude that does *not* decrease with an
increasing number of terms in the partial sums. As the number of terms
increases, the *duration* of the overshoot will indeed decrease, but the
maximum overshoot amplitude will not decrease. And, if you look back at
figures 4.2.3 through 4.2.6, you'll see precisely the behavior that Gibbs
describes in his letter.

Then, without providing any derivation or explanation, Gibbs simply *states* that the maximum overshoot is related to a certain integral: as t approaches π from the left, the maximum amplitude of the Fourier series of a periodically extended $f(t) = t$ is not π but rather is $2 \int_0^\pi (\sin(u)/u \; du)$. The percentage overshoot is thus

$$
\frac{2 \int_0^\pi (\sin(u)/u) \; du - \pi}{\pi} \cdot 100\%
$$

$$
= \frac{2 \cdot 1.851937 - \pi}{\pi} \cdot 100\% = 17.9 \text{ percent}
$$

or, more generally, the Fourier series of *any* discontinuous function will have an overshoot in the neighborhood of the discontinuity of about 8.9 percent of the *total jump* of the discontinuity. This is a number you can find in innumerable modern engineering, mathematics, and physics textbooks. If there is any citation at all in those texts (and there often is not) it is always to Gibbs's second letter in *Nature*, and the overshoot is called the "Gibbs phenomenon."

The Gibbs phenomenon does not occur in the Fourier series of periodic functions that are free of discontinuities. For example, consider the function $f(t) = \sin(t)$ in the interval $0 \le t \le \pi$. This function is equal to zero at both end points, and so if we *extend* it in an *even* way to include $-\pi \le t \le 0$, that is,

$$
f(t) = \begin{cases} \sin(t), & 0 \le t \le \pi, \\ -\sin(t), & -\pi \le t \le 0, \end{cases}
$$

and then periodically extend *this* to $-\infty < t < \infty$, we'll have a periodic function that is continuous everywhere, and when we plot the partial sums of its Fourier series you'll see that there is no Gibbs phenomenon. And, as you'll also see, the Fourier series will consist only of *cosines* (because of the *even* extension), and so we'll have an example of Fourier's claim that $\sin(t)$ (in the interval $0 \le t \le \pi$) can be written in terms of an infinity of cosines. The problem is easy to set up: since $T = 2\pi$ (and $\omega_0 T = 2\pi$, as always) then $\omega_0 = 1$ and so, with $f(t)$ now denoting the periodic function on the entire real line $-\infty < t < \infty$, we have

$$f(t) = \sum_{n=-\infty}^{\infty} c_n e^{int},$$

where

$$c_n = \frac{1}{2\pi}\left[\int_0^{\pi} \sin(t)e^{-int}\,dt - \int_{-\pi}^{0} \sin(t)e^{-int}\,dt\right].$$

As it almost always is, it is convenient now to first write $\sin(t)$ in terms of complex exponentials and then, after a bit of algebra (which I'll let you do), we'll get

$$c_n = \begin{cases} \dfrac{2}{\pi}\cdot\dfrac{1}{1-n^2}, & n \text{ even}, \\ 0, & n \text{ odd}. \end{cases}$$

Inserting this result into the summation for $f(t)$, we have

$$f(t) = \frac{2}{\pi}\cdot\sum_{n=-\infty,\,n\text{ even}}^{\infty} \frac{1}{1-n^2}e^{int} = \frac{2}{\pi} + \frac{2}{\pi}\cdot\sum_{n=-\infty,\,n\text{ even}}^{\infty} \frac{2\cos(nt)}{1-n^2}$$

or, if we write $n = 2k$ where $k = 1, 2, 3, \ldots$ (and so $n = 2, 4, 6, \ldots$), we then have

$$f(t) = \frac{2}{\pi} + \frac{4}{\pi}\cdot\sum_{k=1}^{\infty} \frac{\cos(2kt)}{1-4k^2}.$$

This *should*, if I've made no mistakes, be equal to $\sin(t)$ in the interval $0 \le t \le \pi$. As a check, I've plotted some partial sums of this Fourier series to see what they look like, in figures 4.4.1 through 4.4.4 (for the first one, five, ten, and twenty terms in the summation, respectively). And, sure enough, those plots *do* appear to "look more and more like" $\sin(t)$ for $0 \le t \le \pi$ as more and more terms of the Fourier series are used. And, as you can see, there is no sign of a Gibbs phenomenon (because there are no discontinuities).

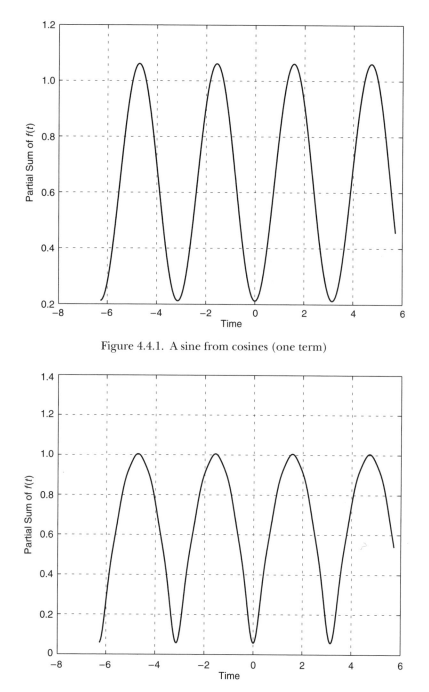

Figure 4.4.1. A sine from cosines (one term)

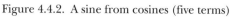

Figure 4.4.2. A sine from cosines (five terms)

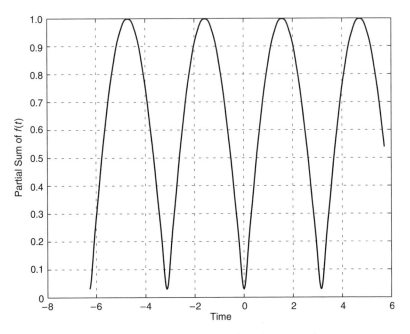

Figure 4.4.3. A sine from cosines (ten terms)

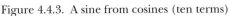
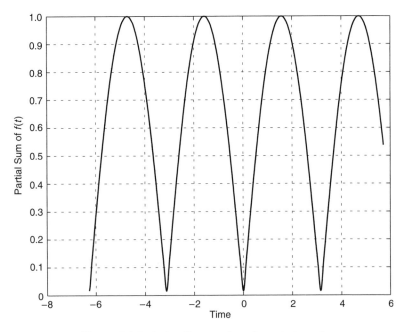

Figure 4.4.4. A sine from cosines (twenty terms)

As a final comment on this particular Fourier series, it has an important place in electronics as well as in mathematics. This periodically extended, everywhere continuous function is called a *full-wave rectified sine wave*. It is, in fact, the signal that is produced by certain electrical circuits that transform the a-c (alternating current) waveform that is present at wall sockets into a pulsating d-c (direct current) waveform that can be further "smoothed" into a *constant* d-c waveform (the waveform that is the usual requirement for the proper operation of electronic circuits, such as radios and computers).

The *Nature* correspondence soon came to an end, first with a testimonial letter from the French mathematician Henri Poincaré (1854–1912) written in support of Michelson (which, curiously, gives the wrong values to Dirichlet's discontinuous integral). And then a final letter came from Love, in response to Poincaré and trying, one last time, to explain matters to Michelson.[17] In that I think Love almost certainly failed, but it is important to understand that Michelson was *not* being dense in this affair. He was really at a mathematical disadvantage that wasn't entirely his own fault—the *mathematicians* of the time didn't understand the full behavior of the Fourier series of discontinuous functions, either. In fact, far from being dense, Michelson was a world-class experimenter; it was the famous "Michelson-Morley experiment," using a fantastic optical interferometer, that showed the absence of a detectable ether effect on the speed of light and thereby hinted at the constancy of the speed of light in all inertial frames of reference (a fundamental axiom of Einstein's special theory of relativity). For that work Michelson received the 1907 Nobel Prize in physics. No, he wasn't dense.

It was the puzzling results from yet another fantastic experimental gadget that he had helped design and build that prompted his initial letter to *Nature*. In a paper published the previous year (1898), Michelson (and his coauthor S. W. Stratton) described this amazing mechanical device; as they wrote in their opening paragraph,[18]

Every one who has had occasion to calculate or to construct graphically the resultant of a large number of simple harmonic motions has felt the need of some simple and fairly accurate machine which would save the considerable time and labor involved in such computations.

Today we all have such machines, of course—our personal computers running some scientific programming language such as MATLAB or Mathematica, but in Michelson's day there had been only one other such machine constructed in the entire world: by William Thomson in the 1880s as an aid in predicting tides.[19] The actual *invention* of the tidal harmonic analyzer was due to William's brother James Thomson, and they had published the conceptual details in the *Proceedings of the Royal Society* of 1876 and 1878. Foreshadowing Michelson and Stratton's opening words, William wrote in 1882 that "The object of this machine [the tidal harmonic analyzer] is to substitute brass for brains." (Since modern computer processing chips are made from silicon, I suppose an up-dating of Thomson's words would say our computers "substitute sand for smarts"!) The tidal analyzer superficially resembles the power train of a modern automobile engine[20] and is most impressive, but, as Michelson and Stratton went on to write in their 1898 paper, the construction details of the ingenious Thomson brothers' machine eventually led to cumulative errors that overwhelmed any benefit obtained by adding ever more terms to the partial sum of a Fourier series (the tidal analyzer could add fifteen terms).

Somewhat mysteriously writing "It occurred to one of us some years ago . . . that most of the imperfections in existing machines [there was only the tidal analyzer] might be practically eliminated" (no clue is given as to which of the two coauthors is the "one"), Michelson and Stratton then stated

> About a year ago [1897] a machine was constructed on [their new principle of adding harmonic functions] with twenty elements [i.e., the number of harmonic functions added] and the results obtained were so encouraging it was decided [to build] the present machine with eighty [!] elements.

The Thomson analyzer used a flexible cord passing through a complicated arrangement of fixed and moveable pulleys, and cord stretching eventually limited the machine's accuracy. The Michelson/Stratton machine,[21] on the other hand, used gears, spiral springs, pinions, and eccentric wheels, all (it seems) much less vulnerable to cumulative error build-up. Indeed, at the end of their paper they asserted "it would be quite feasible to increase the number of elements to several hundred or even to a thousand with a proportional increase [in accuracy]." ("Big machine" physics evidently did not start with the cyclotron builders in the twentieth century–Michelson and Stratton were there first in the nineteenth!)

Both the Thomson and the Michelson/Stratton harmonic analyzers automatically generated ink drawings on paper of the resulting sum-mations, and in their 1898 paper Michelson and Stratton reproduced some of their more spectacular results, some of which (the plots of theoretically discontinuous functions) *distinctly* exhibit the Gibbs phe-nomenon. And now you can appreciate what prompted Michelson's opening line in his first letter to *Nature*–the plots produced by his machine were *not* discontinuous, even with eighty terms. If he wondered at all about the wiggles around the sharp, nearly vertical transitions his ink pen traveled as it moved through the mathematical discontinuities isn't clear—he probably thought they were just the cumulative effect of numerous small but unavoidable deviations from machine perfection. Michelson's letter was the result of *physical experiments* that deviated from what he had been led to expect by *pure mathematics*. This fact is interesting in its own right, because the *Nature* controversy eventually resulted in mathematicians making new studies into just what is meant by a Fourier series expansion of a periodic function. As interesting as all that is, how-ever, I haven't yet told you what I think is the *most* interesting part of the story.

The answer to Michelson's query, and Gibbs's discovery of the curious behavior of a Fourier series near a discontinuity, had been anticipated *and* published *fifty years* before in 1848, by the forgotten English-man Henry Wilbraham (1825–1883), then a twenty-two-year-old student at Trinity College, Cambridge. Wilbraham's paper[22] opens with the following passage:

Fourier, in his Treatise, after discussing the equation

$$y = \cos x - \frac{1}{3}\cos 3x + \frac{1}{5}\cos 5x - \cdots ad\ inf.,$$

says that, if x be taken for abscissa and y for ordinate, it [the equation] will represent a locus composed of separate straight lines, of which each is equal to π [in length], parallel to the axis of x, and at a distance of $\frac{1}{4}\pi$ alternately above and below it, joined by perpendiculars which are themselves part of the locus of the equation. Some writers who have subsequently considered the equation, have stated that the part of the lines perpendicular to the axis of x which satisfies the equation is contained between the limits $\pm\frac{1}{4}\pi$. The following calculation will shew that these limits are erroneous.

You'll recognize the equation cited by Wilbraham as being the square-wave function we analyzed in the last section, and which led to figure 4.3.4.

Then, in his next two pages, Wilbraham *derived* the result that the Fourier series overshoot at the discontinuity is proportional to the integral $\int_0^\pi (\sin(u)/u)\,du$, the same integral given by Gibbs *without* derivation. In addition, Wilbraham's paper contains a number of laboriously calculated, hand-plotted partial sums for the square-wave Fourier series; those plots clearly display the wiggles we see in figure 4.3.4, that is, the Gibbs phenomenon was *in print* when Gibbs was just nine years old. Nevertheless, when the American mathematician Maxime Bôcher (1867–1918), on the faculty at Harvard, finally gave a full mathematical analysis on the behavior of Fourier series at discontinuities in 1906 (all the while being unaware of Wilbraham), he coined the term *Gibbs's phenomenon* for that behavior (now just called *Gibbs phenomenon*), and that's the name that has stuck.

How could such a gross oversight have happened, you might wonder? Wilbraham was clearly a talented mathematician, but that talent did not lead to an academic career and numerous published papers, books, and other forms of professional recognition by which he and his work would have becomewell known. Indeed, a search of the *Royal Society Catalogue of Scientific Papers* shows only seven published papers by Wilbraham, the first of which was his Fourier series paper, and the last a beautifully written 1857 paper on actuarial mathematics that appeared in the *Assurance Magazine*.[23] So, let me end this section by telling you what I have been able to find out about this long-forgotten *almost*-mathematician whose very first published paper was fifty years ahead of its time.[24]

Born the second youngest of five brothers on July 25, 1825, Henry Wilbraham came from a privileged family. His father George (1779–1852) served for ten years in Parliament, his mother Lady Anne (died 1864) was the daughter of the first Earl Fortescue, and they moved in high circles. In October 1832, for example, Henry's parents were in one of the formal carriages that followed the royal procession to the ceremony where the then thirteen-year old Princess Victoria (who would be queen five years later) opened the Grosvenor Bridge (then the largest single-span stone arch bridge in the world). After attending the elite Harrow School Wilbraham was admitted to Trinity College, Cambridge in October 1841, shortly after his sixteenth birthday.[25] He received his BA in 1846 and an MA in 1849. For several years (at least up to 1856) he was a Fellow at Trinity, and then sometime after 1856 he left academia for unknown reasons.

In 1864 he married Mary Jane Marriott (died 1914); they had seven children (one survived until 1954). By 1875 he was not employed as a mathematician at all, but rather held the post of District Registrar of the Chancery Court at Manchester. And that was the only position mentioned (along with being a late Fellow of Trinity) in his will when he died (from tuberculosis) on February 13, 1883, still in his 57th year. He left his family well provided for with, among other valuable possessions, a pension valued in excess of £37,000. And then, with the exception of a few historians, the world forgot Henry Wilbraham.[26]

When the *Nature* debate began, Wilbraham had been dead for more than fifteen years, and all of his work from half a century earlier had to be rediscovered. If his paper had not vanished from view almost immediately after publication, perhaps the state of Fourier analysis might have been greatly accelerated. For example, it wasn't until Bôcher's 1906 paper that it was finally proved that if the periodic function $f(t)$ is discontinuous at $t = t_0$ then at $t = t_0$ the Fourier series for $f(t)$ converges to $1/2[f(t_0-) + f(t_0+)]$, where $f(t_0-)$ and $f(t_0+)$ are the values of $f(t)$ "just before" and "just after" the discontinuity, respectively.

4.5 Dirichlet's evaluation of Gauss's quadratic sum.
In this, the penultimate section of the chapter, I'll show you a beautiful application of Fourier series that most engineers and physicists never

see because its origin is in abstract number theory; even mathematicians might not see it until graduate school. This analysis will be doubly interesting for us because not only does it use Fourier series and Euler's formula, it also involves the solution to a problem that stumped the great Gauss for years. He finally solved it using number theoretic arguments by 1805, just as Fourier was starting his work. Then, thirty years later (1835) Dirichlet (who appeared earlier in sections 1.7 and 1.8) used Fourier's method to produce an entirely new, extremely elegant derivation of Gauss's result. The problem is easy to state.

We wish to evaluate a function called $G(m)$—G for Gauss, of course—defined as

$$G(m) = \sum_{k=0}^{m-1} e^{-i(2\pi/m)k^2}, \quad m > 1.$$

Gauss encountered this expression during his study of regular n-gons (see section 1.6 again). Often you'll see $G(m)$ defined with a positive exponent,

$$G(m) = \sum_{k=0}^{m-1} e^{i(2\pi/m)k^2},$$

but this is a trivial variation. This $G(m)$, and the one I first defined with a minus sign in the exponent, are simply conjugates. That is, once we have the answer for one of the Gs, then the answer for the other G is the conjugate of the first answer; I'll return to this point at the end of the analysis.

$G(m)$ is called a Gaussian *quadratic* sum because of the k^2 in the exponent, and it's the *squaring* of k that makes the problem hard.[27] If it were simply k, then we would have nothing more complicated than a geometric series, which is easy to work with (you should be able to show, for that case, that the sum is zero). Geometrically, too, the zero sum is obvious as $G(m)$ is the sum of $m \geq 2$ equal-length vectors, all originating at the origin and angularly spaced *uniformly* around a circle. The net vector sum is, *by inspection*, zero. But make the k in the exponent *k-squared* and the uniform angular spacing goes out the window. Then the problem gets tough.

There is, of course, no reason to stop with k^2. In fact, mathematicians have generalized the Gaussian sum exponent to k^p, where p is *any* positive integer. And so for $p = 3, 4, 5, 6, \ldots$ we have the Gaussian cubic, quartic, quintic, sextic, ... sums. Mathematicians have studied Gaussian sums up to *at least* (as far as I know) $p = 24$ (the biduodecimic) Gaussian sum.[28] Exercising great and admirable restraint, however, I'll limit myself here to Gauss's original quadratic sum.

Now, to start, let's do a very quick summary of what we've developed in the earlier sections of this chapter, cast into the particular form we'll need here. Suppose $f(t)$ is a periodic function with period one. Then, its complex exponential Fourier series is (remember, $\omega_0 T = 2\pi$ and, with $T = 1$, we have $\omega_0 = 2\pi$)

$$f(t) = \sum_{k=-\infty}^{\infty} c_k e^{ik2\pi t},$$

where the Fourier coefficients are given by

$$c_k = \int_0^1 f(t) e^{-ik2\pi t} \, dt.$$

That's the review—I told you it would be quick! Now, what should we use for $f(t)$?

It was Dirichlet's clever idea to define $f(t)$ as follows: let $g(t)$ be a function defined as

$$g(t) = \sum_{k=0}^{m-1} e^{-i(2\pi/m)(k+t)^2}, \quad 0 \le t < 1,$$

and take this as giving the behavior of one complete period of the periodic function $f(t)$. That is, our *periodic* $f(t)$ is simply the $g(t)$ defined above, repeated over and over as t goes off (in both directions) to infinity. Notice, *carefully*, that

$$G(m) = g(0) = f(0).$$

If you look at the complex exponential series expression for $f(t)$, you can see that if we set $t = 0$ we get

$$f(0) = \sum_{k=-\infty}^{\infty} c_k = G(m).$$

Now all we have to do is find an expression for the Fourier coefficients c_k of $f(t)$, stick that expression into the summation, and calculate the result. This might seem like we've simply traded one sum, our original $G(m)$, for just another sum, but the trade is very much in our favor—the second sum will prove to be far easier to do. So, let's do it.

Inserting the expression for a period of $f(t)$, that is, $g(t)$, into the Fourier coefficient integral, we have

$$c_n = \int_0^1 \left\{ \sum_{k=0}^{m-1} e^{-i(2\pi/m)(k+t)^2} \right\} e^{-i2\pi nt}\, dt,$$

where I've changed the subscript on c to n (from k) because I'm using k as the dummy index of summation on the sum inside the integral. This is "just" notational stuff, but it's *really important* to keep your indices straight. Continuing,

$$c_n = \sum_{k=0}^{m-1} \int_0^1 e^{-i(2\pi/m)(k+t)^2} e^{-i2\pi nt}\, dt = \sum_{k=0}^{m-1} \int_0^1 e^{-i2\pi((k+t)^2+mnt)/m}\, dt.$$

It is easy to confirm (and you should do the *easy* algebra) that

$$(k+t)^2 + mnt = \left[k + t + \frac{1}{2}mn \right]^2 - \left[mnk + \frac{1}{4}m^2 n^2 \right]$$

and so

$$c_n = \sum_{k=0}^{m-1} \int_0^1 e^{-i2\pi([k+t+mn/2]^2-[mnk+m^2 n^2/4])/m}\, dt$$

$$= \sum_{k=0}^{m-1} \int_0^1 e^{-i2\pi[k+t+mn/2]^2/m} e^{i2\pi nk} e^{i2\pi mn^2/4}\, dt$$

or, as $e^{i2\pi\, nk} = 1$ from Euler's formula (because n and k, and so their product, are integers), we have

$$c_n = e^{i2\pi\, mn^2/4} \sum_{k=0}^{m-1} \int_0^1 e^{-i2\pi[k+t+mn/2]^2/m}\, dt.$$

Next, change variables in the integral to $u = k + t + \frac{1}{2}mn$. Then

$$c_n = e^{i2\pi\, mn^2/4} \sum_{k=0}^{m-1} \int_{k+mn/2}^{k+1+mn/2} e^{-i2\pi\, u^2/m}\, du.$$

We can greatly simplify this by making the *almost* trivial observation that

$$\sum_{k=0}^{m-1} \int_{k+mn/2}^{k+1+mn/2} = \int_{mn/2}^{1+mn/2} + \int_{1+mn/2}^{2+mn/2} + \int_{2+mn/2}^{3+mn/2} + \cdots + \int_{m-1+mn/2}^{m+mn/2} = \int_{mn/2}^{m+mn/2}.$$

Therefore,

$$c_n = e^{i2\pi\, mn^2/4} \int_{mn/2}^{m+mn/2} e^{-i2\pi\, u^2/m}\, du$$

and, since as shown earlier we have $G(m) = \sum_{n=-\infty}^{\infty} c_n$,

$$G(m) = \sum_{n=-\infty}^{\infty} e^{i2\pi\, mn^2/4} \int_{mn/2}^{m+mn/2} e^{-i2\pi\, u^2/m}\, du.$$

In this expression, to repeat an obvious but important detail, m and n are both *integers*; $m > 1$ and $-\infty < n < \infty$. Let's now take a look at the behavior of that $mn^2/4$ in the exponent of the first exponential. Specifically, we'll consider the two all-encompassing possibilities of n being even and then odd.

The even case is easy: if n is *even* $(\cdots, -4, -2, 0, 2, 4, \cdots)$, then $mn^2/4$ is an integer and so $e^{i2\pi\, mn^2/4} = 1$.

The odd case is just a bit more subtle: if n is *odd* $(\cdots, -3, -1, 1, 3, \cdots)$, then with l some integer we can write $n = 2l + 1$ and so $mn^2/4 = m(2l+1)^2/4 = m(4l^2 + 4l + 1)/4 = ml^2 + ml + m/4$, which means $mn^2/4$ is an integer plus a fraction that depends *only* on m. This fraction is, of course, one of just four possibilities: 0 (this is the case when m is a multiple of 4), or $\dfrac{1}{4}$, or $\dfrac{2}{4}$, or $\dfrac{3}{4}$. Considering each possibility in turn, we see that

if the fraction is 0, then $e^{i2\pi \, mn^2/4} = 1$,

if the fraction is $\dfrac{1}{4}$, then $e^{i2\pi \, mn^2/4} = i$,

if the fraction is $\dfrac{2}{4}$, then $e^{i2\pi \, mn^2/4} = -1$,

if the fraction is $\dfrac{3}{4}$, then $e^{i2\pi \, mn^2/4} = -i$.

The standard way of writing all of this is with what mathematicians call *congruence arithmetic*, that is, if m, q, and r are all integers, then

$$\frac{m}{4} = q + \frac{r}{4},$$

which is written in shorthand as $m \equiv r \bmod 4$. It is *read* as "m is congruent to r, mod(ulus)4." So, writing $\eta = e^{i2\pi \, mn^2/4}$, then for odd n we have

$$\eta = \begin{cases} 1 & \text{if } m \equiv 0 \quad \bmod 4, \\ i & \text{if } m \equiv 1 \quad \bmod 4, \\ -1 & \text{if } m \equiv 2 \quad \bmod 4, \\ -i & \text{if } m \equiv 3 \quad \bmod 4. \end{cases}$$

Therefore, returning to our last expression for $G(m)$, we see that we can write it as

$$G(m) = \sum_{n=-\infty \, (even)}^{\infty} \int_{mn/2}^{m+mn/2} e^{-i2\pi \, u^2/m} \, du$$

$$+ \sum_{n=-\infty \, (odd)}^{\infty} \eta \int_{mn/2}^{m+mn/2} e^{-i2\pi \, u^2/m} \, du.$$

Now, just as we did before when we showed that $\displaystyle\sum_{k=0}^{m-1} \int_{k+mn/2}^{k+1+mn/2} = \int_{mn/2}^{m+mn/2}$,

it is equally easy to show that $\displaystyle\sum_{n=-\infty\,(\text{even})}^{\infty} \int_{mn/2}^{m+mn/2} = \int_{-\infty}^{\infty}$ and that the

same is true for $\displaystyle\sum_{n=-\infty\,(\text{odd})}^{\infty} \int_{mn/2}^{m+mn/2}$. Thus,

$$G(m) = \int_{-\infty}^{\infty} e^{-i2\pi u^2/m}\,du + \eta \int_{-\infty}^{\infty} e^{-i2\pi u^2/m}\,du = (1+\eta) \int_{-\infty}^{\infty} e^{-i2\pi u^2/m}\,du,$$

or, using Euler's formula in the obvious way,

$$G(m) = (1+\eta)\left[\int_{-\infty}^{\infty} \cos\left(2\pi\frac{u^2}{m}\right)du - i\int_{-\infty}^{\infty} \sin\left(2\pi\frac{u^2}{m}\right)du\right].$$

Both of these integrals are the so-called *Fresnel integrals*, even though it was Euler who first evaluated them in 1781 using complex numbers.[29] For our purposes here, let me simply quote what you'll find in any good table of definite integrals:

$$\int_0^{\infty} \sin(ax^2)\,dx = \int_0^{\infty} \cos(ax^2)\,dx = \frac{1}{2}\sqrt{\frac{\pi}{2a}}.$$

So, with $a = 2\pi/m$, we have (because the integrands are even)

$$\int_{-\infty}^{\infty} \cos\left(2\pi\frac{u^2}{m}\right)du = \int_{-\infty}^{\infty} \sin\left(2\pi\frac{u^2}{m}\right)du = 2\left[\frac{1}{2}\sqrt{\frac{\pi}{2\cdot 2\pi/m}}\right] = \sqrt{\frac{m}{4}} = \frac{1}{2}\sqrt{m}.$$

Thus,

$$G(m) = (1 + \eta)\left(\frac{1}{2}\sqrt{m} - i\frac{1}{2}\sqrt{m}\right) = \frac{1}{2}\sqrt{m}(1 + \eta)(1 - i).$$

So, at last,

$$m \equiv 0 \bmod 4 \implies G(m) = \frac{1}{2}\sqrt{m}(1 + 1)(1 - i) = (1 - i)\sqrt{m},$$

$$m \equiv 1 \bmod 4 \implies G(m) = \frac{1}{2}\sqrt{m}(1 + i)(1 - i) = \sqrt{m},$$

$$m \equiv 2 \bmod 4 \implies G(m) = \frac{1}{2}\sqrt{m}(1 - 1)(1 - i) = 0,$$

$$m \equiv 3 \bmod 4 \implies G(m) = \frac{1}{2}\sqrt{m}(1 - i)(1 - i) = -i\sqrt{m}.$$

For example, if $m = 93$, then $m \equiv 1 \bmod 4$ and so our formula says

$$G(93) = \sum_{k=0}^{92} e^{-i2\pi k^2/93} = \sqrt{93} = 9.64365076099317.$$

It is easy, a matter of a few key strokes, to directly code the summation itself in MATLAB, and when I did that the result was 9.64365076099295. (The difference, starting in the 12th decimal place, is round-off error.) Hurrah for Gauss!

And finally, as mentioned earlier, if we alter our definition of $G(m)$ to its conjugate form, then we have

$$G(m) = \sum_{k=0}^{m-1} e^{i2\pi \frac{k^2}{m}} = \begin{cases} (1 + i)\sqrt{m} & \text{for } m \equiv 0 \bmod 4, \\ \sqrt{m} & \text{for } m \equiv 1 \bmod 4, \\ 0 & \text{for } m \equiv 2 \bmod 4, \\ i\sqrt{m} & \text{for } m \equiv 3 \bmod 4. \end{cases}$$

As you might imagine, Gauss had great admiration for Dirichlet's beautiful new proof of the Master's old problem, as well as for Dirichlet's many other brilliant contributions to mathematics. In a letter dated November 2, 1838, Gauss wrote to Dirichlet (at the Berlin Academy of Sciences) to say "my thanks for sending me your beautiful papers." Beautiful indeed was his work and, as a measure of Dirichlet's eminence in the world of mathematics, he was selected to be Gauss's successor at Göttingen when Gauss died in 1855.

4.6 Hurwitz and the isoperimetric inequality.

The title to the final section of this chapter refers to an ancient problem: how, with a fence of fixed length, to enclose the greatest area. The problem proved extraordinarily frustrating, for literally thousands of years, because its answer (the fence should form a circle) is both so very obvious and so very difficult to actually *prove*. It stumped the brightest of mathematicians for centuries. I've told the story of the isoperimetric inequality in detail elsewhere, including giving a proof using the calculus of variations,[30] and so here I'll simply state the problem and show you how to establish it with an elegant Fourier series analysis. The Fourier proof is a relatively new one (1901), and is due to the German mathematician Adolf Hurwitz (1859–1919).

The isoperimetric inequality has two parts:

(i) the area inside a closed, simple (i.e., non-self-intersecting) curve of given perimeter L cannot exceed the area of a circle with circumference L;

(ii) the circle is the *only* simple, closed curve that actually achieves the maximum possible enclosed area.

If we call the curve C and its enclosed area A, then mathematically the isoperimetric inequality says

$$A \leq \frac{L^2}{4\pi},$$

and equality occurs if and only if C is a circle.

Before starting our proof of this, let me point out that this is what mathematicians call a *scalable* problem. That is, the inequality holds for any value of L we wish to use, and so if we can establish it for a particular

L then we can scale up (or down) to any other value that we wish. To see this, suppose a curve C' has the same shape as C and a perimeter that is l times that of C, (i.e., $L' = lL$). Now, I think I can appeal to your intuition when I say that, if $L' = lL$, then $A' = l^2A$. That is, $L = L'/l$ and $A = A'/l^2$. Then our original statement says

$$\frac{A'}{l^2} \leq \frac{(L'/l)^2}{4\pi} = \frac{L'^2}{l^2 4\pi},$$

or,

$$A' \leq \frac{L'^2}{4\pi}.$$

So, the use of any scale factor $l' > 0$ has no effect on the truth of the inequality, which means we can establish the inequality for any particular value of L that we wish, with no loss of generality. You'll see, in fact, that $L = 2\pi$ is a particularly attractive choice (in that case, the isoperimetric inequality reduces to $A \leq \pi$).

To start the analysis, I'll first show you a preliminary result we'll need about halfway into Hurwitz's proof. Interestingly, this preliminary result also uses Fourier methods—in particular, Parseval's theorem. Suppose $f(t)$ is a real-valued periodic function, with period $T = 2\pi$ *and* an average value of zero. Then, with $\omega_0 = 1$ (since $\omega_0 T = 2\pi$), we can write

$$f(t) = \sum_{k=-\infty}^{\infty} c_k e^{ikt}.$$

Assuming we can differentiate term by term to get the Fourier series for $f'(t) = \dfrac{df}{dt}$, then

$$f'(t) = \sum_{k=-\infty}^{\infty} ikc_k e^{ikt} = \sum_{k=-\infty}^{\infty} c'_k e^{ikt}$$

where $c'_k = ikc_k$. Now, since we have the general coefficient formula

$$c_k = \frac{1}{T} \int_{\text{period}} f(t) e^{-ikt}\, dt,$$

in particular, c_0 is obviously the average value of $f(t)$, which we are given as being zero. (Notice that $c_0' = 0$ even in the case of $c_0 \neq 0$.) Parseval's theorem, from section 4.3, tells us that

$$\frac{1}{2\pi} \int_0^{2\pi} f^2(t)\,dt = c_0^2 + 2\sum_{k=1}^{\infty} |c_k|^2 = 2\sum_{k=1}^{\infty} |c_k|^2$$

and that

$$\frac{1}{2\pi} \int_0^{2\pi} f'^2(t)\,dt = c_0'^2 + 2\sum_{k=1}^{\infty} |c_k'|^2 = 2\sum_{k=1}^{\infty} k^2 |c_k|^2.$$

Obviously,

$$\sum_{k=1}^{\infty} k^2 |c_k|^2 \geq \sum_{k=1}^{\infty} |c_k|^2,$$

and so we have our preliminary result: if the real-valued $f(t)$ is periodic with period 2π, with zero average value, then

$$\boxed{\int_0^{2\pi} f'^2(t)\,dt \geq \int_0^{2\pi} f^2(t)\,dt}\ ,$$

a result called *Wirtinger's inequality* because it is generally attributed to the Austrian mathematician Wilhelm Wirtinger (1865–1945).

Now we can start Hurwitz's proof. Imagine, as illustrated in figure 4.6.1, that we have positioned C on coordinate axes such that P denotes the rightmost x-axis crossing of C. We will take L, the perimeter of C, to be 2π (a choice, as I discussed earlier, that involves no loss of generality). Starting at P, we travel once around C in a *counterclockwise* sense and measure our x and y coordinates as a function of the *counterclockwise* arc-length s that separates us from P. It should then be clear that $x(s)$ and $y(s)$ are each periodic with period 2π (one complete trip around C increases s from 0 to 2π, and $x(0) = x(2\pi)$ and $y(0) = y(2\pi)$)

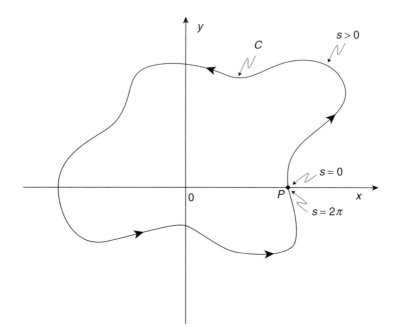

Figure 4.6.1. The geometry of the isoperimetric problem

because when $s = 2\pi$ we have *returned* to P). We know from calculus that the differential arc-length ds satisfies the relation $(ds)^2 = (dx)^2 + (dy)^2$ and so

$$\left(\frac{dx}{ds}\right)^2 + \left(\frac{dy}{ds}\right)^2 = 1 = x'^2 + y'^2.$$

We can express all of this mathematically by writing (where I've changed variable from s to t just to make the equations look more familiar) with, again, $\omega_0 = 1$ as $\omega_0 T = 2\pi$ and $T = 2\pi$,

$$x(t) = \sum_{k=-\infty}^{\infty} x_k e^{ikt}, \quad y(t) = \sum_{k=-\infty}^{\infty} y_k e^{ikt}$$

$$x'^2 + y'^2 = 1.$$

And finally, with no loss in generality, I'll assume it is physically obvious that we can always position our coordinates axes in such a way that the center of gravity of the perimeter of C lies on the y-axis, which means

$$\int_0^{2\pi} x(t)\,dt = 0,$$

that is, $x(t)$ has zero average value. We'll do this, as you'll soon see, so that we can invoke Wirtinger's inequality for $x(t)$ and $x'(t)$.

Now, it is known from calculus[31] that the area enclosed by C is given by

$$A = \int_0^{2\pi} x\frac{dy}{dt}\,dt = \int_0^{2\pi} xy'\,dt.$$

Therefore,

$$2(\pi - A) = 2\pi - 2A = \int_0^{2\pi} dt - 2\int_0^{2\pi} xy'\,dt.$$

Since $x'^2 + y'^2 = 1$,

$$2(\pi - A) = \int_0^{2\pi} (x'^2 + y'^2)\,dt - 2\int_0^{2\pi} xy'\,dt = \int_0^{2\pi} (x'^2 - 2xy' + y'^2)\,dt,$$

or

$$2(\pi - A) = \int_0^{2\pi} (x'^2 - x^2)\,dt + \int_0^{2\pi} (x - y')^2\,dt.$$

Wirtinger's inequality tells us that the first integral is nonnegative, and the second integral (with its *squared* integrand) is obviously non-negative. Thus, $2(\pi - A) \geq 0$ or, $A \leq \pi$, which establishes part (i) of the isoperimetric inequality. Notice that if C is a circle then $A = \pi$. To show part (ii) of the isoperimetric inequality we must show that $A = \pi$ means C *must* be a circle. This is actually not hard to do.

If $A = \pi$, then $2(\pi - A) = 0$ and each of our two nonnegative integrals must individually be zero. In particular, the second integral, with a *squared* integrand, can vanish only if $y' = x$ for every t. If we combine this conclusion with our earlier equation $x'^2 + y'^2 = 1$, then we have

$$\left(\frac{dx}{dt}\right)^2 + \left(\frac{dy}{dt}\right)^2 = 1 \quad \text{and} \quad \frac{dy}{dt} = x.$$

Thus,

$$(dx)^2 + (dy)^2 = (dt)^2 \quad \text{and} \quad dt = \frac{1}{x}dy,$$

and so

$$(dx)^2 + (dy)^2 = \frac{1}{x^2}(dy)^2,$$

or

$$1 + \left(\frac{dy}{dx}\right)^2 = \frac{1}{x^2}\left(\frac{dy}{dx}\right)^2.$$

This is quickly rewritten as

$$\left(\frac{dy}{dx}\right)^2 = \frac{x^2}{1-x^2},$$

which becomes

$$\frac{dy}{dx} = \pm\frac{x}{\sqrt{1-x^2}},$$

and so

$$\int dy = \pm\int \frac{x}{\sqrt{1-x^2}}dx.$$

Integrating, with K as the arbitrary constant of indefinite integration, gives

$$y + K = \pm\sqrt{1-x^2},$$

or

$$(y+K)^2 = 1 - x^2$$

which becomes, at last,

$$x^2 + (y + K)^2 = 1,$$

the equation of a *circle* centered on $x = 0$, $y = -K$, with radius one (such a circle has an enclosed area, of course, of π), and this completes part (ii) of the isoperimetric inequality. As a final comment, you might be wondering why the circle is specifically centered on $x = 0$, while the y-coordinate of the center is arbitrary. Remember, however, that we started our analysis with the *assumption* that the center of gravity of the perimeter of C lies on the y-axis (that allowed us to use Wirtinger's inequality) and for the symmetrical circle that means the center of C must be such that its x-coordinate is zero.

Chapter 5
Fourier Integrals

5.1 Dirac's impulse "function."

In this fairly short introductory section I want to take a break from Fourier and jump ahead a century to Paul Dirac, the English mathematical physicist I mentioned way back in the Preface. His name today is synonymous with the concept of the *impulse function* (often called the *Dirac delta function*), which will be of great use to us—as much as will Euler's formula—in the next section on the *Fourier transform*. The impulse (I'll define it soon) is one of the most important technical tools a physicist or engineer has; Dirac himself was originally trained as an electrical engineer (first-class honors in the 1921 class of the University of Bristol). His Ph.D. was, however, in mathematics—despite the fact that he received the Nobel Prize in *physics*, he was the Lucasian Professor of *Mathematics* at Cambridge (until late in life, when he became a professor of physics at Florida State University in Tallahassee). The Lucasian Professorship is the same position once held by Isaac Newton, and now occupied by the famous mathematical physicist Stephen Hawking.

To trade rigor for clarity (I hope!), an impulse represents something that occurs "all at once." (When a baseball player hits a home run he has delivered an *impulsive force* to the ball.) Dirac's own view of an impulse is shown in figure 5.1.1, which displays a narrow pulse of duration α and height $1/\alpha$, centered on $t = 0$. This pulse, which I'll call $f(t)$, is zero for all $|t| > \alpha/2$. For *any* $\alpha > 0$, it is obvious that $f(t)$ bounds unit area, and it is a perfectly ordinary, well-behaved function. But there's magic hidden in it.

Imagine next that we multiply $f(t)$ by some other nearly arbitrary (we'll demand only that it be *continuous*) function $\phi(t)$, and then

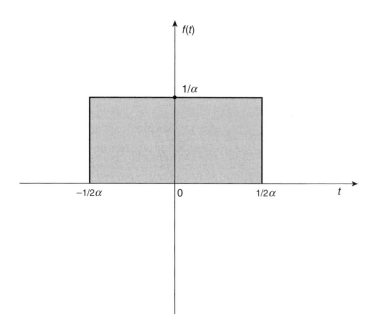

Figure 5.1.1. A nonzero-width pulse with unit area

integrate the product over all t, that is, let's define the integral

$$I = \int_{-\infty}^{\infty} f(t)\phi(t)\,dt = \int_{-\alpha/2}^{\alpha/2} \frac{1}{\alpha}\phi(t)\,dt = \frac{1}{\alpha} \int_{-\alpha/2}^{\alpha/2} \phi(t)\,dt.$$

Now suppose we let $\alpha \to 0$, which means that the height of the pulse $f(t)$ becomes arbitrarily large and the interval of integration (the duration of $f(t)$) becomes arbitrarily small. Since $\phi(t)$ is continuous, I'll take it as physically obvious that $\phi(t)$ cannot change by very much from the start of the integration interval to the end of that interval. Indeed, as $\alpha \to 0$ we can essentially treat $\phi(t)$ as constant over the entire interval, that is, equal to $\phi(0)$, and so we can pull it outside of the integral. That is,

$$\lim_{\alpha \to 0} I = \lim_{\alpha \to 0} \frac{1}{\alpha}\phi(0) \int_{-\alpha/2}^{\alpha/2} dt = \phi(0).$$

The limit of $f(t)$ as $\alpha \to 0$ is shown in figure 5.1.2, which is an attempt to indicate that our perfectly ordinary pulse "becomes" an infinitely

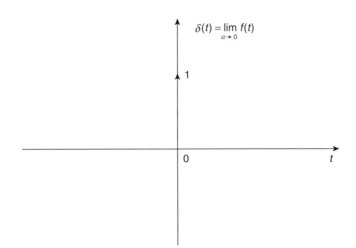

Figure 5.1.2. The Dirac impulse "function," $\delta(t)$

high *spike* of zero duration that bounds unit area. Since I can't draw an infinitely high spike, the figure simply shows an upward pointing arrow with "1" written next to it, to indicate that the spike bounds unit *area* (the spike does *not* have unit height—the height is *infinity!*); the spike is a so-called *unit strength* impulse. More formally, we write the limit of $f(t)$ as

$$\lim_{\alpha \to 0} f(t) = \delta(t),$$

which denotes an impulse that occurs at $t = 0$.

This is a very strange object when it sits naked on the page, all by itself, but in fact it behaves quite nicely *inside integrals*; for instance, as we've just seen, for any continuous $\phi(t)$

$$\int_{-\infty}^{\infty} \delta(t)\phi(t)\,dt = \phi(0).$$

There is, in fact, nothing particularly special about $t = 0$, and we can position the impulse to occur at any time, say $t = t_0$, by simply writing $\delta(t - t_0)$. Then, by the argument above, if $\phi(t)$ is any continuous function we have

$$\int_{-\infty}^{\infty} \delta(t - t_0)\phi(t)\,dt = \phi(t_0),$$

which is called the *sampling* (sometimes the *sifting*) property of the impulse function. I'll show you a wonderful application of this property at the end of this section.

This sort of reasoning is of the type once dismissed by many nineteenth-century analysts as merely "stuff and nonsense." And not without cause, I must admit. In particular, when I pulled the $1/\alpha$ outside of the integral when calculating $I = \int_{-\infty}^{\infty} f(t)\phi(t)\,dt$, leaving just $\phi(t)$ inside, I was treating α as a constant. But then I let $\alpha \to 0$, which certainly means α is *not* a constant. Since the operation of integration, itself, is defined in terms of a limiting operation, what I was really doing was reversing the order of taking two limits. "Well," says a skeptic with a questioning raise of her eyebrows, "How do you know that's mathematically valid?" "Well," I reply, "I don't—and I also realize it often isn't. Being adventurous, however, I won't let that paralyze me into inaction, and I will simply go ahead with it all until (unless) something terrible happens in the math that tells me I have finally gone too far." That was Dirac's attitude as well.[1]

Dirac of course knew he was being nonrigorous with his development of impulses. (To reflect its slightly suspect origin, the impulse is often called an "improper" or "singular" function.) As Dirac wrote in his pioneering 1927 paper that introduced physicists to the impulse function, published while he was still only twenty-five,[2]

> Strictly, of course, $\delta(x)$ is not a proper function of x, but can be regarded only as a limit of a certain sequence of functions. All the same one can use $\delta(x)$ as though it were a proper function for practically all the purposes of quantum mechanics without getting incorrect results. One can also use the [derivatives] of $\delta(x)$, namely $\delta'(x)$, $\delta''(x)$, ..., which are even more discontinuous and less "proper" than $\delta(x)$ itself.

Many years later Dirac gave credit to his youthful *engineering* training for his ability to break free of a too-restrictive loyalty to absolutely pure mathematical rigor[3]:

> I would like to try to explain the effect of this engineering training on me. I did not make any further use of the detailed applications of this work, but it did change my whole outlook to a very large

extent. Previously, I was interested only in exact equations. Well, the engineering training which I received did teach me to tolerate approximations, and I was able to see that even theories based on approximations could sometimes have a considerable amount of beauty in them,... I think that if I had not had this engineering training, I should not have had any success with the kind of work that I did later on,... I continued in my later work to use mostly the non rigorous mathematics of the engineers, and I think that you will find that most of my later writings do involve non rigorous mathematics.... *The pure mathematician who wants to set up all of his work with absolute accuracy is not likely to get very far in physics.* (my emphasis)

None of this is to shove sleazy mathematics under the rug, as the mathematics of impulses *has* been placed on a firm foundation since Dirac's initial use of them. While much of the early work in doing so is due to the Russian mathematician Sergei Sobolev (1908–1989), the central figure in that great achievement is generally considered to be the French mathematician Laurent Schwartz (1915–2002), with the publication of his two books *Theory of Distributions* (1950, 1951). For his work in distribution theory Schwartz received the 1950 Fields Medal, an award often called the "Nobel Prize in mathematics."

Now, at last, we come to the central idea of what an impulse *is*. Using the imagery that an integral of a function is the *area* bounded by that function, it should be clear that

$$\int_{-\infty}^{t} \delta(s)\,ds = \begin{cases} 0, & t < 0, \\ 1, & t > 0, \end{cases}$$

where s is, of course, simply a dummy variable of integration. If $t < 0$ then the impulse (which has unit area) is not in the interval of integration—$\delta(s)$ is located *at* $s = 0$—and so the integral is zero, while if $t > 0$ then the impulse *is* in the interval of integration and so the integral *is the area of the impulse*, that is, one. This behavior, a function that is zero for $t < 0$ and one for $t > 0$, is called the *unit step* function (because, as figure 5.1.3 shows, its graph looks like the cross-section of a *step*). The unit step, written as $u(t)$, is discontinuous at $t = 0$, and I'll

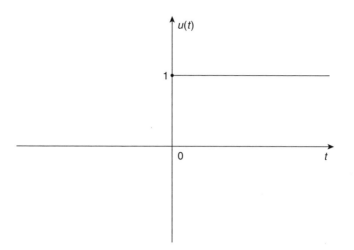

Figure 5.1.3. The unit step function

avoid (until section 5.3) the question what $u(0)$ "equals." Thus,

$$u(t) = \int\limits_{-\infty}^{t} \delta(s)\,ds,$$

and so, differentiating both sides, we formally have the unit area impulse as the derivative of the unit step:

$$\boxed{\delta(t) = \frac{d}{dt}u(t)}\;.$$

This makes some intuitive sense, too, as $u(t)$ is a *constant* everywhere except at $t = 0$, where it makes a jump of one in zero time (and now recall the definition of the derivative). This imagery relating the step and impulse functions was almost certainly suggested to Dirac by his under-graduate electrical engineering training, where he first encountered the step function from the books on electrical circuits and electromag-netic wave theory by the English mathematical electrical engineer Oliver Heaviside (1850–1925). One still occasionally finds the step function called the *Heaviside step* by both electrical engineers and mathematicians (often with the notation $H(t)$ in his honor).

We've actually encountered the step earlier in this book; look back to section 1.8, where we derived Dirichlet's discontinuous integral

$$\int_0^\infty \frac{\sin(\omega x)}{\omega} d\omega = \begin{cases} +\frac{\pi}{2}, & x > 0, \\ -\frac{\pi}{2}, & x < 0. \end{cases}$$

There I used the sgn(x) function to rewrite the right-hand side more compactly, but we could just as easily have used the step. That is,

$$\int_0^\infty \frac{\sin(\omega x)}{\omega} d\omega = \pi \left[u(x) - \frac{1}{2} \right].$$

Suppose we now proceed, in a formal way, to differentiate both sides of this; then

$$\frac{d}{dx} \int_0^\infty \frac{\sin(\omega x)}{\omega} d\omega = \pi \delta(x) = \int_0^\infty \cos(\omega x) d\omega,$$

where the last integral comes from our usual assumption that we can reverse the order of differentiation and integration. Since the cosine is an even function, we double the integral if we extend the integration interval from $(0, \infty)$ to $(-\infty, \infty)$:

$$2\pi \delta(x) = \int_{-\infty}^\infty \cos(\omega x) d\omega.$$

Since the sine is an odd function,

$$\int_{-\infty}^\infty \sin(\omega x) d\omega = 0,$$

and Euler's formula then tells us that

$$2\pi \delta(x) = \int_{-\infty}^\infty \cos(\omega x) d\omega + i \int_{-\infty}^\infty \sin(\omega x) d\omega = \int_{-\infty}^\infty e^{i\omega x} d\omega.$$

That is, we have the absolutely *astonishing* statement that

$$\delta(x) = \frac{1}{2\pi} \int_{-\infty}^{\infty} e^{i\omega x} d\omega .$$

This *is* an astonishing statement because the integral just *doesn't make any sense* if we attempt to actually evaluate it, because $e^{i\omega t}$ does not approach a limit as $|\omega| \to \infty$. The real and imaginary parts of $e^{i\omega t}$ both simply *oscillate* forever, and never approach any final values. The only way we can make any sense out of it, at all, is to interpret the right-hand side of the boxed expression (the integral) as just a collection of printed squiggles that denote the same *concept* that the printed squiggles (the impulse) on the left do. Any time we encounter the right-hand squiggles, we'll just replace them with a δ.

All of this "symbol-pushing" may be leaving you just a bit numb, so let me end this section with an example that I think will convince you that there *is* more than madness here. Consider the periodic function shown in figure 5.1.4, consisting entirely of unit impulses. The spacing between adjacent impulses is unity, that is, $f(t)$ is periodic with period $T = 1$. This particular function is often called a *periodic impulse train* (and you'll see it again when we get to the important *sampling theorem*). The Fourier series

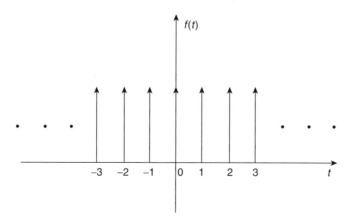

Figure 5.1.4. A periodic impulse train

for $f(t)$ is, formally (with $\omega_0 = 2\pi$ because, remember, $\omega_0 T = 2\pi$),

$$f(t) = \sum_{k=-\infty}^{\infty} c_k e^{ik2\pi t},$$

where

$$c_k = \frac{1}{T} \int_{\text{period}} f(t) e^{-ik2\pi t} dt = \int_{\text{period}} \sum_{k=-\infty}^{\infty} \delta(t-k) e^{-ik2\pi t} dt.$$

The interval of integration (of length $T = 1$) can be anywhere, but I'll pick it to be $-\frac{1}{2} < t < \frac{1}{2}$, to make it easy to see just what the integral equals. This choice of integration interval places just *one* impulse—the one at $t = 0$—right in the middle of the integration interval. If I used the interval 0 to 1, on the other hand, then there would be *two* impulses in the integration interval, one at each end—or would there? Maybe it would be two "half-impulses," whatever that might mean. The $-\frac{1}{2} < t < \frac{1}{2}$ choice smoothly avoids that nasty ambiguity, and we find that

$$c_k = \int_{-1/2}^{1/2} \delta(t) e^{-ik2\pi t} dt = 1$$

because of the sampling property of the impulse. Thus,

$$f(t) = \sum_{k=-\infty}^{\infty} e^{ik2\pi t} = \sum_{k=-\infty}^{\infty} \{\cos(k2\pi t) + i\sin(k2\pi t)\}.$$

For all t the imaginary part of $f(t)$ vanishes (which is *good*, since the $f(t)$ we started with is purely real), that is, writing the imaginary part of the right-hand sum out, term by term in pairs ($k = \pm 1, \pm 2, \pm 3, \ldots$), along with the $k = 0$ term, we see that

$$\sum_{k=-\infty}^{\infty} \sin(k2\pi t) = \sin(0) + \{\sin(2\pi t) + \sin(-2\pi t)\}$$
$$+ \{\sin(4\pi t) + \sin(-4\pi t)\} + \{\sin(6\pi t)$$
$$+ \sin(-6\pi t)\} + \cdots = 0,$$

as each bracketed pair of terms is *identically zero* for all t (and certainly $\sin(0) = 0$). So, our function $f(t)$ is given by

$$f(t) = \sum_{k=-\infty}^{\infty} \delta(t-k) = 1 + 2\sum_{k=1}^{\infty} \cos(k2\pi t)$$

But is this true?

In keeping with previous questions of this type, let's take a pragmatic approach and just calculate and plot the right-hand side and *see what it looks like*. If our boxed result makes any sense at all we should see the partial sums of the right-hand side start to 'look like' a periodic train of impulses. Figures 5.1.5 through 5.1.7 show three partial sums (using the first five, ten, and twenty terms, respectively) of the right-hand side, and

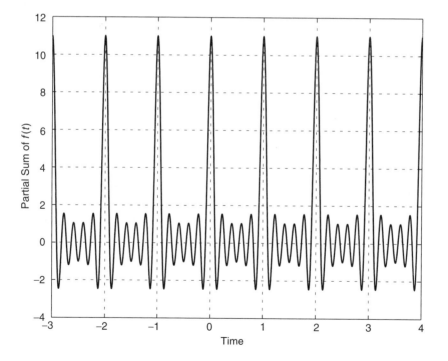

Figure 5.1.5. First five terms of an impulse train Fourier series

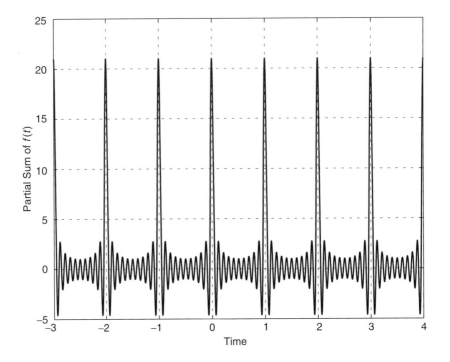

Figure 5.1.6. First ten terms of an impulse train Fourier series

you can indeed see "impulse-building" taking place. What is happening is that the cosine terms are adding together (*constructively interfering*) at integer values of t, and *destructively interfering* at all other values of t. These plots aren't a *proof* of anything, of course, but I think they are quite compelling and give us reason to believe that there is sense to all of the symbol pushing.

One last remark about impulses. Figure 5.1.1 shows us that the pulse-like function Dirac used in his original conceptualization of the impulse is even. We can show, formally, that $\delta(t)$ itself has that property, too. It is *almost* trivial to show this, in fact. If $\phi(t)$ is, as usual, any continuous function, then

$$\int_{-\infty}^{\infty} \delta(t)\phi(t)\,dt = \phi(0)$$

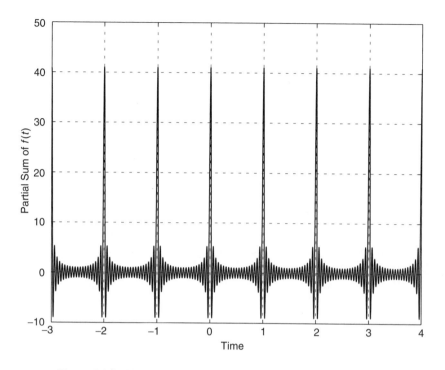

Figure 5.1.7. First twenty terms of an impulse train Fourier series

by the sampling property. But, if we make the change of variable to $s = -t$, then

$$\int\limits_{-\infty}^{\infty} \delta(-t)\phi(t)\,dt = \int\limits_{\infty}^{-\infty} \delta(s)\phi(-s)(-ds)$$

$$= \int\limits_{-\infty}^{\infty} \delta(s)\phi(-s)\,ds = \phi(-0) = \phi(0)$$

too. So, we say $\delta(t)$ and $\delta(-t)$ are *equivalent* (more loosely, equal) because they produce the same result when applied to *any* continuous $\phi(t)$ *inside an integral*. So, $\delta(t) = \delta(-t)$ and $\delta(t)$ is even. We'll use this important result at the end of the next section.

5.2 Fourier's integral theorem.

Fourier series are a mathematical description of a *periodic* function $f(t)$ (as usual, we'll think of t as representing *time*). But what if we have a function $f(t)$, defined on $-\infty < t < \infty$, that is *not* periodic? Since the function definition already "uses up" the entire infinite t-line, we can't use our usual trick of periodically extending the function—there's nowhere *left* to extend $f(t)$ into. So, Fourier *series* are out. But there is yet another devilishly clever trick available for us to play. What if we simply think of our nonperiodic $f(t)$ as having an *infinitely long* period, and so it *is* periodic (we are simply seeing, as t goes from $-\infty$ to $+\infty$, the period we "just happen" to be living in)? This *is* clever—no, I'll be honest, it's actually pretty *outrageous*—but in fact it is the standard approach used in just about every engineering textbook I have seen. That approach takes the Fourier series equations

$$f(t) = \sum_{k=-\infty}^{\infty} c_k e^{ik\omega_0 t},$$

$$c_k = \frac{1}{T} \int_{\text{period}} f(t) e^{-ik\omega_0 t}\, dt, \quad \omega_0 T = 2\pi$$

and explores what happens to them "mathematically" as $T \to \infty$.

Perhaps "explore" is too gentle a word—what we'll do is *play* with the two Fourier series equations in a pretty rough-and-ready way with little (if any) regard to justifying the manipulations. But—and this is important to understand—once we are done *it won't matter*. Once we have the mathematical result that is our goal (i.e., the answer to "what happens as $T \to \infty$?") we can *forget* how we got it and simply treat it as a *definition*. The reason we can do this is because that result—called the *Fourier integral*—has deep *physical significance*, which is Nature's way of telling us that, although we may have been a bit "casual" in getting to the result, the result is still a *good* result. Indeed, many books on the Fourier integral, written by *mathematicians*, take precisely this course of action.

Okay, let's get started on seeing what happens as $T \to \infty$. Notice first that the $k\omega_0$ in the exponent of the c_k integral changes by ω_0 as k increases from one integer to the next. If we call this *change* $\Delta\omega$, then

$\Delta\omega = \omega_0$. Now, since $\omega_0 = 2\pi/T$, as $T \to \infty$ we see that $\omega_0 \to 0$, that is, ω_0 becomes arbitrarily small, and so we should write $d\omega$ and not $\Delta\omega$ as $T \to \infty$, that is, as $T \to \infty$ the change in the fundamental frequency becomes a *differential* change. Thus, for our first result, we have

$$\lim_{T\to\infty} \Delta\omega = \lim_{T\to\infty} \omega_0 = \lim_{T\to\infty} \frac{2\pi}{T} = d\omega.$$

In addition, it therefore follows that, as $T \to \infty$, $k(2\pi/T) \to kd\omega$. Since $d\omega$ is, by definition, infinitesimally small, as k varies from $-\infty$ to $+\infty$, $kd\omega$ "should behave" like a *continuous* variable which we will call ω, that is,

$$\lim_{T\to\infty} k\omega_0 = \omega.$$

Now, if you accept all of that, then let's continue by rewriting the c_k expression (with $T \to \infty$ and integrating over the period symmetrical around $t = 0$) as

$$\lim_{T\to\infty} c_k = \lim_{T\to\infty} \frac{1}{T} \int_{-T/2}^{T/2} f(t)e^{-ik\omega_0 t}\,dt = \lim_{T\to\infty} \frac{1}{2\pi}\cdot\frac{2\pi}{T} \int_{-T/2}^{T/2} f(t)e^{-ik\omega_0 t}\,dt$$

$$= \lim_{T\to\infty} \frac{1}{2\pi}\left[\int_{-T/2}^{T/2} f(t)e^{-ik\omega_0 t}\,dt\right]\frac{2\pi}{T} = \frac{1}{2\pi}\left[\int_{-T/2}^{T/2} f(t)e^{-i\omega t}\,dt\right]d\omega.$$

Or, if we *define* the integral in the brackets as the so-called *Fourier transform* of $f(t)$, written as $F(\omega)$, then we have

$$\lim_{T\to\infty} c_k = \frac{1}{2\pi}F(\omega)\,d\omega,$$

where

$$F(\omega) = \int_{-\infty}^{\infty} f(t)e^{-i\omega t}\,dt \quad.$$

The Fourier transform has many wonderful mathematical properties and a highly useful physical interpretation, all of which I'll develop as we move further into this chapter.

Finally, to complete this line of "reasoning," if we now insert our result for $\lim_{T \to \infty} c_k$ back into our formula for $f(t)$, we get

$$f(t) = \sum_{k=-\infty}^{\infty} \left\{ \frac{1}{2\pi} F(\omega) \, d\omega \right\} e^{ik\omega_0 t} = \frac{1}{2\pi} \sum_{k=-\infty}^{\infty} F(\omega) e^{ik\omega_0 t} \, d\omega.$$

This formula is, of course, a bit of a mixed bag of notation, since parts of it are for $T \to \infty$ and other parts of it are written as if T is still finite. In fact, so goes the final step of this argument, as $T \to \infty$ the summation becomes an integral (and "of course" $k\omega_0 \to \omega$), and so we arrive at the *inverse* Fourier transform that goes in the reverse direction from the Fourier transform itself, that is, the inverse transform gives $f(t)$ from $F(\omega)$,

$$f(t) = \frac{1}{2\pi} \int_{-\infty}^{\infty} F(\omega) e^{i\omega t} \, d\omega \quad .$$

The two expressions in the boxes are called the *Fourier transform integral pair*. Unlike Fourier series, which were actually used long before Fourier, the Fourier transform is his alone, appearing for the first time in Fourier's writings. Symbolically, we'll write the one-to-one correspondence of $f(t)$ and its Fourier transform $F(\omega)$ as

$$f(t) \leftrightarrow F(\omega).$$

The doubleheaded arrow indicates that each side is *uniquely* determined by the other. The convention is to always write the time function on the left, and the frequency function—the transform—on the right. Written this way, $f(t)$ and $F(\omega)$ are simply called a *Fourier pair*.

Many "pure" analysts are truly aghast as the preceding "derivation," and I must admit, although I have introduced the Fourier transform pair to my own students in just that way for the past thirty years, I've always felt just a tiny bit guilty about it. Not enough to stop me, of course—I'll reverse the order of integration on a double integral as fast as you can snap your fingers. I pin my fate (perhaps foolishly) on the hope that, as one mathematician recently put it,[4] "In mathematics, as in life, virtue is not always rewarded, *nor vice always punished*" (my emphasis). Still, since it's always good to know how to do something in more than one way, let me give you an alternative approach to the Fourier transform pair that perhaps is just a little easier to swallow.

Recall, from the previous section, our "astonishing expression" for $\delta(x)$, written as an integral:

$$\delta(x) = \frac{1}{2\pi} \int_{-\infty}^{\infty} e^{i\omega x}\, d\omega.$$

From this it immediately follows that

$$\delta(x-y) = \frac{1}{2\pi} \int_{-\infty}^{\infty} e^{i\omega(x-y)}\, d\omega.$$

Now, suppose we take an arbitrary function $h(x)$ and use the sampling property of the impulse to write

$$h(x) = \int_{-\infty}^{\infty} \delta(x-y) h(y)\, dy.$$

Thus,

$$h(x) = \int_{-\infty}^{\infty} h(y) \left\{ \frac{1}{2\pi} \int_{-\infty}^{\infty} e^{i\omega(x-y)}\, d\omega \right\} dy$$

$$= \frac{1}{2\pi} \int_{-\infty}^{\infty} e^{i\omega x} \left\{ \int_{-\infty}^{\infty} h(y) e^{-i\omega y}\, dy \right\} d\omega.$$

The last step is surely now obvious—we write this as a *pair* of integrals; the interior integral is

$$H(\omega) = \int_{-\infty}^{\infty} h(y) e^{-i\omega y}\, dy = \int_{-\infty}^{\infty} h(x) e^{-i\omega x}\, dx$$

and the exterior integral is

$$h(x) = \frac{1}{2\pi} \int_{-\infty}^{\infty} H(\omega) e^{i\omega x}\, d\omega.$$

But this is just the Fourier transform pair again (with $h(x)$ instead of $f(t)$). That's it!

The Fourier transform of a real-valued function $f(t)$ is generally a complex-valued quantity, but it can't just be *anything*. There *are* constraints. It is easy to show, for example, that $|F(\omega)|^2$ is always *even*. The reason we are going to be interested in this is that the quantity $|F(\omega)|^2$ is intimately related to the *physical* interpretation of $F(\omega)$—$(1/2\pi)|F(\omega)|^2$ is called the *energy spectrum* of $f(t)$, and you'll see why in the next section—but for now let me just demonstrate the evenness property of $|F(\omega)|^2$ for any real-valued $f(t)$. Using Euler's formula, we can write

$$F(\omega) = \int_{-\infty}^{\infty} f(t)e^{-i\omega t}\, dt = \int_{-\infty}^{\infty} f(t)\cos(\omega t)\, dt - i\int_{-\infty}^{\infty} f(t)\sin(\omega t)\, dt.$$

Writing $F(\omega)$ explicitly as a complex-valued quantity, that is, writing

$$F(\omega) = R(\omega) + iX(\omega),$$

gives

$$R(\omega) = \int_{-\infty}^{\infty} f(t)\cos(\omega t)\, dt$$

and

$$X(\omega) = -\int_{-\infty}^{\infty} f(t)\sin(\omega t)\, dt,$$

where of course both $R(\omega)$ and $X(\omega)$ are real-valued functions of ω (because $f(t)$ is real).

Since $\cos(\omega t)$ and $\sin(\omega t)$ are even and odd, respectively,

$$R(-\omega) = \int_{-\infty}^{\infty} f(t)\cos(-\omega t)\, dt = R(\omega),$$

$$X(-\omega) = -\int_{-\infty}^{\infty} f(t)\sin(-\omega t)\, dt = -X(\omega).$$

That is, $R(\omega)$ is even, while $X(\omega)$ is odd, so both $R^2(\omega)$ and $X^2(\omega)$ are even and, since

$$|F(\omega)|^2 = R^2(\omega) + X^2(\omega),$$

$|F(\omega)|^2$ (and so $|F(\omega)|$, too) is even. If—remember—$f(t)$ is *real*-valued.

If we impose further requirements on $f(t)$ beyond being simply real, then we can say even more about $F(\omega)$. For example, if $f(t)$ is even (odd) then it should be clear from the $R(\omega)$ and $X(\omega)$ integrals that $F(\omega)$ is purely real (imaginary). (I'll use this observation at the end of this section.) And in chapter 6 I'll show you that if $f(t)$ is a *causal* function (defined as any $f(t)$ with the property that $f(t) = 0$ for $t < 0$) then $R(\omega)$ and $X(\omega)$ are so closely related that each *completely* determines the other.

As a final observation in this section on the Fourier transform, the transform is in a certain sense more general than the Fourier series, even though you'll recall that we "derived" the transform *from* the series. By this I mean that nonperiodic functions do not have a Fourier series, but *all* "well-behaved" functions have Fourier transforms, even periodic ones (which have Fourier series, too, of course). That is, if we take the Fourier series of a periodic function,

$$f(t) = \sum_{k=-\infty}^{\infty} c_k e^{ik\omega_0 t},$$

and insert it into the Fourier transform integral, we get

$$f(t) \longleftrightarrow F(\omega) = \int_{-\infty}^{\infty} \left\{ \sum_{k=-\infty}^{\infty} c_k e^{ik\omega_0 t} \right\} e^{-i\omega t}\, dt = \sum_{k=-\infty}^{\infty} c_k \int_{-\infty}^{\infty} e^{i(k\omega_0 - \omega)t}\, dt.$$

That integral ought to look vaguely familiar—if you remember our astonishing statement from the previous section—

$$\delta(x) = \frac{1}{2\pi} \int_{-\infty}^{\infty} e^{i\omega x}\, d\omega,$$

which if we replace x with t becomes

$$\delta(t) = \frac{1}{2\pi} \int_{-\infty}^{\infty} e^{i\omega t}\, d\omega.$$

Now, let me show you a little "trick": replace every t with ω and every ω with t! This "trick" is based on the observation that the particular squiggles we use in our equations are all historical accidents. The *only absolute constraint* that we *must* follow in writing equations is to be consistent; if we do something on one side of an equation then we must do the same thing on the other side. Swapping t for ω and ω for t *on both sides* does *nothing* to alter the mathematical truth of the original statement. So,

$$\delta(\omega) = \frac{1}{2\pi} \int_{-\infty}^{\infty} e^{i\omega t}\, dt,$$

which gives us an integral representation for an impulse in the *frequency* domain that says

$$\int_{-\infty}^{\infty} e^{i(k\omega_0 - \omega)t}\, dt = 2\pi\,\delta(k\omega_0 - \omega) = 2\pi\,\delta(\omega - k\omega_0),$$

because the impulse is even. Thus, a periodic $f(t)$ with period $2\pi/\omega_0$ in the t-domain has the Fourier transform

$$F(\omega) = 2\pi \sum_{k=-\infty}^{\infty} c_k\delta(\omega - k\omega_0),$$

which is an impulse train in the ω-domain (the impulses are regularly spaced at intervals of ω_0 but the train itself is *not* necessarily periodic as the c_k are generally all different).

5.3 Rayleigh's energy formula, convolution, and the autocorrelation function.

The Fourier transform has a beautiful physical interpretation, an *energy* property analogous to Parseval's *power* formula for Fourier series. The total energy of the nonperiodic real function $f(t)$ is defined to be

$$W = \int_{-\infty}^{\infty} f^2(t)\,dt = \int_{-\infty}^{\infty} f(t)f(t)\,dt.$$

(The total energy of a *periodic* function is, of course, infinite, which is why we used the energy *per period*—the power—when working with periodic functions in chapter 4.) Writing one of the $f(t)$ factors in the last integrand above in terms of the inverse Fourier transform,

$$W = \int\limits_{-\infty}^{\infty} f(t) \left\{ \frac{1}{2\pi} \int\limits_{-\infty}^{\infty} F(\omega) e^{i\omega t} \, d\omega \right\} dt$$

or, reversing the order of integration,

$$W = \int\limits_{-\infty}^{\infty} \frac{1}{2\pi} F(\omega) \left\{ \int\limits_{-\infty}^{\infty} f(t) e^{i\omega t} \, dt \right\} d\omega.$$

Since $f(t)$ is real, then the second integral is the *conjugate* of $F(\omega)$, that is,

$$\int\limits_{-\infty}^{\infty} f(t) e^{i\omega t} \, dt = F^*(\omega),$$

and so

$$W = \int\limits_{-\infty}^{\infty} \frac{1}{2\pi} F(\omega) F^*(\omega) \, d\omega = \boxed{\int\limits_{-\infty}^{\infty} \frac{1}{2\pi} |F(\omega)|^2 \, d\omega = \int\limits_{-\infty}^{\infty} f^2(t) \, dt} \,.$$

The formula in the box is the Fourier transform equivalent of Parseval's formula in Fourier series, and is often called *Rayleigh's energy formula*, after the great English mathematical physicist John William Strutt (1842–1919), better known as Lord Rayleigh, who published it in 1889. The energy *spectrum* of the function $f(t)$ is defined to be $(1/2\pi) \, | \, F(\omega) \, |^2$ because that quantity describes how the energy of $f(t)$ is *distributed* over frequency (ω). That is, integration of $(1/2\pi) \, | \, F(\omega) \, |^2$ over the interval $\omega_1 < \omega < \omega_2$ gives the energy of $f(t)$ in that frequency interval $(\omega_1 = -\infty$ and $\omega_2 = \infty$ gives, of course, the *total* energy of $f(t))$. For this reason, $(1/2\pi) \, | \, F(\omega) \, |^2$ is called the *energy spectral density* *(ESD)*.

A little aside. We could, and many analysts do, eliminate the sometimes annoying $1/2\pi$ (annoying because it's just one more thing to have to carry along in the equations). This is done by expressing the interval of integration in the frequency *domain in terms of hertz (look back at section 1.4) instead of radians per second, that is, by using v instead of ω $(= 2\pi v)$. Thus, $d\omega = 2\pi dv$ and so*

$$W = \int_{-\infty}^{\infty} f^2(t)\,dt = \int_{-\infty}^{\infty} |F(v)|^2\,dv.$$

The energy formula is very useful as a pure mathematical tool, in addition to its physical energy interpretation. Let me give you just three examples of this right now (more later).

Example 1. Consider the very simple function

$$f(t) = \begin{cases} 1, & |t| < \frac{\tau}{2}, \\ 0, & \text{otherwise}, \end{cases}$$

a pulse of duration τ centered on the origin ($\tau = 0$). Its Fourier transform is

$$F(\omega) = \int_{-\infty}^{\infty} f(t)e^{-i\omega t}\,dt = \int_{-\tau/2}^{\tau/2} e^{-i\omega t}\,dt = \left(\frac{e^{-i\omega t}}{-i\omega}\right)\Big|_{-\tau/2}^{\tau/2}$$

$$= \frac{e^{-\omega\tau/2} - e^{\omega\tau/2}}{-i\omega} = \frac{-i2\sin(\omega\tau/2)}{-i\omega} = 2\frac{\sin(\omega\tau/2)}{\omega} = \tau\frac{\sin(\omega\tau/2)}{(\omega\tau/2)}.$$

Rayleigh's energy formula then tells us that

$$\int_{-\infty}^{\infty} f^2(t)\,dt = \int_{-\tau/2}^{\tau/2} dt = \tau = \frac{1}{2\pi}\int_{-\infty}^{\infty} \tau^2\frac{\sin^2(\omega\tau/2)}{(\omega\tau/2)^2}\,d\omega.$$

Or, if we change variable to $x = \omega\tau/2$ (and so $d\omega = (2/\tau)dx$), we have

$$\tau = \frac{1}{2\pi}\int_{-\infty}^{\infty} \tau^2\frac{\sin^2(\omega\tau/2)}{x^2}\cdot\frac{2}{\tau}\,dx.$$

That is,

$$\int_{-\infty}^{\infty} \frac{\sin^2(x)}{x^2}\,dx = \pi,$$

or, since the integrand is even,

$$\int_0^\infty \frac{\sin^2(x)}{x^2}\,dx = \frac{\pi}{2}.$$

This definite integral—which is similar to the result we derived back in section 1.8, $\int_0^\infty (\sin(u)/u)\,du = (\pi/2)$—occurs often in advanced mathematics, physics, and engineering analysis, and is *not* easily derived by other means.

Example 2. Suppose now that $f(t) = e^{-\sigma t}u(t)$, where σ is any positive ($\sigma > 0$) constant and $u(t)$ is the step function introduced in the previous section. The Fourier transform of $f(t)$ is

$$F(\omega) = \int_{-\infty}^\infty f(t)e^{-i\omega t}\,dt = \int_0^\infty e^{-\sigma t}e^{-i\omega t}\,dt$$

$$= \int_0^\infty e^{-(\sigma+i\omega)t}\,dt = \left(\frac{e^{-(\sigma+i\omega)t}}{-(\sigma+i\omega)}\right)\bigg|_0^\infty = \frac{1}{\sigma+i\omega};$$

that is, we have the Fourier transform pair

$$e^{-\sigma t}u(t) \leftrightarrow \frac{1}{\sigma+i\omega}, \sigma > 0.$$

We thus have $|F(\omega)|^2 = 1/(\sigma^2+\omega^2)$, and Rayleigh's energy formula then tells us that

$$\int_{-\infty}^\infty f^2(t)\,dt = \int_0^\infty e^{-2\sigma t}\,dt = \left(\frac{e^{-2\sigma t}}{-2\sigma}\right)\bigg|_0^\infty = \frac{1}{2\sigma} = \frac{1}{2\pi}\int_{-\infty}^\infty |F(\omega)|^2\,d\omega$$

$$= \frac{1}{2\pi}\int_{-\infty}^\infty \frac{d\omega}{\sigma^2+\omega^2},$$

or

$$\int_{-\infty}^\infty \frac{d\omega}{\sigma^2+\omega^2} = \frac{\pi}{\sigma}, \sigma > 0.$$

We can write this as

$$\int_{-\infty}^{\infty} \frac{d\omega}{\sigma^2(1 + (\omega^2/\sigma^2))} = \frac{\pi}{\sigma},$$

or

$$\int_{-\infty}^{\infty} \frac{d\omega}{(1 + (\omega^2/\sigma^2))} = \pi\sigma.$$

Changing variables to $x = \omega/\sigma$ (and so $d\omega = \sigma \, dx$), we have

$$\int_{-\infty}^{\infty} \frac{\sigma dx}{1 + x^2} = \pi\sigma,$$

or, at last,

$$\boxed{\int_{-\infty}^{\infty} \frac{dx}{1 + x^2} = \pi}.$$

This is, of course, just a special case of the general integration formula $\int (dx/1 + x^2) = \tan^{-1}(x)$. But, it is useful for us, here, to use Fourier transform theory to get this result because now we can use our transform pair to answer a puzzle I mentioned back in section 5.1—what is the "value" of $u(0)$, the step function *at* $t = 0$? We have left that "value" undefined up to now, but in fact Fourier transform theory says that $u(0)$ *cannot* just be anything. To see this, we'll use the inverse Fourier transform to write

$$e^{-\sigma t} u(t) = \frac{1}{2\pi} \int_{-\infty}^{\infty} \frac{1}{\sigma + i\omega} e^{i\omega t} \, d\omega.$$

Then, setting $t = 0$,

$$u(0) = \frac{1}{2\pi} \int_{-\infty}^{\infty} \frac{1}{\sigma + i\omega} d\omega = \frac{1}{2\pi} \int_{-\infty}^{\infty} \frac{\sigma - i\omega}{\sigma^2 + \omega^2} d\omega$$

$$= \frac{\sigma}{2\pi} \int_{-\infty}^{\infty} \frac{d\omega}{\sigma^2 + \omega^2} - i \frac{1}{2\pi} \int_{-\infty}^{\infty} \frac{\omega}{\sigma^2 + \omega^2} d\omega.$$

The second integral is zero because its integrand is an odd function of ω, and so $u(0)$ has the purely real value (no surprise with that!) of

$$u(0) = \frac{\sigma}{2\pi} \int_{-\infty}^{\infty} \frac{d\omega}{\sigma^2 + \omega^2}.$$

Just a few steps back, however, we showed that this integral equals (π/σ), and so we have $u(0) = \frac{1}{2}$. In retrospect, from how Fourier series behave at a discontinuity, this is just what we'd expect; $u(0)$ is the *average* of the values of $u(t)$ on each side of the discontinuity *at $t = 0$*.

Example 3. Suppose now that

$$f(t) = \begin{cases} e^{-at}, & 0 \le t \le m, \\ 0, & \text{otherwise.} \end{cases}$$

Then,

$$F(\omega) = \int_{-\infty}^{\infty} f(t) e^{-i\omega t}\, dt = \int_0^m e^{-at} e^{-i\omega t}\, dt = \int_0^m e^{-(a+i\omega)t}\, dt$$

$$= \left(\frac{e^{-(a+i\omega)t}}{-(a+i\omega)} \right)\Bigg|_0^m = \frac{e^{-(a+i\omega)m} - 1}{-(a+i\omega)} = \frac{e^{-ma}\{\cos(m\omega) - i\sin(m\omega)\} - 1}{-(a+i\omega)}$$

$$= \frac{1 - e^{-ma}\cos(m\omega) + ie^{-ma}\sin(m\omega)}{a + i\omega}.$$

Therefore,

$$|F(\omega)|^2 = \frac{\{1 - e^{-ma}\cos(m\omega)\}^2 + \{e^{-ma}\sin(m\omega)\}^2}{\omega^2 + a^2}$$

$$= \frac{1 - 2e^{-ma}\cos(m\omega) + e^{-2ma}\cos^2(m\omega) + e^{-2ma}\sin^2(m\omega)}{\omega^2 + a^2}$$

$$= \frac{1 + e^{-2ma} - 2e^{-ma}\cos(m\omega)}{\omega^2 + a^2}.$$

So, by Rayleigh's energy formula, we have

$$\frac{1}{2\pi} \int_{-\infty}^{\infty} \frac{1 + e^{-2ma} - 2e^{-ma}\cos(m\omega)}{\omega^2 + a^2} \, d\omega$$

$$= \int_{-\infty}^{\infty} f^2(t) \, dt = \int_{0}^{m} e^{-2at} \, dt = \left(\frac{e^{-2at}}{-2a} \Big|_{0}^{m} \right) = \frac{1 - e^{-2ma}}{2a}.$$

Or,

$$\int_{-\infty}^{\infty} \frac{1 + e^{-2ma} - 2e^{-ma}\cos(m\omega)}{\omega^2 + a^2} \, d\omega = \frac{\pi}{a}(1 - e^{-2ma}).$$

Or,

$$2e^{-ma} \int_{-\infty}^{\infty} \frac{\cos(m\omega)}{\omega^2 + a^2} \, d\omega = (1 + e^{-2ma}) \int_{-\infty}^{\infty} \frac{d\omega}{\omega^2 + a^2} - \frac{\pi}{a}(1 - e^{-2ma}).$$

In the integral on the right, change variables to $x = \omega/a$, $(d\omega = a \, dx)$, and so

$$\int_{-\infty}^{\infty} \frac{d\omega}{\omega^2 + a^2} = \frac{1}{a^2} \int_{-\infty}^{\infty} \frac{d\omega}{(\omega/a)^2 + 1} = \frac{1}{a^2} \int_{-\infty}^{\infty} \frac{a \, dx}{x^2 + 1} = \frac{1}{a} \int_{-\infty}^{\infty} \frac{dx}{x^2 + 1},$$

or, using our result from example 2,

$$\int_{-\infty}^{\infty} \frac{d\omega}{\omega^2 + a^2} = \frac{\pi}{a}.$$

Thus,

$$2e^{-ma} \int_{-\infty}^{\infty} \frac{\cos(m\omega)}{\omega^2 + a^2} \, d\omega = (1 + e^{-2ma})\frac{\pi}{a} - \frac{\pi}{a}(1 - e^{-2ma}) = \frac{2\pi}{a} e^{-2ma},$$

or

$$\int_{-\infty}^{\infty} \frac{\cos(m\omega)}{\omega^2 + a^2} \, d\omega = \frac{\pi}{a} e^{-ma},$$

a very pretty result that would, without the aid of Rayleigh's energy formula, be much more difficult to derive. (Notice that it reduces to our earlier, simpler integral for the case of $m = 0$.)

As I've derived this last result, we have explicitly taken $m > 0$, that is, we started the derivation with an $f(t)$ defined on the interval $0 < t < m$. I'll leave it for you to verify, but if one does this calculation for $f(t) = e^{at}$ for $m < t < 0$ (which of course says $m < 0$), then the result is

$$\int_{-\infty}^{\infty} \frac{\cos(m\omega)}{\omega^2 + a^2} \, d\omega = \frac{\pi}{a} e^{ma}, \quad m < 0.$$

We can write one expression that covers both possibilities for m as follows:

$$\boxed{\int_{-\infty}^{\infty} \frac{\cos(m\omega)}{\omega^2 + a^2} \, d\omega = \frac{\pi}{a} e^{-|m|a}}\,.$$

The Rayleigh energy formula has an immediate *mathematical* implication that has a quite interesting *physical* interpretation. If $f(t)$ has finite energy (a constraint that certainly includes almost all time functions of engineering interest—the impulse and step are exceptions, as I'll discuss soon), then Rayleigh's energy formula says

$$\int_{-\infty}^{\infty} f^2(t) \, dt = \frac{1}{2\pi} \int_{-\infty}^{\infty} |F(\omega)|^2 \, d\omega < \infty.$$

But this can be true only if $\lim_{|\omega| \to \infty} |F(\omega)|^2 = 0$, as otherwise the ω-integral would blow up. (I've noticed that sometimes students have to think a bit about this: suppose instead that $\lim_{|\omega| \to \infty} |F(\omega)|^2 = \epsilon > 0$, where ϵ is as small as you like *but not zero*; since the ω-integral is from $-\infty$ to $+\infty$, $|F(\omega)|^2$ would bound *infinite* area above the ω-axis.) In fact, $|F(\omega)|^2$ not only must vanish as $|\omega| \to \infty$, but must do so at a sufficiently rapid rate, that is, it must vanish faster than does $1/|\omega|$ for the ω-integral to exist ($|F(\omega)|^2$ vanishing only as fast as $1/|\omega|$ still allows the ω-integral to diverge logarithmically). That is, for the energy of $f(t)$ to be finite $F(\omega)$ must vanish fast enough that $\lim_{|\omega| \to \infty} \omega |F(\omega)|^2 = 0$.

Mathematicians call this the *Riemann-Lebesgue lemma*, after the French mathematician Henri Lebesgue (1875–1941) and, of course, Riemann. For an electrical engineer it means that the energy spectral density of any "real-world" time function must "roll off to zero" at a fairly fast rate as one goes ever higher in frequency. Notice that, for each of our three examples, the Riemann-Lebesgue lemma is indeed satisfied. The same sort of argument, by the way, used with Parseval's theorem for a periodic function $f(t)$ with *finite* power, shows that $\lim_{|n|\to\infty} |c_n|^2 = 0$, where the c_n are the coefficients in the complex Fourier series expansion of $f(t)$. In that case an electrical engineer would say the *power* spectrum of $f(t)$ "rolls off to zero" as one goes ever higher in frequency.

The Rayleigh energy formula is actually a special case of a much more general result that we can find by asking the following question: if $m(t)$ and $g(t)$ are two time functions with Fourier transforms $M(\omega)$ and $G(\omega)$, respectively, then what is the Fourier transform of $m(t)g(t)$? When we get to chapter 6, you'll see how forming the product of two time functions is essential to the operation of speech scramblers and radios. For now, this is purely of *mathematical* interest. The transform of $m(t)g(t)$ is, by definition,

$$\int_{-\infty}^{\infty} m(t)g(t)e^{-i\omega t}\,dt = \int_{-\infty}^{\infty} m(t)\left\{\frac{1}{2\pi}\int_{-\infty}^{\infty} G(u)e^{iut}\,du\right\}e^{-i\omega t}\,dt,$$

where $g(t)$ has been written in the form of an inverse Fourier transform (I've used u as the dummy variable of integration in the inner integral, rather than ω, to avoid confusion with the outer ω). So, continuing, if we reverse the order of integration we have the transform of $m(t)g(t)$ as

$$\int_{-\infty}^{\infty} \frac{1}{2\pi}G(u)\left\{\int_{-\infty}^{\infty} m(t)e^{iut}e^{-i\omega t}\,dt\right\}du$$

$$= \frac{1}{2\pi}\int_{-\infty}^{\infty} G(u)\left\{\int_{-\infty}^{\infty} m(t)e^{-i(\omega-u)t}\,dt\right\}du,$$

or, as the inner integral is just $M(\omega - u)$, we have the Fourier transform pair

$$m(t)g(t) \longleftrightarrow \frac{1}{2\pi} \int\limits_{-\infty}^{\infty} G(u)M(\omega - u)\,du \quad.$$

The integral on the right occurs so often in mathematics and engineering that it has its own name; the *convolution integral*. In general, if one has any two functions $x(t)$ and $y(t)$ that are "combined" as $\int_{-\infty}^{\infty} x(\tau)y(t - \tau)\,d\tau$, we say that $x(t)$ and $y(t)$ are *convolved*, and write it in shorthand as $x(t) * y(t)$. (You'll see a lot more on convolution, and some of its applications in electronic technology, in the next chapter.) So, our new pair is simply

$$m(t)g(t) \longleftrightarrow \frac{1}{2\pi} G(\omega) * M(\omega) \quad.$$

Note carefully: the $*$ symbol denotes *conjugation* when used as a super-script, and *convolution* when used in-line in a formula. And since it is arbitrary which function we call $m(t)$ and which we call $g(t)$, then in fact convolution is commutative (this is easy to prove formally, as well—just make the obvious change of variable in the convolution integral) and so $m(t)g(t) \longleftrightarrow (1/2\pi)M(\omega) * G(\omega)$, too.

For the special case of $m(t) = g(t)$ we have

$$g^2(t) \longleftrightarrow \frac{1}{2\pi} G(\omega) * G(\omega) \quad,$$

and we can use this special case as a nice example of how a purely mathematical result can also give us a lot of insight in the physical world. Suppose $g(t)$ is what electrical engineers call a *bandlimited baseband signal*, by which they mean that $g(t)$ has all of its energy confined to the finite frequency interval $|\omega| \leq \omega_m$ (that's the *bandlimited* part). (*Baseband* means the interval is centered on $\omega = 0$.) This means, of course, that $G(\omega) = 0$ for $|\omega| > \omega_m$. The Fourier transform of $g^2(t)$ (which tells us where the energy of $g^2(t)$ is located) is, as just derived,

$$\frac{1}{2\pi} G(\omega) * G(\omega) = \int_{-\infty}^{\infty} G(u)G(\omega - u)\,du,$$

which will certainly be zero if the integrand is zero, that is, if ω is so positive (or so negative) that $G(u)G(\omega-u) = 0$. Now, obviously $G(u) \neq 0$ only if $-\omega_m \leq u \leq \omega_m$, by definition (see the upper hashed interval in figure 5.3.1). And $G(\omega - u) \neq 0$ only if $-\omega_m \leq \omega - u \leq \omega_m$, that is, if $-\omega_m - \omega \leq -u \leq \omega_m - \omega$, that is, if $\omega_m + \omega \geq u \geq -\omega_m + \omega$, that is, if $-\omega_m + \omega \leq u \leq \omega_m + \omega$ (see the lower hashed interval in figure 5.3.1). Now, imagine that we *increase* ω, which shifts the lower hashed interval to the right. We will have an overlap of the two hashed intervals (and so a nonzero integrand) as long as $-\omega_m + \omega \leq \omega_m$, that is, as long as $\omega \leq 2\omega_m$. And, if we *decrease* ω (which shifts the lower hashed interval to the left)

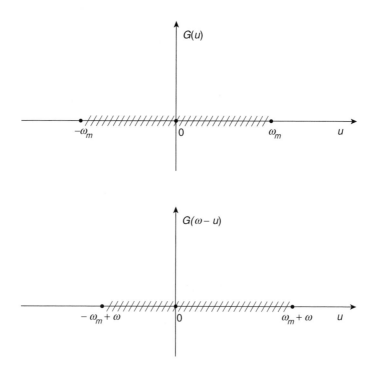

Figure 5.3.1. The integrand of a convolution integral

we will have an overlap of the two hashed intervals (and so a nonzero integrand) as long as $-\omega_m \le \omega_m + \omega$, that is, as long as $-2\omega_m \le \omega$. Thus, we have a nonzero integrand as long as $|\omega| \le 2\omega_m$. This tells us that if $g(t)$ has all of its energy confined to the frequency interval $|\omega| \le \omega_m$ then $g^2(t)$ has all of *its* energy confined to the frequency interval $|\omega| \le 2\omega_m$. You'll see, in chapter 6, how useful this conclusion is to the construction of a quite interesting electronic gadget.

In the last example we avoided the actual details of evaluating a convolution integral. Let me now show you an example of "doing the details" of a calculation, which we'll return to in the last section of this chapter. Suppose we have the time function (defined for all t)

$$f(t) = e^{-\alpha|t|}\cos(t),$$

where $\alpha > 0$. We could determine $F(\omega)$ by simply stuffing $f(t)$ into the defining integral for the Fourier transform, but our convolution result provides an elegant (and *short*) alternative. If we write $g(t) = e^{-\alpha|t|}$ and $h(t) = \cos(t)$, then we know we can write

$$F(\omega) = \frac{1}{2\pi}H(\omega) * G(\omega) = \frac{1}{2\pi}\int_{-\infty}^{\infty} H(\tau)G(\omega - \tau)d\tau.$$

Now, since $|t| = t$ for $t > 0$ and $|t| = -t$ for $t < 0$,

$$G(\omega) = \int_{-\infty}^{\infty} e^{-\alpha|t|}e^{-i\omega t}dt = \int_{-\infty}^{0} e^{\alpha t}e^{-i\omega t}dt + \int_{0}^{\infty} e^{-\alpha t}e^{-i\omega t}dt$$

$$= \int_{-\infty}^{0} e^{(\alpha - i\omega)t}dt + \int_{0}^{\infty} e^{-(\alpha + i\omega)t}dt = \left(\frac{e^{(\alpha - i\omega)t}}{\alpha - i\omega}\right)\Bigg|_{-\infty}^{0} + \left(\frac{e^{-(\alpha - i\omega)t}}{-(\alpha + i\omega)}\right)\Bigg|_{0}^{\infty}$$

$$= \frac{1}{\alpha - i\omega} + \frac{1}{\alpha + i\omega} = \frac{2\alpha}{\alpha^2 + \omega^2}.$$

Also,

$$H(\omega) = \int_{-\infty}^{\infty} \cos(t) e^{-i\omega t}\, dt = \int_{-\infty}^{\infty} \frac{e^{it} + e^{-it}}{2} e^{-i\omega t}\, dt$$

$$= \frac{1}{2}\left[\int_{-\infty}^{\infty} e^{i(1-\omega)t}\, dt + \int_{-\infty}^{\infty} e^{-i(1+\omega)t}\, dt \right].$$

At the end of section 5.2 you'll recall that we derived the result

$$\delta(\omega) = \frac{1}{2\pi} \int_{-\infty}^{\infty} e^{i\omega t}\, dt.$$

Thus,

$$\int_{-\infty}^{\infty} e^{i(1-\omega)t}\, dt = 2\pi\, \delta(1 - \omega)$$

and (because, don't forget, the impulse is *even*)

$$\int_{-\infty}^{\infty} e^{-i(1+\omega)t}\, dt = 2\pi\, \delta(1 + \omega).$$

Therefore,

$$H(\omega) = \pi\, \delta(1 - \omega) + \pi\, \delta(1 + \omega),$$

and so we immediately have

$$F(\omega) = \frac{1}{2\pi} \int_{-\infty}^{\infty} \pi\{\delta(1 - \tau) + \delta(1 + \tau)\} \frac{2\alpha}{\alpha^2 + (\omega - \tau)^2}\, d\tau,$$

which, despite its formidable appearance, is trivial to evaluate because of the sampling property of impulse functions that appear inside integrals. That is, we have the pair

$$e^{-\alpha|t|} \cos(t) \longleftrightarrow F(\omega) = \alpha\left[\frac{1}{\alpha^2 + (\omega - 1)^2} + \frac{1}{\alpha^2 + (\omega + 1)^2} \right].$$

That's it! You'll see this particular result again in section 5.6, as an illustration of the famous *uncertainty principle*.

Since we have been convolving ω-functions in the previous calculations, these last results are often referred to as *frequency convolutions*. You'll notice that ω is an arbitrary variable in our result for the general transform pair for $m(t)g(t)$ and, as a special case, if we set $\omega = 0$, we get

$$\int_{-\infty}^{\infty} m(t)g(t)e^{-i\omega t}\,dt \mid_{\omega=0} = \frac{1}{2\pi}\int_{-\infty}^{\infty} G(u)M(\omega - u)\,du \mid_{\omega=0}$$

or

$$\int_{-\infty}^{\infty} m(t)g(t)\,dt = \frac{1}{2\pi}\int_{-\infty}^{\infty} G(u)M(-u)\,du.$$

Now, since

$$M(u) = \int_{-\infty}^{\infty} m(t)e^{-iut}\,dt,$$

it is clear that $M(-u) = M^*(u)$. Thus, we have

$$\boxed{\int_{-\infty}^{\infty} m(t)g(t)\,dt = \frac{1}{2\pi}\int_{-\infty}^{\infty} G(u)M^*(u)\,du}\;.$$

And finally, if we again specialize this to the case of $m(t) = g(t)$, then $M(\omega) = G(\omega)$ and so

$$\int_{-\infty}^{\infty} g^2(t)\,dt = \frac{1}{2\pi}\int_{-\infty}^{\infty} G(u)G^*(u)\,du = \frac{1}{2\pi}\int_{-\infty}^{\infty} |\,G(\omega)\,|^2\,d\omega,$$

where in the last integral I changed the dummy variable of integration from u to ω to make it more familar—this is, of course, just Rayleigh's energy formula.

To follow up briefly on our result that $m(t)g(t) \longleftrightarrow M(\omega) * G(\omega)$, before returning to the Rayleigh energy formula itself, you might be wondering what would be the "reverse" of this pair. That is, what is the transform of $m(t) * g(t)$? If you like symmetry, you might argue that since multiplication in time pairs with convolution in frequency, then convolution in time should pair with multiplication in frequency. And you'd be right! This is actually quite easy (and most important, too) to prove. Since

$$m(t) * g(t) = \int_{-\infty}^{\infty} m(u)g(t - u)du,$$

the transform of $m(t) * g(t)$ is

$$\int_{-\infty}^{\infty} \left\{ \int_{-\infty}^{\infty} m(u)g(t - u)du \right\} e^{-i\omega t}dt,$$

or, reversing the order of integration, the transform is equal to

$$\int_{-\infty}^{\infty} m(u) \left\{ \int_{-\infty}^{\infty} g(t - u)e^{-i\omega t}dt \right\} du.$$

Now, in the inner integral let $\tau = t - u$ $(d\tau = dt)$ and our transform becomes

$$\int_{-\infty}^{\infty} m(u) \left\{ \int_{-\infty}^{\infty} g(\tau)e^{-i\omega(\tau+u)}d\tau \right\} du$$

$$= \int_{-\infty}^{\infty} m(u)e^{-i\omega u} \left\{ \int_{-\infty}^{\infty} g(\tau)e^{-i\omega\tau}d\tau \right\} du$$

$$= \int_{-\infty}^{\infty} m(u)e^{-i\omega u} G(\omega)du = G(\omega) \int_{-\infty}^{\infty} m(u)e^{-i\omega u}du = G(\omega)M(\omega).$$

That is, as claimed, we have

$$\boxed{m(t) * g(t) \longleftrightarrow M(\omega)G(\omega)} \ .$$

And once again, if we look at the special case of $m(t) = g(t)$, we have

$$\boxed{m(t) * m(t) \longleftrightarrow M^2(\omega)} \ .$$

For the last topic of this section, let me show you a result related to this last special case; this new result, in turn, is intimately related to one of the most celebrated theorems in Fourier analysis. And to *really* get your interest up, let me alert you to the little-known fact that it wasn't a mathematician who discovered it, but rather the *physicist* Albert Einstein.

I'll start by defining what is called the *autocorrelation* of the real-valued function $f(t)$:

$$R_f(\tau) = \int_{-\infty}^{\infty} f(t)f(t - \tau)\,dt$$.

Notice, *carefully*, that $R_f(\tau)$ is a function of τ, not t. I say this because the above integral bears a superficial resemblence to a convolution integral,

$$f(t) * f(t) = \int_{-\infty}^{\infty} f(\tau)f(t - \tau)\,d\tau,$$

which is a function of t, not τ. (These two integrals *do* look a lot alike—we are used to thinking of t as time, and dimensionally τ has the units of time, too, but what *is* τ?) But there is a profound difference, which I'll develop shortly. You may have your doubts about this. After all, if you remember our symbol swapping trick, you might well argue that we can make both expressions functions of the same variable—for example, in the $R_f(\tau)$ equation let's simply write t for τ and τ for t. The truth of the equality is unaffected, and so now we have

$$R_f(t) = \int_{-\infty}^{\infty} f(\tau)f(\tau - t)\,d\tau.$$

Now the only difference in the two expressions is quite clear: in $R_f(t)$ the integrand has the factor $f(\tau - t)$ while in $f(t) * f(t)$ the integrand has the factor $f(t - \tau)$. How much difference can a simple reversal in the argument of f really make, you ask? Well, you'll see that it makes a *lot* of difference.

What is $R_f(\tau)$, *physically*? $R_f(\tau)$ is a measure of the similarity of $f(t)$ with a *shifted* (τ is a time *shift!*) version of itself; hence the name *correlation*. The *auto*, of course, comes from $f(t)$ being measured against itself. While I'm not going to pursue it here, $R_f(\tau)$ can be generalized to measure $f(t)$ against *any* function $g(t)$ (also real-valued) as

$$R_{fg}(\tau) = \int_{-\infty}^{\infty} f(t)g(t-\tau)dt.$$

Then, if $g = f$ we see that $R_{fg}(\tau) = R_{ff}(\tau)$, which is just $R_f(t)$. $R_{fg}(\tau)$ is called the *cross*-correlation, and that's all I'll say about it in this book. And, while I won't pursue the technological uses of the autocorrelation function either, I see no reason not to tell you at least that it has enormous application in the construction of electronic signal processing circuitry that can "extract" an information-bearing signal that is literally *buried* in random noise. To pursue this matter at any depth would require us to plunge into the theory of stochastic processes, which would be getting pretty far beyond the scope of this book.

I'll show you a specific example of an $R_f(\tau)$ calculation soon, once we've established the following three (there are actually lots more) fundamental *general* properties of *any* $R_f(\tau)$.

(i) $R_f(0) \geq 0$. This follows immediately by simply inserting $\tau = 0$ into the defining integral and arriving at $R_f(0) = \int_{-\infty}^{\infty} f^2(t)dt$, which is certainly never negative. Indeed, $R_f(0)$ is the *energy* of $f(t)$.

(ii) $R_f(\tau)$ is even, that is, $R_f(-\tau) = R_f(\tau)$. This follows by first writing $R_f(-\tau) = \int_{-\infty}^{\infty} f(t)f(t+\tau)dt$ and then changing variables to $s = t + \tau$ ($ds = dt$). Then, $R_f(-\tau) = \int_{-\infty}^{\infty} f(s-\tau)$ $f(s)ds$, or, making the trivial notational change in the dummy variable of integration from s to t, $R_f(-\tau) = \int_{-\infty}^{\infty} f(t-\tau)f(t)$ $dt = R_f(\tau)$.

(iii) $R_f(0) \geq |R_f(\tau)|$. That is, $R_f(\tau)$ has its *maximum* value for zero time shift.[5] To establish this, first write $\int_{-\infty}^{\infty}\{f(t)\pm f(t-\tau)\}^2$ $dt \geq 0$, which should be obviously true (the integral of *anything* real, squared, is nonnegative. Then, expanding, we have

$$\int\limits_{-\infty}^{\infty} f^2(t)\,dt \pm 2\int\limits_{-\infty}^{\infty} f(t)f(t-\tau)\,dt + \int\limits_{-\infty}^{\infty} f^2(t-\tau)\,dt \geq 0.$$

Or, $R_f(0) \pm 2R_f(\tau) + R_f(0) \geq 0$ or, $R_f(0) \geq \pm R_f(\tau)$, from which $R_f(0) \geq |R_f(\tau)|$ immediately follows.

As an example to illustrate these properties, suppose that $f(t) = e^{-t}u(t)$, where $u(t)$ is the unit step function. That is, $f(t) = 0$ for $t < 0$ and $f(t) = e^{-t}$ for $t > 0$. Then,

$$R_f(\tau) = \int\limits_{-\infty}^{\infty} e^{-t}u(t)e^{-(t-\tau)}u(t-\tau)\,dt.$$

To understand the next step in the analysis, remember how the step function "works": since $u(t) = 0$ for $t < 0$ and $u(t) = 1$ for $t > 0$, $u(t-\tau) = 0$ for $t < \tau$ and $u(t-\tau) = 1$ for $t > \tau$. So, since we need *both* steps to be 1 to have a nonzero integrand, our expression for $R_f(\tau)$ takes on two different forms, depending on whether $\tau < 0$ or $\tau > 0$. Specifically,

$$R_f(\tau) = \begin{cases} \displaystyle\int\limits_{\tau}^{\infty} e^{-t}e^{-(t-\tau)}\,dt, & \tau > 0, \\[2ex] \displaystyle\int\limits_{0}^{\infty} e^{-t}e^{-(t-\tau)}\,dt, & \tau < 0. \end{cases}$$

Both of these integrals are quite easy to do. So,

If $\tau > 0$, $R_f(\tau) = e^{\tau}\displaystyle\int\limits_{\tau}^{\infty} e^{-2t}\,dt = e^{\tau}\left(\dfrac{e^{-2t}}{-2}\Bigg|_{\tau}^{\infty}\right) = e^{\tau}\dfrac{e^{-2\tau}}{2} = \dfrac{1}{2}e^{-\tau}$.

If $\tau < 0$, $R_f(\tau) = e^{\tau}\displaystyle\int\limits_{0}^{\infty} e^{-2t}\,dt = e^{\tau}\left(\dfrac{e^{-2t}}{-2}\Bigg|_{0}^{\infty}\right) = \dfrac{1}{2}e^{\tau}$.

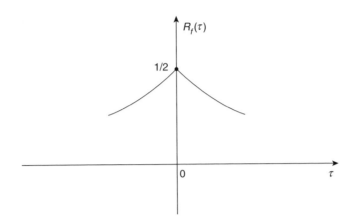

Figure 5.3.2. An autocorrelation function

We can combine these two expressions into a single formula, valid for all τ, as

$$R_f(\tau) = \frac{1}{2}e^{-|\tau|}$$,

which is shown in figure 5.3.2. That plot exhibits all three of the general properties developed earlier. Notice that $\lim_{\tau \to \pm\infty} R_f(\tau) = 0$, which means that $e^{-t}u(t)$ "looks less and less like itself" as the time shift τ increases, a conclusion I think you should find obviously true.

Now, let me end this topic by demonstrating the great difference between autocorrelation and autoconvolution, using the Fourier transform. You'll recall that earlier in this section we derived the time-*convolution* pair

$$f(t) * f(t) \longleftrightarrow F^2(\omega).$$

Let's now derive the Fourier transform of the autocorrelation (where now I'll use R_f written as a function of t). By definition, the transform of $R_f(t)$ is

$$
\int\limits_{-\infty}^{\infty} R_f(t)e^{-i\omega t}\,dt = \int\limits_{-\infty}^{\infty}\left\{\int\limits_{-\infty}^{\infty} f(\tau)f(\tau-t)\,d\tau\right\}e^{-i\omega t}\,dt
$$

$$
= \int\limits_{-\infty}^{\infty} f(\tau)\left\{\int\limits_{-\infty}^{\infty} f(\tau-t)e^{-i\omega t}\,dt\right\}d\tau
$$

where I've reversed the order of integration. Now, in the inner integral make the change of variables to $s = \tau - t$ (so $ds = -dt$). Thus, the transform of $R_f(t)$ is

$$
\int\limits_{-\infty}^{\infty} f(\tau)\left\{\int\limits_{\infty}^{-\infty} f(s)e^{-i\omega(\tau-s)}(-ds)\right\}d\tau = \int\limits_{-\infty}^{\infty} f(\tau)e^{-i\omega\tau}\left\{\int\limits_{-\infty}^{\infty} f(s)e^{i\omega s}\,ds\right\}d\tau.
$$

The inner integral is simply $F^*(\omega)$, and so the transform of $R_f(t)$ is

$$
\int\limits_{-\infty}^{\infty} f(\tau)e^{-i\omega\tau}F^*(\omega)\,d\tau = F^*(\omega)\int\limits_{-\infty}^{\infty} f(\tau)e^{-i\omega\tau}\,d\tau = F^*(\omega)F(\omega) = |F(\omega)|^2 .
$$

That is, we have the time-*correlation* pair

$$
R_f(t) \longleftrightarrow |F(\omega)|^2,
$$

which is *very different* from the $F^2(\omega)$ we got for time-convolution. $|F(\omega)|^2$ is the magnitude squared of $F(\omega)$, while $F^2(\omega)$ is the direct square of $F(\omega)$ itself. $|F(\omega)|^2$ is purely real, *always*, while $F^2(\omega)$ is generally complex.

Now, I'll end with a really big surprise. Notice that, to within a factor of 2π, $R_f(t)$ and the energy spectral density are a Fourier transform pair! This is a remarkable result, so remarkable that it has its own name: the *Wiener-Khinchin theorem*, named after the American mathematician Norbert Wiener (1894–1964), who discovered it in 1930, and the Russian mathematician Aleksandr Khinchin (1894–1959), who independently discovered it in 1934. In fact, however, both men were following in the footsteps of the theoretical physicist Albert Einstein (1879–1955), who discovered it in 1914 (in a paper delivered at the Swiss Physical Society's meeting in Basel in February of that year).[6] Like Wilbraham's 1848

paper, however, the 1914 paper by Einstein was overlooked by mathematicians until *sixty-five* years later, in 1979, the year of the Einstein
centenary. Today we remember Einstein for his revolutionary gravitational physics, but, as his 1914 *math* discovery shows, in his youth just
about *everything* technical that Einstein touched turned to gold.

5.4 Some curious spectra.

All through our discussions on the energy of a time function I've avoided
the question of *existence*, the question of whether there actually *is* a
finite value to $\int_{-\infty}^{\infty} f^2(t)\,dt$. In fact, a very useful function, the unit step,
obviously has infinite energy because

$$\int_{-\infty}^{\infty} u^2(t)\,dt = \int_{0}^{\infty} dt = \infty.$$

What does this mean with respect to the Fourier transform of $u(t)$? To
be consistent in our notation, I'll write that transform as $U(\omega)$, that
is, $u(t) \longleftrightarrow U(\omega)$. But—does $U(\omega)$ even exist? After all, if we simply
substitute $u(t)$ into the definition of $U(\omega)$, we get

$$U(\omega) = \int_{-\infty}^{\infty} u(t)e^{-i\omega t}\,dt = \int_{0}^{\infty} e^{-i\omega t}\,dt = \left(\frac{e^{-i\omega t}}{-i\omega}\Big|_{0}^{\infty}\right) = \frac{e^{-i\infty}-1}{-i\omega} = ?$$

This seems to be a first-class puzzle, as what in the world could $e^{-i\infty}$ possibly mean? And how about the impulse function, which seems to be even
more mysterious? After all, what meaning can we attach to $\int_{-\infty}^{\infty} \delta^2(t)\,dt$?
You might argue that, from the sampling property of the impulse, we
can write

$$\int_{-\infty}^{\infty} \delta^2(t)\,dt = \int_{-\infty}^{\infty} \delta(t)\delta(t)\,dt = \delta(0) = \infty,$$

but that's assuming our $\phi(t)$ function (introduced at the beginning of
section 5.1) can be set equal to $\delta(t)$, which goes far beyond what I claimed
earlier for the mathematical nature of $\phi(t)$, for example, that it is *continuous* at $t = 0$ (which $\delta(t)$ surely is not!). Oddly enough, however, it is the
impulse function and not the benign-appearing step that has the more

easily calculated Fourier transform. We can do the impulse calculation as follows.

The Fourier transform of $\delta(t - t_0)$, an impulse located at $t = t_0$, is by definition (and the sampling property) given by

$$\delta(t - t_0) \longleftrightarrow \int_{-\infty}^{\infty} \delta(t - t_0)e^{-i\omega t}\,dt = e^{-i\omega t}\,|_{t=t_0} = e^{-i\omega t_0}.$$

In particular, for $t_0 = 0$ we have the astonishingly simple Fourier transform pair

$$\boxed{\delta(t) \longleftrightarrow 1}\ .$$

And for t_0 *anything* we have the *magnitude squared* of the Fourier transform of $\delta(t - t_0)$ as $|e^{-i\omega t_0}|^2$, which Euler's formula tells us is equal to 1, a constant over all frequencies. The energy spectral density of $\delta(t - t_0)$, for *any* t_0, is simply $1/2\pi$, $-\infty < \omega < \infty$. An impulse in *time* has its energy *uniformly* distributed over all frequencies. It is clear now that the total energy of an impulse is infinite, as Rayleigh's energy formula tells us that we can express that energy as

$$\int_{-\infty}^{\infty} \frac{1}{2\pi}\,d\omega = \frac{1}{2\pi} \int_{-\infty}^{\infty} d\omega = \infty.$$

We've encountered impulses *in time* a number of times now, and that may have prompted a related question in your mind: what *time* function goes with $\delta(\omega)$, an impulse *in frequency*? All we need to do to answer that question is to substitute $\delta(\omega)$ into the inverse Fourier transform integral to get the pair

$$\frac{1}{2\pi} \int_{-\infty}^{\infty} \delta(\omega)e^{i\omega t}\,d\omega \longleftrightarrow \delta(\omega),$$

which, from the sampling property of the impulse, gives us the important result that

$$\boxed{\frac{1}{2\pi} \longleftrightarrow \delta(\omega)}\ .$$

This purely mathematical result has a certain physical plausibility to it, too. The time function $1/2\pi$, *a constant over all time,* is what electrical engineers call a *constant d c-signal.* And another way to think of a *constant* d c-signal is that it is a signal of *zero* frequency, that is, it doesn't *vary* with time. And that means *all* of the energy of such a signal is concentrated at the *single* frequency $\omega = 0$. And what does that remind you of? An *impulse,* that's what, an impulse located at $\omega = 0$. An impulse at zero frequency is associated with a time function that has all of its energy uniformly distributed over all time, from minus infinity to plus infinity. It clearly has infinite energy, as

$$\int_{-\infty}^{\infty} 1^2 \, dt = \int_{-\infty}^{\infty} dt = \infty.$$

Well, you might be thinking at this point, the impulse wasn't too hard to handle. Maybe it won't be so hard after all to calculate $U(\omega)$, the Fourier transform of the unit step $u(t)$, either. You might suggest, for example, looking back at the earlier calculation we did for $f(t) = e^{-\sigma t} u(t)$ (example 2 in the previous section), where we found its Fourier transform to be $F(\omega) = 1/(\sigma + i\omega)$. Thus, you might argue that since

$$\lim_{\sigma \to 0} f(t) = \lim_{\sigma \to 0} e^{-\sigma t} u(t) = u(t),$$

it should be the case that

$$U(\omega) = \lim_{\sigma \to 0} F(\omega) = \lim_{\sigma \to 0} \frac{1}{\sigma + i\omega} = \frac{1}{i\omega}.$$

This would say that

$$|U(\omega)|^2 = \frac{1}{\omega^2},$$

and so Rayleigh's energy formula would say the energy of the unit step is

$$\int_{-\infty}^{\infty} \frac{1}{2\pi} |U(\omega)|^2 \, d\omega = \frac{1}{2\pi} 2 \int_{0}^{\infty} \frac{d\omega}{\omega^2} = \infty,$$

which is just what we calculated at the start of this section. *But,* this calculation can *not* be right. Here's why.

$U(\omega) = 1/i\omega$ *obviously* says that $U(\omega)$ is purely imaginary, which in turn says (as we showed back in section 5.2) that $u(t)$ "must be" an odd function of t—which $u(t)$ is clearly *not*. We have an apparent *deep* inconsistency here and we seem to be right back where we started, that is, *nowhere*, with calculating $U(\omega)$—what is going on? What is going on is that $U(\omega) = 1/i\omega$ is *almost* right, but it's also "missing something." "What 'something'?" you ask, and the surprising answer is—an *impulse*. Here's how to show that.

The key idea is to write $u(t)$ in terms of the sgn(t) function, which we first encountered back in section 1.8. There we wrote

$$\text{sgn}(t) = \begin{cases} +1, & t > 0, \\ -1, & t < 0. \end{cases}$$

Now, clearly (as I've already done in section 5.1 in the discussion on Dirichlet's discontinuous integral) we can write

$$u(t) = \frac{1}{2} + \frac{1}{2}\text{sgn}(t).$$

Thus, $U(\omega)$, the transform of $u(t)$, is the sum of the transforms of $\frac{1}{2}$ and $\frac{1}{2}\text{sgn}(t)$. As we've seen, we have the pair $1/2\pi \longleftrightarrow \delta(\omega)$, and so the transform of $\frac{1}{2}$ is $\pi\delta(\omega)$. And the transform for sgn(t) is $2/i\omega$, a result we can verify by inserting $2/i\omega$ into the inverse Fourier transform integral:

$$\frac{1}{2\pi} \int\limits_{-\infty}^{\infty} \frac{2}{i\omega} e^{i\omega t}\, d\omega = \frac{1}{\pi i} \int\limits_{-\infty}^{\infty} \frac{e^{i\omega t}}{\omega}\, d\omega.$$

This integral, as shown at the very end of section 1.8, is $\pi i \cdot \text{sgn}(t)$. So, the time function that is paired with $2/i\omega$ is $1/\pi i[\pi i \cdot \text{sgn}(t)] = \text{sgn}(t)$. That's it. Thus,

$$U(\omega) = \pi\delta(\omega) + \frac{1}{2}\left(\frac{2}{i\omega}\right).$$

That is, we at last have the pair that we are after:

$$\boxed{u(t) \leftrightarrow U(\omega) = \pi\delta(\omega) + \frac{1}{i\omega}}.$$

It's worth a little time here to stop and understand *why* the $\delta(\omega)$ is needed. Intuitively, $u(t)$ has an "average" value of $\frac{1}{2}$ (0 for $-\infty < t < 0$ and $+1$ for $0 < t < \infty$), and that's the source of the $\pi\delta(\omega)$. We do *not* have a $\delta(\omega)$ in the transform of sgn(t) since it has an average value of zero (-1 for $-\infty < t < 0$ and $+1$ for $0 < t < \infty$). The purely imaginary $2/i\omega$ is just fine, all by itself, for the *odd* sgn(t). For both $u(t)$ and sgn(t), however, the $1/i\omega$ and $2/i\omega$ are needed, respectively, because of the *sudden jumps* at $t = 0$, that is, both functions need energy at arbitrarily high frequencies to support a *discontinuous* change in time. Notice that the form of these terms is $1/i\omega$, times the magnitude of the jump (one for $u(t)$ and two for sgn(t)).

We can now use our result for $U(\omega)$, curious in its own right, to derive what might be considered an even more curious Fourier transform pair. That is, we can now answer the following intriguing *mathematical* question (one that you'll see in chapter 6 has important *practical* value in the theory of radio): what time function pairs with a *step in frequency?* That is, what's on the left in the pair? $\longleftrightarrow u(\omega)$? Be crystal-clear on the notation: $u(\omega) \neq U(\omega)$. $U(\omega)$ is the transform of the unit step *in time*, that is, $u(t) \longleftrightarrow U(\omega)$, while $u(\omega)$ is the unit step *in frequency* ($u(\omega) = 1$ for $\omega > 0$, and $u(\omega) = 0$ for $\omega < 0$). Such an ω-function is called a *single-sided spectrum*, with all of its energy at positive frequencies only. You shouldn't be surprised, then, when we find that the time function that goes with such a spectrum is quite unusual indeed.

To answer our question, I need to establish one more theoretical result. Suppose we have the pair $g(t) \longleftrightarrow G(\omega)$. Then, from the inverse Fourier transform integral we have

$$g(t) = \frac{1}{2\pi} \int_{-\infty}^{\infty} G(\omega)e^{i\omega t}\,d\omega.$$

If we now replace t with $-t$ on both sides (which retains the truth of the equality), we have

$$g(-t) = \frac{1}{2\pi} \int_{-\infty}^{\infty} G(\omega)e^{-i\omega t}\,d\omega.$$

And then if we use again our symbol swapping trick that I used in the last section, that is, if we replace every ω with t and every t with ω, we have

$$g(-\omega) = \frac{1}{2\pi} \int_{-\infty}^{\infty} G(t)e^{-it\omega}\, dt,$$

or

$$2\pi g(-\omega) = \int_{-\infty}^{\infty} G(t)e^{-i\omega t}\, dt.$$

But the integral is simply the Fourier transform of $G(t)$, and so we have the following wonderful result, known as the *duality theorem*:

$$
\begin{array}{ll}
\text{if} & g(t) \longleftrightarrow G(\omega) \\
\text{then} & G(t) \longleftrightarrow 2\pi g(-\omega)
\end{array}
$$

Now we are all set to go. According to the duality theorem, since we have already established that

$$u(t) \longleftrightarrow \pi\delta(\omega) + \frac{1}{i\omega},$$

we have

$$\pi\delta(t) + \frac{1}{it} \longleftrightarrow 2\pi u(-\omega).$$

This isn't *quite* what we're after, of course, which is the time function that goes with $u(\omega)$, rather than with the $u(-\omega)$ we have here. But we are *almost* done, because all we need to do to complete the analysis is to make one last, easy observation. If $f(t)$ is any function in general, then of course we have (assuming the integral exists)

$$f(t) \longleftrightarrow \int_{-\infty}^{\infty} f(t)e^{-i\omega t}\, dt = F(\omega).$$

Thus,

$$F(-\omega) = \int_{-\infty}^{\infty} f(t)e^{-i(-\omega)t}\, dt = \int_{-\infty}^{\infty} f(t)e^{i\omega t}\, dt.$$

Now, make the change of variable to $v = -t$. Then,

$$F(-\omega) = \int_{\infty}^{-\infty} f(-v)e^{i\omega(-v)}(-dv) = \int_{-\infty}^{\infty} f(-v)e^{-i\omega v}\,dv = \int_{-\infty}^{\infty} f(-t)e^{-i\omega t}\,dt,$$

which is just the Fourier transform of $f(-t)$—the last integral follows by making the trivial change in the dummy variable of integration from v to t. That is, we have the general pair $f(-t) \longleftrightarrow F(-\omega)$.

So, returning to our "almost" result for $u(-\omega)$, we have (once we replace ω with $-\omega$ and t with $-t$)

$$\pi\delta(-t) + \frac{1}{i(-t)} \longleftrightarrow 2\pi\,u(\omega).$$

Or, remembering that $\delta(t)$ is even, we have the exotic (this is *not* too strong a word to use) pair shown in the box below as the answer to our original question of what time function pairs with the unit step in the frequency domain?

$$\boxed{\frac{1}{2}\delta(t) + i\frac{1}{2\pi t} \longleftrightarrow u(\omega)} \quad.$$

This pair, I think, involving a *complex* time function, is *really nonobvious!*

This section is titled "curious spectra" (and we've certainly seen some pretty curious examples) and to end it I want to show you analyses of two *really curious* signals. The first example is rather fanciful. It isn't of much practical interest, but it does present a maze of fascinating mathematical puzzles which Fourier theory will see us through. The second and concluding example is, on the other hand, of absolutely enormous practical interest to electrical engineers.

For our first example, the function we are going to study is $g(t) = |t|$, which I first asked you to think about back in section 1.8 (where I asked you to convince yourself that $\operatorname{sgn}(t) = (d/dt)|t|$). Our question here is: how is the energy of $g(t)$ distributed in frequency? Now $g(t) = |t|$ is obviously an infinite energy signal, but you'll see that it is *very unlike* the other infinite energy signals we've already encountered (the step and the impulse). Our first step is, of course, to calculate the Fourier transform of $g(t)$, which immediately gets us into trouble. That is, it does if we simply plug $|t|$ directly into the Fourier transform integral and expand with Euler's formula:

$$G(\omega) = \int_{-\infty}^{\infty} |t| e^{-i\omega t} dt = \int_{-\infty}^{\infty} |t| \cos(\omega t) dt + i \int_{-\infty}^{\infty} |t| \sin(\omega t) dt,$$

or, because the first integrand on the right is even and the second integrand is odd, we have

$$G(\omega) = 2 \int_{0}^{\infty} t \cos(\omega t) dt = \, ?$$

So, let's try another approach. It doesn't completely work, either, but it *almost* does. And with just one *more* try, we'll be able to patch it up.

We know the Fourier transform of $\text{sgn}(t)$, and so

$$\text{sgn}(t) = \frac{d}{dt}|t| \longleftrightarrow \frac{2}{i\omega}.$$

Can we use this to get our hands on $G(\omega)$, the Fourier transform of $|t|$? Well, as I said above, *almost*. Here's how, using what is called the *time differentiation theorem*. If we have the pair $f(t) \longleftrightarrow F(\omega)$, then the inverse Fourier transform integral says

$$f(t) = \frac{1}{2\pi} \int_{-\infty}^{\infty} F(\omega) e^{i\omega t} d\omega.$$

Differentiating with respect to t gives

$$\frac{df}{dt} = \frac{1}{2\pi} \int_{-\infty}^{\infty} F(\omega) i\omega e^{i\omega t} d\omega$$

which says that if $f(t) \longleftrightarrow F(\omega)$ then $df/dt \longleftrightarrow i\omega F(\omega)$. (We'll use this theorem again in section 5.6, in the discussion of the uncertainty principle.) This assumes, of course, that $f(t)$ *is* differentiable and, since the derivative of $|t|$ does *not* exist everywhere (i.e., at $t = 0$) then we might not be too surprised if we run into at least a little trouble. Well, let's tighten our seatbelts and see what happens.

To help keep things straight and obvious, let's write **T** as the Fourier transform *integral operator* on $f(t)$:

$$\mathbf{T}\{f(t)\} = F(\omega).$$

Our time differentiation theorem, for example, is simply[7]

$$\mathbf{T}\left\{\frac{df}{dt}\right\} = i\omega\,\mathbf{T}\{f(t)\}.$$

So,

$$\mathbf{T}\{\text{sgn}(t)\} = \mathbf{T}\left\{\frac{d}{dt}|t|\right\} = i\omega\,\mathbf{T}\{|t|\} = i\omega G(\omega),$$

or

$$G(\omega) = \frac{1}{i\omega}\mathbf{T}\{\text{sgn}(t)\} = \frac{1}{i\omega}\cdot\frac{2}{i\omega} = -\frac{2}{\omega^2},$$

that is, we have the pair $|t| \longleftrightarrow -2/\omega^2$. We seem to have solved our problem. But this *cannot* be right! Here's why. The Fourier transform of $g(t)$ is, of course,

$$G(\omega) = \int_{-\infty}^{\infty} g(t)e^{-i\omega t}\,dt,$$

and so for $\omega = 0$ we have

$$G(0) = \int_{-\infty}^{\infty} g(t)\,dt = \int_{-\infty}^{\infty} |t|\,dt = +\infty.$$

But our result says

$$G(0) = -\frac{2}{\omega^2}\,|_{\omega=0} = -\infty.$$

Alas, we have the wrong sign on the infinity. The $-2/\omega^2$ is, in fact, correct for all ω *except* at $\omega = 0$, where, just as with our first attempt at transforming the step function $u(t)$, we are still "missing something." So, let's try yet another approach to find that "something."

You'll recall that back in section 1.8 I also asked you to think about the following formula:

$$|t| = \int_{0}^{t} \text{sgn}(s)\,ds.$$

This is actually quite easy to establish, if we consider the two cases $t > 0$ and $t < 0$ separately. The first case is nearly obvious: if $t > 0$ then $s > 0$ over the entire interval of integration and we have

$$\int_0^t \operatorname{sgn}(s)\,ds = \int_0^t 1 \cdot ds = t = |t| \text{ because } t > 0.$$

The case $t < 0$ is just slightly more subtle. Let $l = -t$ (where, of course, $l > 0$); then

$$\int_0^t \operatorname{sgn}(s)\,ds = \int_0^{-l} \operatorname{sgn}(s)\,ds = -\int_{-l}^0 \operatorname{sgn}(s)\,ds.$$

Over the entire interval of integration s is negative, and so

$$\int_0^t \operatorname{sgn}(s)\,ds = -\int_{-l}^0 (-1)\,ds = \int_{-l}^0 ds = s\Big|_{-l}^0$$

$$= 0 - (-l) = l = -t = |t| \text{ because } t < 0.$$

Thus, as claimed, for all t we have

$$|t| = \int_0^t \operatorname{sgn}(s)\,ds.$$

Now, recall our result from section 1.8 that resulted in Dirichlet's discontinuous integral:

$$\int_{-\infty}^\infty \frac{\sin(\omega t)}{\omega}\,d\omega = \pi \operatorname{sgn}(t).$$

Combining this with our last result for $|t|$, we have

$$|t| = \int_0^t \left\{ \frac{1}{\pi} \int_{-\infty}^\infty \frac{\sin(\omega s)}{\omega}\,d\omega \right\} ds.$$

Reversing the order of integration, we have

$$|t| = \frac{1}{\pi} \int\limits_{-\infty}^{\infty} \frac{1}{\omega} \left\{ \int\limits_{0}^{t} \sin(\omega s)\,ds \right\} d\omega = \frac{1}{\pi} \int\limits_{-\infty}^{\infty} \frac{1}{\omega} \left\{ \frac{-\cos(\omega s)}{\omega} \right\} \Bigg|_{0}^{t} d\omega$$

$$= \frac{1}{\pi} \int\limits_{-\infty}^{\infty} \frac{1 - \cos(\omega t)}{\omega^2}\,d\omega.$$

Now, to return to our original problem of calculating the Fourier transform of $|t|$, let's stick this integral representation[8] for $|t|$ into the Fourier transform integral. That gives us

$$G(\omega) = \int\limits_{-\infty}^{\infty} \left\{ \frac{1}{\pi} \int\limits_{-\infty}^{\infty} \frac{1 - \cos(\alpha t)}{\alpha^2}\,d\alpha \right\} e^{-i\omega t}\,dt,$$

where I've changed the dummy variable of integration in the inner integral from ω to α to avoid confussion with the independent variable ω in the *outer* integral. Reversing the order of integration gives

$$\boxed{\; G(\omega) = \int\limits_{-\infty}^{\infty} \frac{1}{\pi \alpha^2} \left\{ \int\limits_{-\infty}^{\infty} \{1 - \cos(\alpha t)\} e^{-i\omega t}\,dt \right\} d\alpha \;}.$$

We can write the inner integral as

$$\int\limits_{-\infty}^{\infty} \{1 - \cos(\alpha t)\} e^{-i\omega t}\,dt = \int\limits_{-\infty}^{\infty} e^{-i\omega t}\,dt - \int\limits_{-\infty}^{\infty} \cos(\alpha t) e^{-i\omega t}\,dt.$$

The first integral on the right is just our "astonishing statement" from section 5.1, that is, from there we have

$$\delta(x) = \frac{1}{2\pi} \int\limits_{-\infty}^{\infty} e^{i\omega x}\,d\omega = \frac{1}{2\pi} \int\limits_{-\infty}^{\infty} e^{isx}\,ds.$$

Replacing x with ω gives

$$\delta(\omega) = \frac{1}{2\pi} \int\limits_{-\infty}^{\infty} e^{is\omega}\,ds,$$

and so, replacing ω with $-\omega$,

$$\int_{-\infty}^{\infty} e^{is(-\omega)}\,ds = 2\pi\,\delta(-\omega) = 2\pi\,\delta(\omega) = \int_{-\infty}^{\infty} e^{-i\omega s}\,ds = \int_{-\infty}^{\infty} e^{-i\omega t}\,dt.$$

Thus, our inner integral is

$$\int_{-\infty}^{\infty} \{1 - \cos(\alpha t)\} e^{-i\omega t}\,dt = 2\pi\,\delta(\omega) - \int_{-\infty}^{\infty} \cos(\alpha t) e^{-i\omega t}\,dt.$$

The remaining integral on the right is just the Fourier transform of $\cos(\alpha t)$, and by now this should be a routine Euler's formula calculation for you:

$$\int_{-\infty}^{\infty} \cos(\alpha t) e^{-i\omega t}\,dt = \int_{-\infty}^{\infty} \frac{e^{i\alpha t} + e^{-i\alpha t}}{2} e^{-i\omega t}\,dt$$

$$= \frac{1}{2}\int_{-\infty}^{\infty} e^{it[-(\omega-\alpha)]}\,dt + \frac{1}{2}\int_{-\infty}^{\infty} e^{it[-(\omega+\alpha)]}\,dt$$

$$= \frac{1}{2}2\pi\,\delta(-\{\omega-\alpha\}) + \frac{1}{2}2\pi\,\delta(-\{\omega+\alpha\})$$

$$= \pi\,\delta(\omega-\alpha) + \pi\,\delta(\omega+\alpha).$$

Thus, finally, our inner integral is

$$\int_{-\infty}^{\infty} \{1 - \cos(\alpha t)\} e^{-i\omega t}\,dt = 2\pi\,\delta(\omega) - \pi\,\delta(\omega-\alpha) - \pi\,\delta(\omega+\alpha),$$

and if we insert this result into our boxed expression for $G(\omega)$ we arrive at

$$G(\omega) = \int_{-\infty}^{\infty} \frac{1}{\pi\alpha^2}\{2\pi\,\delta(\omega) - \pi\,\delta(\omega-\alpha) - \pi\,\delta(\omega+\alpha)\}\,d\alpha$$

$$= 2\delta(\omega)\int_{-\infty}^{\infty} \frac{d\alpha}{\alpha^2} - \int_{-\infty}^{\infty} \frac{1}{\alpha^2}\delta(\omega-\alpha)\,d\alpha - \int_{-\infty}^{\infty} \frac{1}{\alpha^2}\delta(\omega+\alpha)\,d\alpha.$$

The first term is an impulse at $\omega = 0$ with *infinite strength!* Remember, all impulses have infinite *height*, all by themselves—the factor 2 $\int_{-\infty}^{\infty} d\alpha/\alpha^2 = 4\int_0^{\infty} d\alpha/\alpha^2 = \infty$ is the *area* or *strength* of the impulse. This is one **mega-powerful** impulse! As it well should be, of course, since you will recall from our earlier analyses of sgn(t) and $u(t)$ how $\delta(\omega)$ played its role in *their* transforms—the strength factors there that multiplied $\delta(\omega)$ (0 and $\frac{1}{2}$, respectively), were the *average value* of the time function. And what is the "average value" of $|t|$ over all time?—*infinity*. This huge impulse is the "missing something" we've been after. Located at $\omega = 0$, it is just what we need to turn the $-\infty$ we got for $G(0)$ from the *almost* correct $G(\omega) = -2/\omega^2$ into the $+\infty$ $G(0)$ actually equals.

The remaining two integrals in our last expression for $G(\omega)$ are equal to the $-2/\omega^2$ we calculated before. That is, by the sampling property of the impulse we have

$$-\int_{-\infty}^{\infty} \frac{1}{\alpha^2}\delta(\omega - \alpha)\,d\alpha - \int_{-\infty}^{\infty} \frac{1}{\alpha^2}\delta(\omega + \alpha)\,d\alpha = -\frac{1}{(\omega)^2} - \frac{1}{(-\omega)^2} = -\frac{2}{\omega^2}.$$

So, *at last*, we have the pair

$$\boxed{\, |t| \longleftrightarrow 4\delta(\omega)\int_0^{\infty} \frac{d\alpha}{\alpha^2} - \frac{2}{\omega^2} \,},$$

which is indeed one curious spectrum.

The energy of $|t|$ is, as mentioned at the start of this analysis, infinite, but the above curious pair tells us that all that energy is distributed over frequency in an equally curious way. If we stay away from $\omega = 0$, then the energy spectral density is given by $4/\omega^4$ and so, for *any* $\omega_1 > 0$, no matter how small (just not zero), the total energy over all ω (excluding the arbitrarily small "hole" at $-\omega_1 < \omega < \omega_1$) is, by Rayleigh's energy formula,

$$\frac{1}{2\pi}\left[\int_{-\infty}^{-\omega_1} \frac{4}{\omega^4}\,d\omega + \int_{\omega_1}^{\infty} \frac{4}{\omega^4}\,d\omega\right] = \frac{4}{2\pi}2\int_{\omega_1}^{\infty} \frac{d\omega}{\omega^4} = \frac{4}{\pi}\left(-\frac{1}{3\omega^3}\Big|_{\omega_1}^{\infty}\right) = \frac{4}{3\pi\omega_1^3} < \infty.$$

That is, there is just a *finite* amount of energy in any frequency interval (even of infinite width) that does not include $\omega = 0$. The infinite energy of $|t|$ is, therefore, packed into the infinitesimally tiny frequency interval around $\omega = 0$, which is in marked contrast to the infinite energy of a "mere" impulse, which we found earlier in this section is *uniformly* distributed over all frequencies.

For the second and final example of our pair of *really curious* spectra, a very interesting problem occurs in radio engineering with the analysis of the elementary signal $s(t) = \sin(\omega_c t)$, $-\infty < t < \infty$. When you listen to a radio you are hearing the *information signal* (voice or music) that has been "placed onto" what is called the *carrier signal*, which is our sinusoidal function $s(t)$ at frequency ω_c. (In chapter 6 I'll elaborate on what "placed onto" means, and you'll see that Fourier theory and complex numbers will be very helpful with that, too.) In commercial AM broadcast radio $v_c = \omega_c/2\pi$ is called the *carrier frequency*, and its value (measured in hertz) is different for each transmitting station in the same listening area in order to avoid mutual interference at receivers. If a station's carrier frequency is 1.27 MHz (megahertz), for example, which is 1,270 kHz (kilohertz), then the station's announcer will often say something like "you're listening to twelve-seventy on your radio dial." A very natural question to ask, here, is where is the energy of a carrier signal? This is not a mystery of the ages, of course—the energy is at the carrier frequency. But the mathematics of it is not so obvious. With a clever use of Fourier theory, however, this question will be easy for us to answer.

We have from Euler's formula that

$$s(t) = \sin(\omega_c t) = \frac{e^{i\omega_c t} - e^{-i\omega_c t}}{2i},$$

and so the Fourier transform of $s(t)$ is

$$S(\omega) = \int_{-\infty}^{\infty} \frac{e^{i\omega_c t} - e^{-i\omega_c t}}{2i} e^{-i\omega t}\, dt$$

$$= \frac{1}{2i}\left[\int_{-\infty}^{\infty} e^{i(\omega_c - \omega)t}\, dt - \int_{-\infty}^{\infty} e^{i[-(\omega_c + \omega)]t}\, dt \right].$$

We have from our previous work that

$$\delta(\omega) = \frac{1}{2\pi} \int\limits_{-\infty}^{\infty} e^{i\omega t}\, dt,$$

and, since the impulse is even, we have

$$\int\limits_{-\infty}^{\infty} e^{i(\omega_c - \omega)t}\, dt = 2\pi\, \delta(\omega_c - \omega) = 2\pi\, \delta(\omega - \omega_c)$$

and

$$\int\limits_{-\infty}^{\infty} e^{i[-(\omega_c + \omega)]t}\, dt = 2\pi\, \delta[-(\omega_c + \omega)] = 2\pi\, \delta(\omega + \omega_c).$$

That is,

$$S(\omega) = \frac{1}{2i}[2\pi\, \delta(\omega - \omega_c) - 2\pi\, \delta(\omega + \omega_c)]$$

$$= -\pi\, i[\delta(\omega - \omega_c) - \delta(\omega + \omega_c)].$$

The Fourier transform of $\sin(\omega_c t)$ is purely imaginary because the real-valued time function $\sin(\omega_c t)$ is odd, and the transform consists of just two impulses, one at $\omega = \omega_c$ and the other at $\omega = -\omega_c$.

The formal answer to our initial question, then, of "where's the energy of the carrier?" is given by looking at the energy spectral density $(1/2\pi) \mid S(\omega) \mid^2$. Admittedly, that's a bit difficult to visualize since $S(\omega)$ is impulsive. What does the square of an impulse look like? Since the impulses themselves are at $\pm\omega_c$, then it seems clear that their "squares" (and so their energies) are located at $\pm\omega_c$ as well. But what does the energy spectral density *look like?*

Here's a really clever way to sneak up on the answer. Instead of calculating the transform of a sinusoidal carrier that's been "on" for *all* time (that's nearly as unrealistic as $|t|$—don't you agree?—and in fact it is this very point that results in those two impulses), let's imagine that we have a sinusoid that's been "on" for exactly N complete cycles. The transform of such a signal is *non*-impulsive. *Then* we can study where the energy is, and also what happens as $N \to \infty$. Such a signal is called a *sinusoidal burst,* and figure 5.4.1 shows such a burst for $N = 6$. Such

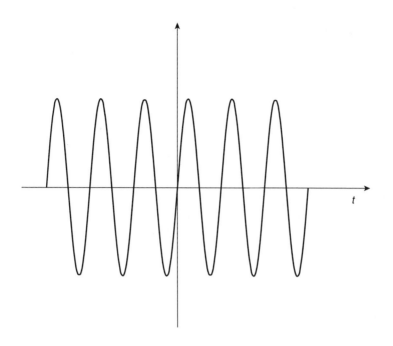

Figure 5.4.1. A sinusoidal burst of six cycles

burst signals are intentionally produced by what are called *pulse radars*, which transmit a very short burst of electromagnetic energy at a very high frequency, and then listen for the echos that indicate the presence of and range to (proportional to the time delay between pulse transmission and echo reception) a target. By studying successive echos such radars can *track* targets. If a pulse radar operates at 10 GHz (gigahertz), for example—this is a so-called "3-centimeter radar," as three centimeters is the wavelength of 10 GHz electromagnetic waves—and if a pulse lasts one microsecond, then $N = 10,000$ cycles.

We set our analysis up mathematically as follows, with the duration of each complete cycle as $2\pi/\omega_c$:

$$s_b(t) = \begin{cases} \sin(\omega_c t), & -\dfrac{2\pi}{\omega_c} \cdot \dfrac{N}{2} \le t \le \dfrac{2\pi}{\omega_c} \cdot \dfrac{N}{2}, \\ 0 & , \quad \text{otherwise}, \end{cases}$$

that is,

$$s_b(t) = \begin{cases} \sin(\omega_c t), & -\dfrac{N\pi}{\omega_c} \le t \le \dfrac{N\pi}{\omega_c}, \\ 0 & , \quad \text{otherwise}. \end{cases}$$

The Fourier transform of $s_b(t)$ is (once again, using Euler's formula)

$$S_b(\omega) = \int_{-\infty}^{\infty} s_b(t)e^{-i\omega t}\,dt = \int_{-N\pi/\omega_c}^{N\pi/\omega_c} \sin(\omega_c t)e^{-i\omega t}\,dt$$

$$= \int_{-N\pi/\omega_c}^{N\pi/\omega_c} \frac{e^{i\omega_c t} - e^{-i\omega_c t}}{2i} e^{-i\omega t}\,dt$$

$$= \frac{1}{2i}\left[\int_{-N\pi/\omega_c}^{N\pi/\omega_c} e^{-i(\omega-\omega_c)t}\,dt - \int_{-N\pi/\omega_c}^{N\pi/\omega_c} e^{-i(\omega+\omega_c)t}\,dt\right]$$

$$= \frac{1}{2i}\left[\left(\frac{e^{-i(\omega-\omega_c)t}}{-i(\omega - \omega_c)}\Big|_{-N\pi/\omega_c}^{N\pi/\omega_c}\right) - \left(\frac{e^{-i(\omega+\omega_c)t}}{-i(\omega + \omega_c)}\Big|_{-N\pi/\omega_c}^{N\pi/\omega_c}\right)\right]$$

$$= \frac{1}{2}\left[\left\{\frac{e^{-i(\omega-\omega_c)(N\pi/\omega_c)} - e^{i(\omega-\omega_c)(N\pi/\omega_c)}}{\omega - \omega_c}\right\}\right.$$
$$\left. - \left\{\frac{e^{-i(\omega+\omega_c)(N\pi/\omega_c)} - e^{i(\omega+\omega_c)(N\pi/\omega_c)}}{\omega + \omega_c}\right\}\right]$$

$$= \frac{1}{2}\left[\left\{\frac{e^{-iN\pi(\omega/\omega_c)}e^{iN\pi} - e^{iN\pi(\omega/\omega_c)}e^{-iN\pi}}{\omega - \omega_c}\right\}\right.$$
$$\left. - \left\{\frac{e^{-iN\pi(\omega/\omega_c)}e^{-iN\pi} - e^{iN\pi(\omega/\omega_c)}e^{iN\pi}}{\omega + \omega_c}\right\}\right].$$

We now consider the two cases of N first an even integer, and then an odd integer. If N is even, then Euler's formula says $e^{iN\pi} = e^{-iN\pi} = 1$, and so

$$S_b(\omega) = \frac{1}{2}\left[\left\{\frac{e^{-iN\pi(\omega/\omega_c)} - e^{iN\pi(\omega/\omega_c)}}{\omega - \omega_c}\right\} - \left\{\frac{e^{-iN\pi(\omega/\omega_c)} - e^{iN\pi(\omega/\omega_c)}}{\omega + \omega_c}\right\}\right]$$

$$= \frac{1}{2}\left[\frac{-2i\sin(N\pi(\omega/\omega_c))}{\omega - \omega_c} - \frac{-2i\sin(N\pi(\omega/\omega_c))}{\omega + \omega_c}\right]$$

$$= -i\sin(N\pi(\omega/\omega_c))\left[\frac{1}{\omega - \omega_c} - \frac{1}{\omega + \omega_c}\right],$$

or,

$$S_b(\omega) = -i\frac{2\omega_c \sin(N\pi\omega/\omega_c)}{\omega^2 - \omega_c^2}, \; N \; even \; .$$

On the other hand, if N is odd, then $e^{iN\pi} = e^{-iN\pi} = -1$, and so

$$S_b(\omega) = \frac{1}{2}\left[\left\{\frac{-e^{-iN\pi(\omega/\omega_c)} + e^{iN\pi(\omega/\omega_c)}}{\omega - \omega_c}\right\} - \left\{\frac{-e^{-iN\pi(\omega/\omega_c)} + e^{iN\pi(\omega/\omega_c)}}{\omega + \omega_c}\right\}\right]$$

$$= \frac{1}{2}\left[\frac{2i\sin(N\pi(\omega/\omega_c))}{\omega - \omega_c} - \frac{2i\sin(N\pi(\omega/\omega_c))}{\omega + \omega_c}\right]$$

$$= i\sin\left(N\pi\frac{\omega}{\omega_c}\right)\left[\frac{1}{\omega - \omega_c} - \frac{1}{\omega + \omega_c}\right],$$

or,

$$S_b(\omega) = i\frac{2\omega_c \sin(N\pi\omega/\omega_c)}{\omega^2 - \omega_c^2}, \; N \; odd \; .$$

Thus, for *any* nonnegative integer N the energy spectral density (ESD) of an N-cycle sinusoidal burst is

$$\frac{1}{2\pi}\mid S_b(\omega)\mid^2 = \frac{2}{\omega_c^2\pi} \cdot \frac{\sin^2(N\pi(\omega/\omega_c))}{[(\frac{\omega}{\omega_c})^2 - 1]^2}.$$

Figures 5.4.2 through 5.4.5 show the ESD (without the constant scale factor $2/\omega_c^2\pi$) for $N = 1, 2, 5,$ and 10, respectively, over the normalized frequency interval $-2 \le \omega/\omega_c \le 2$. (Notice that the plots are *even*, as they should be for a *real* $s_b(t)$.) It is clear that as N increases the ESD is, indeed, looking more and more like two impulses at $(\omega/\omega_c) = \pm1$. It is surprising, I think, just how quickly the ESD does approach two impulses; after all, $N = 10$ is a "pretty small" number and, in the illustration I mentioned earlier of the pulse radar, with $N = 10,000$, the ESD would be virtually indistinguishable from the impulsive ESD of a true (but "unrealistic") sinusoid that has been (and *will be*) "forever on."

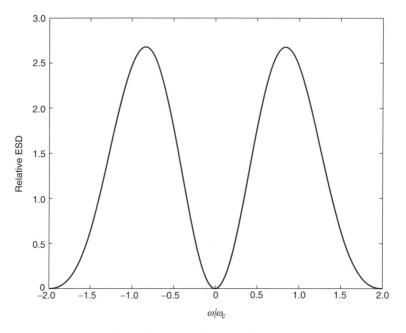

Figure 5.4.2. 1-cycle sinusoidal burst

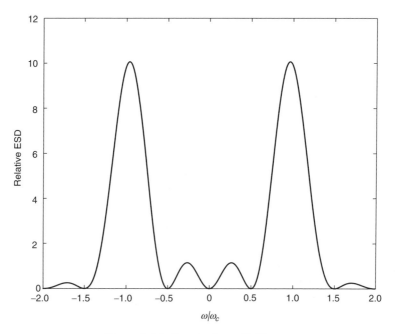

Figure 5.4.3. 2-cycle sinusoidal burst

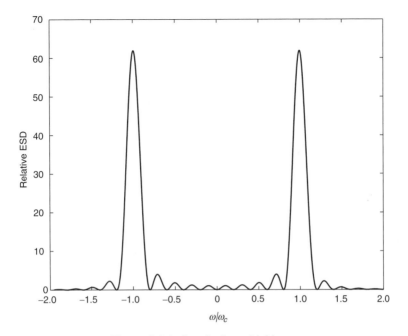

Figure 5.4.4. 5-cycle sinusoidal burst

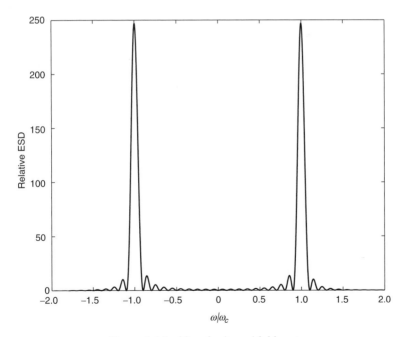

Figure 5.4.5. 10-cycle sinusoidal burst

5.5 Poisson summation.

In this short section I'll show you a beautiful, *surprising* connection between a time function and its Fourier transform. This result, called *Poisson's summation formula*, has an ironic aspect—it is named for the French mathematician Siméon-Denis Poisson (1781–1840) who was one of Fourier's most severe critics. Suppose we have a function $f(t)$ defined over the entire real line, $-\infty < t < \infty$. From this $f(t)$ we then construct another function $g(t)$, defined as

$$g(t) = \sum_{k=-\infty}^{\infty} f(t+k).$$

If you look at this definition for just a bit, you should be able to see that $g(t)$ is *periodic*, with period $T = 1$. We can easily show this formally by writing

$$g(t+1) = \sum_{k=-\infty}^{\infty} f(t+1+k),$$

which becomes, with a change in the summation index to $n = k+1$,

$$g(t+1) = \sum_{n=-\infty}^{\infty} f(t+n) = g(t).$$

This works, of course, because mapping the infinite set of all the integers into itself by simply adding one to each integer changes nothing.

Now, since $g(t)$ is periodic, it can be written as a Fourier series with (as you'll recall from Chapter 4) $\omega_0 = 2\pi$ since $T = 1$:

$$g(t) = \sum_{n=-\infty}^{\infty} c_n e^{in2\pi t},$$

where

$$c_n = \frac{1}{T} \int_{period} g(t)e^{-in2\pi t}\,dt = \int_0^1 \sum_{k=-\infty}^{\infty} f(t+k)e^{-in2\pi t}\,dt$$

$$= \sum_{k=-\infty}^{\infty} \int_0^1 f(t+k)e^{-in2\pi t}\,dt.$$

If we change variables (in the integral) to $s = t + k$, then $ds = dt$ and

$$c_n = \sum_{k=-\infty}^{\infty} \int_k^{k+1} f(s)e^{-in2\pi(s-k)}\,ds = \sum_{k=-\infty}^{\infty} \int_k^{k+1} f(s)e^{-in2\pi s}e^{ink2\pi}\,ds$$

$$= \sum_{k=-\infty}^{\infty} \int_k^{k+1} f(s)e^{-in2\pi s}\,ds$$

as, since n and k are integers, $e^{ink2\pi} = 1$. Then, since

$$\sum_{k=-\infty}^{\infty} \int_k^{k+1} = \int_{-\infty}^{\infty},$$

we have

$$c_n = \int_{-\infty}^{\infty} f(s)e^{-in2\pi s}\,ds = \int_{-\infty}^{\infty} f(t)e^{-in2\pi t}\,dt.$$

Since $f(t) \longleftrightarrow F(\omega)$, where

$$F(\omega) = \int_{-\infty}^{\infty} f(t)e^{-i\omega t}\,dt,$$

we see that

$$c_n = F(2\pi n),$$

and so

$$g(t) = \sum_{n=-\infty}^{\infty} F(2\pi n)e^{in2\pi t} = \sum_{k=-\infty}^{\infty} f(t+k).$$

This is an identity in t and, in particular, for $t = 0$ we get the remarkable Poisson's summation formula, derived[9] by him in 1827,

$$\boxed{\sum_{k=-\infty}^{\infty} f(k) = \sum_{n=-\infty}^{\infty} F(2\pi n)}.$$

This may look pretty benign, but let me now show you three examples that will convince you that such an initial impression is wrong.

Example 1. Suppose $f(t)$ is defined for all t as follows, where α is any positive number:

$$f(t) = e^{-\alpha|t|}, \quad \alpha > 0.$$

Then

$$F(\omega) = \int_{-\infty}^{\infty} f(t) e^{-i\omega t} \, dt$$

$$= \int_{-\infty}^{0} e^{\alpha t} e^{-i\omega t} \, dt + \int_{0}^{\infty} e^{-\alpha t} e^{-i\omega t} \, dt$$

$$= \int_{-\infty}^{0} e^{(\alpha - i\omega)t} \, dt + \int_{0}^{\infty} e^{-(\alpha + i\omega)t} \, dt$$

$$= \left(\frac{e^{(\alpha - i\omega)t}}{\alpha - i\omega} \bigg|_{-\infty}^{0} \right) + \left(\frac{e^{-(\alpha + i\omega)t}}{-(\alpha + i\omega)} \bigg|_{0}^{\infty} \right)$$

$$= \frac{1}{\alpha - i\omega} + \frac{1}{\alpha + i\omega} = \frac{2\alpha}{\alpha^2 + \omega^2}.$$

Poisson's summation formula tells us, therefore, that it must be true that

$$\sum_{k=-\infty}^{\infty} e^{-\alpha|k|} = \sum_{n=-\infty}^{\infty} \frac{2\alpha}{\alpha^2 + (2\pi n)^2}.$$

The sum on the left is, when written out, is

$$\sum_{k=-\infty}^{\infty} e^{-\alpha|k|} = 1 + 2 \sum_{k=1}^{\infty} e^{-\alpha k} = 1 + 2(e^{-\alpha} + e^{-2\alpha} + e^{-3\alpha} + \cdots)$$

and the expression in the parentheses is simply a geometric series which is easily summed to give

$$\sum_{k=-\infty}^{\infty} e^{-\alpha|k|} = 1 + 2 \frac{e^{-\alpha}}{1 - e^{-\alpha}} = \frac{1 + e^{-\alpha}}{1 - e^{-\alpha}}.$$

Thus,

$$\sum_{n=-\infty}^{\infty} \frac{2\alpha}{\alpha^2 + (2\pi n)^2} = \frac{1 + e^{-\alpha}}{1 - e^{-\alpha}},$$

or, with a couple of quick algebraic steps (that I'll let *you* do), we arrive at

$$\sum_{n=-\infty}^{\infty} \frac{1}{(\alpha/2\pi)^2 + n^2} = \pi \left(\frac{2\pi}{\alpha}\right) \frac{1+e^{-\alpha}}{1-e^{-\alpha}}.$$

This is a generalization of a result we derived back in section 4.3:

$$\sum_{n=-\infty}^{\infty} \frac{1}{1+n^2} = \pi \frac{1+e^{-2\pi}}{1-e^{-2\pi}},$$

which is our new result in the above box for the special case of $\alpha = 2\pi$. Our new result has an *infinity* of such special cases, of course, one for any value of $\alpha > 0$ we care to use. For example, if $\alpha = \pi$ then

$$\sum_{n=-\infty}^{\infty} \frac{1}{1/4+n^2} = 2\pi \frac{1+e^{-\pi}}{1-e^{-\pi}}.$$

The expression on the right equals 6.850754, which is easily "checked" by direct calculation of the sum, itself (using $-10,000 \le n \le 10,000$ in the sum gives 6.850734). Who could have even made up such a wonderful formula?

Example 2. A very pretty extension of our special result from section 4.3 is now possible with help from our generalized formula—specifically, we can now calculate the value of

$$\sum_{n=-\infty}^{\infty} \frac{(-1)^n}{1+n^2},$$

our original result from section 4.3 written now with *alternating* signs. To do this, notice that

$$\sum_{n=-\infty}^{\infty} \frac{(-1)^n}{1+n^2} = \sum_{n \text{ even}} \frac{1}{1+n^2} - \sum_{n \text{ odd}} \frac{1}{1+n^2}$$

and that

$$\sum_{n \text{ even}} \frac{1}{1+n^2} = \sum_{n=-\infty}^{\infty} \frac{1}{1+(2n)^2} = \sum_{n=-\infty}^{\infty} \frac{1}{1+4n^2} = \frac{1}{4} \sum_{n=-\infty}^{\infty} \frac{1}{1/4+n^2}.$$

Also,

$$\sum_{n \text{ odd}} \frac{1}{1+n^2} = \sum_{n=-\infty}^{\infty} \frac{1}{1+n^2} - \sum_{n \text{ even}} \frac{1}{1+n^2}$$

$$= \sum_{n=-\infty}^{\infty} \frac{1}{1+n^2} - \frac{1}{4} \sum_{n=-\infty}^{\infty} \frac{1}{1/4+n^2}.$$

Thus,

$$\sum_{n=-\infty}^{\infty} \frac{(-1)^n}{1+n^2} = \frac{1}{4} \sum_{n=-\infty}^{\infty} \frac{1}{1/4+n^2}$$

$$- \left\{ \sum_{n=-\infty}^{\infty} \frac{1}{1+n^2} - \frac{1}{4} \sum_{n=-\infty}^{\infty} \frac{1}{1/4+n^2} \right\}$$

$$= \frac{1}{2} \sum_{n=-\infty}^{\infty} \frac{1}{1/4+n^2} - \sum_{n=-\infty}^{\infty} \frac{1}{1+n^2}$$

$$= \pi \frac{1+e^{-\pi}}{1-e^{-\pi}} - \pi \frac{1+e^{-2\pi}}{1-e^{-2\pi}};$$

with a final couple of quick algebraic steps, we arrive at the very pretty

$$\boxed{\sum_{n=-\infty}^{\infty} \frac{(-1)^n}{1+n^2} = \frac{2\pi}{e^\pi - e^{-\pi}}.}$$

"Confirmation" comes (as usual) from direct calculation; the right-hand side is 0.27202905498213, while using $-100{,}000 \le n \le 100{,}000$ for the sum on the left gives 0.27202905508215.

Example 3. For my third and final example of this section, I'll start by deriving the Fourier transform of the so-called *Gaussian pulse*, that is, the transform of

$$f(t) = e^{-\alpha t^2}, \quad \alpha > 0, \ |t| < \infty.$$

That is, $f(t)$ is an exponential with a *quadratic* exponent. By definition,

$$F(\omega) = \int_{-\infty}^{\infty} e^{-\alpha t^2} e^{-i\omega t} \, dt.$$

This may look like a pretty tough integral to calculate, but it yields to the following clever attack.

Differentiating with respect to ω gives

$$\frac{dF}{d\omega} = -i \int_{-\infty}^{\infty} t e^{-\alpha t^2} e^{-i\omega t} \, dt.$$

If we then integrate by parts on the right, that is, in the classic formula from calculus

$$\int_{-\infty}^{\infty} u \, dv = \left(uv \Big|_{-\infty}^{\infty} - \int_{-\infty}^{\infty} v \, du \right),$$

we let $u = e^{-i\omega t}$ and $dv = t e^{-\alpha t^2} \, dt$, then $du = -i\omega e^{-i\omega t} \, dt$ and $v = -(1/2\alpha)e^{-\alpha t^2}$, and so

$$\int_{-\infty}^{\infty} t e^{-\alpha t^2} e^{-i\omega t} \, dt = \left(-\frac{1}{2\alpha} e^{-\alpha t^2} e^{-i\omega t} \Big|_{-\infty}^{\infty} - i\frac{\omega}{2\alpha} \int_{-\infty}^{\infty} e^{-\alpha t^2} e^{-i\omega t} \, dt \right.$$

$$= -i\frac{\omega}{2\alpha} \int_{-\infty}^{\infty} e^{-\alpha t^2} e^{-i\omega t} \, dt,$$

since $\lim_{|t|\to\infty} e^{-\alpha t^2} e^{-i\omega t} = 0$. But, this last integral is $F(\omega)$ and so we have a simple first order differential equation for $F(\omega)$:

$$\frac{dF}{d\omega} = -i \left[-i\frac{\omega}{2\alpha} F(\omega) \right] = -\frac{\omega}{2\alpha} F(\omega),$$

or

$$\frac{dF}{F} = -\frac{\omega}{2\alpha} \, d\omega.$$

With $\ln(C)$ as the constant of indefinite integration, integrating both sides gives

$$\ln[F(\omega)] = -\frac{\omega^2}{4\alpha} + \ln(C),$$

or

$$F(\omega) = Ce^{-\omega^2/4\alpha}.$$

To evaluate the constant C, notice that $C = F(0)$, that is,

$$C = \int_{-\infty}^{\infty} e^{-\alpha t^2}\, dt.$$

This particular definite integral can be evaluated by elementary (but *very* clever) means[10]; it is equal to $\sqrt{\pi/\alpha}$. So, we have the quite interesting pair

$$\boxed{\; f(t) = e^{-\alpha t^2} \longleftrightarrow F(\omega) = \sqrt{\tfrac{\pi}{\alpha}}\, e^{-\omega^2/4\alpha} \;}\;,$$

which says that a Gaussian pulse in time has a Fourier transform of the same form in frequency, that is, an exponential with an exponent of quadratic variation in ω.

A brief aside: While not the central point of this example (that's the *next* paragraph), notice a pretty little result we now have. Our pair tells us that

$$\int_{-\infty}^{\infty} e^{-\alpha t^2} e^{-i\omega t}\, dt = \sqrt{\tfrac{\pi}{\alpha}}\, e^{-\omega^2/4\alpha},$$

and then Euler's formula tells us that, since the integral is purely real because the right-hand side is purely real,

$$\boxed{\; \int_{-\infty}^{\infty} e^{-\alpha t^2} \cos(\omega t)\, dt = \sqrt{\tfrac{\pi}{\alpha}}\, e^{-\omega^2/4\alpha} \;}\;.$$

This, I think, would not be easy to directly derive. The same argument also tells us, of course, that $\int_{-\infty}^{\infty} e^{-\alpha t^2} \sin(\omega t)\, dt = 0$, but we *already* knew that since the integrand of this integral is odd. Notice, too, that for the special case of $\alpha = \tfrac{1}{2}$ our pair reduces to

$$e^{-t^2/2} \longleftrightarrow \sqrt{2\pi}\, e^{-\omega^2/2},$$

which says that, to within a factor of $\sqrt{2\pi}$, $e^{-(t^2/2)}$ is its *own* Fourier transform.[11]

Now, with the pair in the box in hand, let's apply the Poisson summation formula to it. The result is

$$\sum_{k=-\infty}^{\infty} e^{-\alpha k^2} = \sum_{n=-\infty}^{\infty} \sqrt{\frac{\pi}{\alpha}} e^{-4\pi^2 n^2/4\alpha},$$

or

$$\boxed{\sum_{k=-\infty}^{\infty} e^{-\alpha k^2} = \sqrt{\frac{\pi}{\alpha}} \sum_{n=-\infty}^{\infty} e^{-\pi^2 n^2/\alpha}}.$$

Both of these summations should look at least somewhat familar to you—they have the form of Gauss's quadratic sum that we studied in section 4.5. I'll not pursue that connection any further here, but in fact both of the sums in the box are special cases of the far more general topic of *theta functions,* first studied in a systematic manner by the German mathematician Carl Jacobi (1804–1851) in his 1829 masterpiece *Fundamenta nova theoriae functionum ellipticarum* (*New Foundations of the Theory of Elliptic Functions*). Theta functions[12] are intimately connected with many deep problems in analytic number theory. One purely numerical computational use of our result in the box is that one sum is preferable to the other for calculation purposes, depending on the value of α. If α is "small" then the sum on the right will converge faster than does the sum on the left, while if α is "big" the reverse is true.

5.6 Reciprocal spreading and the uncertainty principle.

In 1927 the German theoretical physicist Werner Heisenberg (1901–1976) published his famous *uncertainty principle* in quantum mechanics (hereafter written as QM). Now, we are not going to take a long side jaunt into QM in this book, but a few philosophical words will perhaps motivate the mathematics that follows. QM is probabilistic physics. In pre-QM times "classical" theoretical physics predicted what *would* (or *would not*) happen in a given situation. QM, on the other hand, accepts the observed fact that, in a given situation in the microworld, several (perhaps many) different things *could* happen, and so QM provides only the *probabilities* of these various possible outcomes as actually being the *one* that occurs. The mathematics of QM is, therefore, as you might expect, probability theory.

In probability theory one quickly learns the following terminology and notation: if **X** (called a *random variable*) denotes a quantity whose value can be measured but is not deterministic (every time you measure **X** you can get a different value), then associated with **X** is a function $f_{\mathbf{X}}(x)$ called the *probability density function* (or, simply, the "pdf") such that

(i) $f_{\mathbf{X}}(x) \geq 0, -\infty < x < \infty$,
(ii) $\int_a^b f_{\mathbf{X}}(x)\,dx =$ probability **X** has a value in the interval
   $a \leq x \leq b$.

As a consequence of (ii), we have the so-called *normalization condition*

(iii) $\int_{-\infty}^{\infty} f_{\mathbf{X}}(x)\,dx = 1$, which simply says the obvious: **X** has a value
   somewhere in the interval $-\infty < x < \infty$.

In probability theory one learns all sorts of clever ways to calculate $f_{\mathbf{X}}(x)$ from other information that is given about the nature of **X**, but here we'll simply assume that such calculations have already been done and $f_{\mathbf{X}}(x)$ is known. Once $f_{\mathbf{X}}(x)$ is known, one uses it to calculate various other quantities that are descriptive of **X**. For example, two such quantities are the *mean* and the *mean square* of **X**, written as $\widehat{\mathbf{X}}$ and $\widehat{\mathbf{X}^2}$, respectively. $\widehat{\mathbf{X}}$ is more commonly known as the *average* of **X**, and is calculated as

$$\widehat{\mathbf{X}} = \int_{-\infty}^{\infty} x f_{\mathbf{X}}(x)\,dx.$$

Without any loss of generality we can assume $\widehat{\mathbf{X}}$ is zero (instead of working with $\widehat{\mathbf{X}}$ itself, imagine we are working with $\mathbf{X}-\widehat{\mathbf{X}}$, which *by construction* has a zero mean because (remember, $\widehat{\mathbf{X}}$ is a *number* and not a function of x)

$$\widehat{\mathbf{X}-\widehat{\mathbf{X}}} = \int_{-\infty}^{\infty} (x - \widehat{\mathbf{X}}) f_{\mathbf{X}}(x)\,dx = \int_{-\infty}^{\infty} x f_{\mathbf{X}}(x)\,dx - \widehat{\mathbf{X}} \int_{-\infty}^{\infty} f_{\mathbf{X}}(x)\,dx$$

$$= \widehat{\mathbf{X}} - \widehat{\mathbf{X}} = 0.$$

If **X** doesn't take on values that deviate much from $\widehat{\mathbf{X}}$, then $\widehat{\mathbf{X}}$ is a good measure of the "typical" value of **X** when many measurements of **X** are made. If **X** does vary a lot around its average value, then $\widehat{\mathbf{X}}$ is not such a good measure. A useful quantity, which measures how good a measure $\widehat{\mathbf{X}}$

is, is the so-called *variance* of $\mathbf{X}$, written as $\sigma_{\mathbf{X}}^2 = \widehat{(\mathbf{X} - \widehat{\mathbf{X}})^2}$, the average of the *square* of the *variation* of $\mathbf{X}$ around $\widehat{\mathbf{X}}$ (the squaring prevents positive and negative variations from negating each other). That is, with our assumption that $\widehat{\mathbf{X}} = 0$ we have

$$\sigma_{\mathbf{X}} = \sqrt{\int_{-\infty}^{\infty} x^2 f_{\mathbf{X}}(x)\, dx}.$$

Notice, carefully, that $\sigma_{\mathbf{X}} \geq 0$, *always*.

Okay, here's where all of this is going. Another way to think of $\sigma_{\mathbf{X}}$ is as a measure of the uncertainty we have in the value of $\mathbf{X}$ (notice that $\sigma_{\mathbf{X}}$ has the same units as $\mathbf{X}$). We think of $\widehat{\mathbf{X}}$ (equal to zero) as being the average value of $\mathbf{X}$. A "small" $\sigma_{\mathbf{X}}$ would mean that it's "probably" a pretty good estimate for $\mathbf{X}$ (i.e., we have "low" uncertainty in the value of $\mathbf{X}$), while a "large" $\sigma_{\mathbf{X}}$ means we would have a "high" uncertainty as to the actual value of $\mathbf{X}$, that is, with significant probability $\mathbf{X}$ could have a value considerably different from its average value of zero.

In QM there are many *coupled pairs* of probabilistic quantities that one encounters, and Heisenberg's uncertainty principle states that the product of the uncertainties of the two variables in such a pair is *at least* equal to some positive constant. The classic example of such a pair, commonly used in physics texts, is the location and the momentum of a particle; if we call $\sigma_{\mathbf{X}}$ the uncertainty in the location and $\sigma_{\mathbf{Y}}$ the uncertainty in the momentum then, for $c > 0$ some constant,

$$\sigma_{\mathbf{X}}\sigma_{\mathbf{Y}} \geq c.$$

This is popularly stated as "it is impossible simultaneously to measure both the position and the momentum of a particle with perfect certainty." When measuring the values of a pair, one trades off more certainty in one quantity with less certainty in the other.

We can do the same sort of thing with the coupled Fourier pair $g(t) \longleftrightarrow G(\omega)$, that is, we can relate the "uncertainty" we have in the location in time of $g(t)$ and the "uncertainty" we have in the location in frequency of $G(\omega)$ (which is a measure of where $g(t)$'s *energy* is in frequency). By "location in time" of $g(t)$ we simply mean that the "significant" part of $g(t)$ generally occurs over an *interval* of time (perhaps even infinite in extent), and similarly for the "significant" part of $G(\omega)$ in the frequency domain. It is generally true that the two intervals vary *inversely* with each other, that is, a signal that occurs in a very narrow interval of time must have a Fourier transform that exhibits significant presence over a very wide interval of frequency. The extreme example of this behavior is the impulse function, $\delta(t)$, which occupies *zero* time and has a transform with uniform amplitude over the entire *infinite* ω-axis. This inverse relationship is called *reciprocal spreading*. Figure 5.6.1 shows an example of reciprocal spreading, using a Fourier transform pair we derived back in section 5.3:

$$e^{-\alpha|t|}\cos(t) \longleftrightarrow \alpha\left[\frac{1}{\alpha^2+(\omega-1)^2}+\frac{1}{\alpha^2+(\omega+1)^2}\right].$$

The top two plots show the time function and its transform for $\alpha = 0.2$ (a "fast" decay in time), while the bottom two plots display the same for $\alpha = 0.05$ (a "slow" decay in time). As the time function becomes "less localized" its transform becomes "more localized."

The very next year (1928) after the appearance of Heisenberg's paper, the German mathematician Hermann Weyl (1885–1955) published an elegant mathematical derivation of the uncertainty principle in his book *The Theory of Groups and Quantum Mechanics* (published by Dover in English in 1931). The Fourier integral derivation of the uncertainty principle that I'll show you next, which uses the Cauchy-Schwarz inequality we derived in section 1.5, is based on Weyl's presentation—although in his book Weyl seems to credit the Austrian physicist Wolfgang Pauli (1900–1958) as the actual originator of this approach.

Suppose $g(t)$ is any real-valued (nonperiodic) time function, where $g(t) \longleftrightarrow G(\omega)$. Then, with $W = \int_{-\infty}^{\infty} g^2(t)\,dt$ as the energy of $g(t)$, we see that $\int_{-\infty}^{\infty} g^2(t)/W\,dt = 1$ and that $g^2(t)/W \geq 0$ for all t. That is, $g^2(t)/W$ "behaves" just like the pdf of a random variable. Taking this analogy even further, let's then define the uncertainty in the time of $g(t)$ as

$$\sigma_t = \sqrt{\int_{-\infty}^{\infty} t^2 \frac{g^2(t)}{W}\,dt}.$$

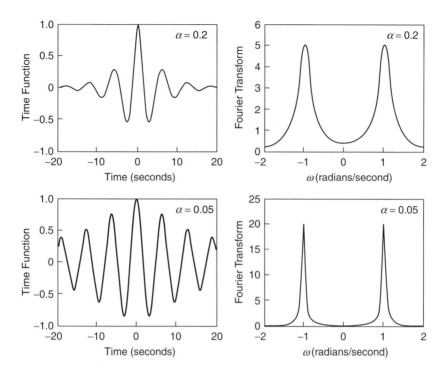

Figure 5.6.1. Reciprocal spreading

That is, we are treating time as a random variable which takes on its values from the interval $-\infty < t < \infty$—and so has an "average value" of zero!—with pdf $(g^2(t)/W$. (Remember, this is an *analogy*.)

In the same way, from Rayleigh's energy formula, we have $W = 1/2\pi \int_{-\infty}^{\infty} |G(\omega)|^2 d\omega$ and so $\int_{-\infty}^{\infty} |G(\omega)|^2/(2\pi W) d\omega = 1$, where $|G(\omega)|^2/2\pi W \geq 0$ for all ω. That is, $|G(\omega)|^2/(2\pi W)$ "behaves" just like the pdf of a random variable that takes on its values from the interval $-\infty < \omega < \infty$—and so has an "average value" of zero. So, as with σ_t, let's write

$$\sigma_\omega = \sqrt{\int_{-\infty}^{\infty} \omega^2 \frac{|G(\omega)|^2}{2\pi W} d\omega}$$

as the uncertainty in frequency. Our question, now, is, simply this: what can we say about the product $\sigma_t \sigma_\omega$?

To answer that question, let me first remind you of the Cauchy-Schwarz inequality from section 1.5: if $h(t)$ and $s(t)$ are two real functions of time then, assuming all the integrals actually exist, we have

$$\left\{\int_{-\infty}^{\infty} s(t)h(t)\,dt\right\}^2 \leq \left\{\int_{-\infty}^{\infty} s^2(t)\,dt\right\}\left\{\int_{-\infty}^{\infty} h^2(t)\,dt\right\}.$$

To have *equality* implies that $h(t)$ is proportional to $s(t)$, that is, $h(t) = ks(t)$, where k is some constant; this condition obviously reduces both sides of the inequality (before canceling the k's) to $k^2\{\int_{-\infty}^{\infty} s^2(t)\,dt\}^2$. If you look back at our derivation of the inequality, you'll see that *equality* meant, in our original notation in section 1.5, that

$$\int_{L}^{U} \{f(t) + \lambda g(t)\}^2\,dt = 0,$$

which can only be true if the *squared* integrand is *identically* zero, that is, that $f(t) = -\lambda g(t)$, that is, the two functions $f(t)$ and $g(t)$ are proportional. I'll put this observation to good use at the end of the following analysis.

Now, to start, let's make the following definitions:

$$s(t) = tg(t), \quad h(t) = \frac{dg}{dt}.$$

Then the Cauchy-Schwarz inequality says

$$\left\{\int_{-\infty}^{\infty} tg(t)\frac{dg}{dt}\,dt\right\}^2 \leq \left\{\int_{-\infty}^{\infty} t^2 g^2(t)\,dt\right\}\left\{\int_{-\infty}^{\infty} \left(\frac{dg}{dt}\right)^2\,dt\right\}.$$

Looking at the integral on the left in the box, we can write

$$\int_{-\infty}^{\infty} tg(t)\frac{dg}{dt}\,dt = \int_{-\infty}^{\infty} t\frac{d(g^2(t)/2)}{dt}\,dt,$$

and we can now evaluate the rewritten integral (on the right) by parts. That is, in

$$\int\limits_{-\infty}^{\infty} u\, dv = (uv)\Big|_{-\infty}^{\infty} - \int\limits_{-\infty}^{\infty} v\, du,$$

let $u = t$ and $dv = d(g^2(t)/2)dt/dt$. Then, $du = dt$ and $v = \frac{1}{2}g^2(t)$, and so

$$\int\limits_{-\infty}^{\infty} t\, \frac{d(g^2(t)/2)}{dt}\, dt = \left(\frac{1}{2} tg^2(t)\right)\Big|_{-\infty}^{\infty} - \int\limits_{-\infty}^{\infty} \frac{1}{2} g^2(t)\, dt.$$

If we make the *assumption* that, as $|t| \rightarrow \infty$, we have $g(t) \rightarrow 0$ faster than $1/\sqrt{t}$ (which is actually *not* very fast), then $\lim_{|t| \to \infty} tg^2(t) = 0$ and so

$$\int\limits_{-\infty}^{\infty} tg(t)\, \frac{dg}{dt}\, dt = -\frac{1}{2} \int\limits_{-\infty}^{\infty} g^2(t)\, dt = -\frac{1}{2} W.$$

That takes care of the left-hand side of the Cauchy-Schwarz inequality in the above box. Let's now look at the two integrals on the right in the box.

The first one is easy: literally by definition we have

$$\int\limits_{-\infty}^{\infty} t^2 g^2(t)\, dt = W\sigma_t^2.$$

And the second integral is almost as easy. Recalling the time differentiation transform pair we derived back in section 5.5,

if $g(t) \longleftrightarrow G(\omega)$,
then $\frac{dg}{dt} \longleftrightarrow i\omega G(\omega)$,

and so Rayleigh's energy formula tells us that

$$\int\limits_{-\infty}^{\infty} \left(\frac{dg}{dt}\right)^2 dt = \frac{1}{2\pi} \int\limits_{-\infty}^{\infty} |i\omega G(\omega)|^2 d\omega = \frac{1}{2\pi} \int\limits_{-\infty}^{\infty} \omega^2 |G(\omega)|^2 d\omega = W\sigma_\omega^2.$$

Our result here says $\lim_{|\omega| \to \infty} \omega^2 |G(\omega)|^2 = 0$ must be true for the ω-integral to exist, which is a stronger requirement than just $\lim_{|\omega| \to \infty} \omega |G(\omega)|^2 = 0$ (which is the requirement of the Riemann-Lebesgue lemma from section 5.3). Example 1 in section 5.3, for example, *fails* this requirement. There we derived the Fourier transform pair of the pulse

$$f(t) = \begin{cases} 1, & |t| < \frac{\tau}{2}, \\ 0, & \text{otherwise,} \end{cases}$$

and found that

$$F(\omega) = \tau \frac{\sin(\omega\tau/2)}{\omega\tau/2}.$$

Thus, since the energy of $f(t)$ is $W = \tau$, we have

$$\sigma_\omega^2 = \int_{-\infty}^{\infty} \omega^2 \tau^2 \frac{\sin^2(\omega\tau/2)}{(\omega\tau/2)^2 2\pi\tau} d\omega = \frac{2}{\pi\tau} \int_{-\infty}^{\infty} \sin^2\left(\frac{\omega\tau}{2}\right) d\omega = \infty.$$

This result for σ_ω^2 is, of course, due to the fact that

$$\lim_{|\omega|\to\infty} \omega^2 |F(\omega)|^2 \neq 0.$$

If we take our three results for the integrals in the Cauchy-Schwarz inequality and substitute them into the inequality we get

$$\left(-\frac{1}{2}W\right)^2 = \frac{1}{4}W^2 \leq (W\sigma_t^2)(W\sigma_\omega^2) = W^2\sigma_t^2\sigma_\omega^2,$$

or

$$\frac{1}{4} \leq \sigma_t^2\sigma_\omega^2$$

or, at last, we have the uncertainty principle for Fourier integral theory:

$$\boxed{\sigma_t\sigma_\omega \geq \tfrac{1}{2}}.$$

This is a result, I think, that is *totally* nonobvious.

A natural question to ask now is, how good a lower bound is $\frac{1}{2}$ on the $\sigma_t\sigma_\omega$ product? Let's do a specific example and see what happens. Suppose that

$$g(t) = e^{-|t|/T},$$

where $T > 0$ and $-\infty < t < \infty$. The energy of $g(t)$ is

$$W = \int_{-\infty}^{\infty} g^2(t)\,dt = \int_{-\infty}^{0} e^{2t/T}\,dt + \int_{0}^{\infty} e^{-2t/T}\,dt$$

$$= \left(\frac{e^{2t/T}}{2/T}\bigg|_{-\infty}^{0}\right) + \left(\frac{e^{-2t/T}}{-2/T}\bigg|_{0}^{\infty}\right)$$

$$= \frac{1}{2/T} + \frac{1}{2/T} = T.$$

Thus,

$$\sigma_t^2 = \frac{1}{W}\int_{-\infty}^{\infty} t^2 g^2(t)\,dt = \frac{1}{T}\left[\int_{-\infty}^{0} t^2 e^{2t/T}\,dt + \int_{0}^{\infty} t^2 e^{-2t/T}\,dt\right],$$

which reduces (I'll let *you* do the integrals by parts or, even easier, look them up in tables) to $T^2/2$,

$$\sigma_t = \frac{T}{\sqrt{2}}.$$

To calculate σ_ω we *could* just substitute into

$$\sigma_\omega = \sqrt{\int_{-\infty}^{\infty} \omega^2 \frac{|G(\omega)|^2}{2\pi W}\,d\omega}$$

once we have $G(\omega)$. Recall our alternative formulation of σ_ω^2, however, which is much easier to use in this case. If you look back to the use of the time differentiation pair to derive this formula for σ_ω^2 you'll see that we also had the expression

$$\sigma_\omega^2 = \frac{1}{W}\int_{-\infty}^{\infty}\left(\frac{dg}{dt}\right)^2 dt.$$

For our problem this is an easy integral to do. We have

$$\frac{dg}{dt} = \begin{cases} \frac{1}{T}e^{t/T}, & t < 0, \\[2mm] -\frac{1}{T}e^{-t/T}, & t > 0, \end{cases}$$

and so (as $W = T$)

$$\sigma_\omega^2 = \frac{1}{T}\left[\frac{1}{T^2}\int_{-\infty}^{0} e^{2t/T}\,dt + \frac{1}{T^2}\int_{0}^{\infty} e^{-2t/T}\,dt\right]$$

$$= \frac{1}{T^3}\left[\left(\frac{e^{2t/T}}{2/T}\Big|_{-\infty}^{0}\right) + \left(\frac{e^{-2t/T}}{-2/T}\Big|_{0}^{\infty}\right)\right] = \frac{1}{2T^2}[1+1] = \frac{1}{T^2}.$$

That is,

$$\sigma_\omega = \frac{1}{T}.$$

Thus,

$$\sigma_t\sigma_\omega = \frac{T}{\sqrt{2}}\cdot\frac{1}{T} = \frac{1}{\sqrt{2}} = 0.707,$$

which is, indeed, greater than our theoretical lower bound of 0.5 (*significantly* greater, in fact, by more than 41%).

This no doubt prompts the obvious question in your mind: is it possible for $\sigma_t\sigma_\omega$ ever to *equal* 0.5? The answer is *yes*. For $\sigma_t\sigma_\omega = \frac{1}{2}$, the Cauchy-Schwarz *in*equality must actually be an equality, which we observed earlier implies (in our earlier notation) that $h(t) = ks(t)$ with k some constant,

$$\frac{dg}{dt} = ktg(t).$$

That is,

$$\frac{dg}{g(t)} = ktdt$$

or, with C the constant of indefinite integration,

$$\ln\{g(t)\} - \ln(C) = \frac{1}{2}kt^2.$$

Now, with $C = 1$ (this choice involves no loss of generality since we can always redefine our amplitude scale) we have

$$\ln\{g(t)\} = \frac{1}{2}kt^2$$

or, simply writing a new constant k for the "old" $\frac{1}{2}k$,

$$g(t) = e^{kt^2},$$

where it is clear that $k < 0$ for $g(t)$ actually to have a Fourier transform. That is, a time function that *achieves* the minimum product of the time and frequency uncertainties is the Gaussian-pulse signal we considered in Example 3 of section 5.5.

The concept of reciprocal spreading, once confined to the worlds of theoretical physics and estoteric mathematics, has today found its way into popular fiction; for example, in Carl Sagan's 1985 novel *Contact*. At one point, his heroine, a radio astronomer searching for radio messages from extraterrestrials, is troubled by the apparent lack of such signals. As she runs through possible explanations for the failure to detect intelligent communications from deep space, her train of thought is as follows; maybe the aliens are

> fast talkers, manic little creatures perhaps moving with quick and jerky motions, who transmitted a complete radio message—the equivalent of hundreds of pages of English text—in a nanosecond. Of course, if you had a very narrow bandpass to your receiver, so you were listening only to a tiny range of frequencies, you were forced to accept the long time-constant. [This is the reciprocal spreading.] You would never be able to detect a rapid modulation. It was a simple consequence of the Fourier Integral Theorem, and closely related to the Heisenberg Uncertainty Principle. So, for example, if you had a bandpass of a kilohertz, you couldn't make out a signal that was modulated at faster than a millisecond. It would be a kind of sonic blur. [Her radio receivers'] bandpasses were narrower than a hertz, so to be detected the [alien] transmitters must be modulating very slowly, slower than one bit of information a second. (pp. 65–66)

5.7 Hardy and Schuster, and their optical integral.

Fourier transform theory is such a beautiful subject that it is easy to get so caught up in the symbols that you lose any connection with physical reality. So, every now and then, I think it's a good idea to see Fourier theory in action in support of *science*. The story I'm going to tell you in this section has, as far as I know, not been told before.

In 1925 the German-born English physicist Arthur Schuster (1851–1934), who became a naturalized British subject in 1875 (and was knighted in 1920) published a paper on the theory of light.[13] In that paper he encountered a definite integral that appears, at first sight, to be particularly challenging. (A second look will *not* change that opinion!) In his notation, that integral is

$$\int_0^\infty \left[\left\{ \frac{1}{2} - C(v) \right\}^2 + \left\{ \frac{1}{2} - S(v) \right\}^2 \right] dv,$$

where

$$C(v) = \int_0^v \cos\left(\frac{\pi}{2} v^2\right) dv, \ S(v) = \int_0^v \sin\left(\frac{\pi}{2} v^2\right) dv.$$

Schuster has, of course, committed a freshman calculus blunder here by the use of the same symbol for the upper limits on the C and S integrals as for the dummy variable of integration. He should have written the C and S integrals as, for example,

$$C(y) = \int_0^y \cos\left(\frac{\pi}{2} v^2\right) dv, \ S(y) = \int_0^y \sin\left(\frac{\pi}{2} v^2\right) dv, \ y \geq 0,$$

and then his integral would become

$$\int_0^\infty \left[\left\{ \frac{1}{2} - C(y) \right\}^2 + \left\{ \frac{1}{2} - S(y) \right\}^2 \right] dy.$$

In any case, awkward notation aside, Schuster was unable to evaulate this integral. At one point, however, he concludes that the *physics* of the problem would be satisfied "if it could be proved that

$$\int_0^\infty \left[\left(\frac{1}{2} - C \right)^2 + \left(\frac{1}{2} - S \right)^2 \right] dv = \pi^{-1}.$$

Alas, he couldn't prove this and that is where he left matters.

Schuster's paper soon came to the attention of the great British mathematician G. H. Hardy, whom I've mentioned several times before in this book (he will appear again in the next chapter). It wasn't Schuster's physics that interested him, however. Indeed, the author of one of his obituary notices wrote that "Hardy had singularly little appreciation of science."[14] I mentioned this curious attitude of Hardy's in the Introduction, and in fact he was such a purist that even applied mathematics had no place in his world. As he wrote in his interesting (and a bit sad) *A Mathematician's Apology*, "It is true that there are branches of applied mathematics, such as ballistics and aerodynamics, which have been developed deliberately for war ... but none of them has any claim to rank as 'real.' They are indeed repulsively ugly and intolerably dull." The physics of light was obviously of no interest to anyone who could write that. Rather, it was Schuster's *definite integral itself* that attracted Hardy's mathematician's eye.

Indeed, displaying an unevaluated definite integral to Hardy was very much like waving a red flag in front of a bull. Hardy's papers[15] are full of such calculations, and one cannot fail to be impressed at the force of intellect he could focus to achieve success in attacking some pretty nasty, downright scary-looking integrals. (As the previously quoted obituary notice put it, "There was hardly a branch of analysis to which he did not make some contribution. He came to be recognized generally as the leading English mathematician of his time.") By comparison, Schuster's integral is much less terrifying than others laid low by Hardy, but, it was still enough of a challenge for the mathematician temporarily to push aside whatever was on his plate at the time. A *physicist* needed mathematical help, and Hardy was just the fellow to come to the rescue. Both men were clearly top-notch in their respective fields; each received the Royal Society's highest honor, the Copley Medal (Schuster in 1931, and Hardy in 1947). But *this* problem was on *Hardy's* turf, and here was an irresistible opportunity to show the brilliance of pure mathematical thought over mere experimental grubbing.

Hardy's first efforts were in simply expressing the problem in cleaner fashion. Hardy opened his paper,[16] with no preamble at all, with

The integral in question is

$$J = \int_0^\infty (C^2 + S^2)\, dx$$

where

$$C = \int_x^\infty \cos t^2 dt, \quad S = \int_x^\infty \sin t^2 dt.$$

Sir Arthur Schuster... suggests that the value of the integral is $\frac{1}{2}\sqrt{\frac{1}{2}\pi}$.

In a footnote he admitted to the reader "I have altered the notation slightly." Indeed!

Here's how I've reconstructed what Hardy probably did to get to his restatement of Schuster's integral. The correctly written C and S integrals are actually the well-known Fresnel integrals that we first encountered in section 4.5, in the discussion on Gauss's quadratic sum; they each have the property that

$$\lim_{y\to\infty} C(y) = \lim_{y\to\infty} S(y) = \frac{1}{2}.$$

Thus, if we redefine the C and S integrals as

$$C(y) = \int_y^\infty \cos\left(\frac{\pi}{2}v^2\right) dv, \quad S(y) = \int_y^\infty \sin\left(\frac{\pi}{2}v^2\right) dv, \quad y \geq 0,$$

then Schuster's integral becomes (because $\int_y^\infty = \int_0^\infty - \int_0^y$ and $\int_0^\infty = \frac{1}{2}$)

$$\int_0^\infty \{C^2(y) + S^2(y)\} dy,$$

and his *conjecture* becomes

$$\int_0^\infty \{C^2(y) + S^2(y)\} dy = \frac{1}{\pi}.$$

If we next make the change of variable to $t = v\sqrt{\pi/2}$, then

$$C(y) = \sqrt{\frac{2}{\pi}} \int_{y\sqrt{\pi/2}}^\infty \cos(t^2) dt, \quad S(y) = \sqrt{\frac{2}{\pi}} \int_{y\sqrt{\pi/2}}^\infty \sin(t^2) dt, \quad y \geq 0,$$

and the Schuster conjecture becomes

$$\int_0^\infty \left\{ \left[\int_{y\sqrt{\pi/2}}^\infty \cos(t^2)\,dt \right]^2 + \left[\int_{y\sqrt{\pi/2}}^\infty \sin(t^2)\,dt \right]^2 \right\} dy = \frac{1}{2}, \ y \geq 0.$$

And then, with one more change of variable to $x = y\sqrt{\pi/2}$, the Schuster conjecture becomes

$$J = \int_0^\infty \left\{ \left[\int_x^\infty \cos(t^2)\,dt \right]^2 + \left[\int_x^\infty \sin(t^2)\,dt \right]^2 \right\} dx = \frac{1}{2}\sqrt{\frac{\pi}{2}}, \ x \geq 0,$$

which is, as mentioned above, where Hardy's paper *begins*. We have, for the rest of this section, Hardy's new and *final* definitions for $C(x)$ and $S(x)$:

$$C(x) = \int_x^\infty \cos(t^2)\,dt, \ S(x) = \int_x^\infty \sin(t^2)\,dt.$$

Notice, *carefully*, that with Hardy's redefinitions we now have $\lim_{x\to\infty} C(x) = \lim_{x\to\infty} S(x) = 0$.

With his opening words Hardy's paper is already nearly one-third done—and then, in less than two dozen more lines (mostly prose), he calculates Schuster's integral (which I will from now on refer to as the *Hardy-Schuster integral*) not once, but *twice*! It is his second solution, using Fourier theory, that is the central point of this section, but let me first say just a few words about his initial solution. It's pretty clever, too, and Euler's formula plays a role in it as well. Hardy begins with a direct frontal assualt on the integral by writing "It is plain that

$$C^2 + S^2 = \int_x^\infty \int_x^\infty \cos(t^2 - u^2)\,dt\,du."$$

Intelligent people might argue about just how "plain" this is, but here's one way to see it without too much work. From Euler's formula we have

$$e^{it^2} = \cos(t^2) + i\sin(t^2),$$

which allows us to write

$$C(x) + iS(x) = \int_x^\infty e^{it^2} dt.$$

Taking the conjugate of both sides gives

$$C(x) - iS(x) = \int_x^\infty e^{-it^2} dt = \int_x^\infty e^{-iu^2} du.$$

Multiplying these last two expressions together gives

$$C^2(x) + S^2(x) = \int_x^\infty e^{it^2} dt \int_x^\infty e^{-iu^2} du$$

$$= \int_x^\infty \int_x^\infty e^{i(t^2 - u^2)} dt\, du$$

$$= \int_x^\infty \int_x^\infty \cos(t^2 - u^2) dt\, du + i \int_x^\infty \int_x^\infty \sin(t^2 - u^2) dt\, du.$$

Now, since $C^2(x)$ and $S^2(x)$ are each purely real, the imaginary part of this expression must vanish and we immediately have Hardy's "plain" result that

$$C^2(x) + S^2(x) = \int_x^\infty \int_x^\infty \cos(t^2 - u^2) dt\, du.$$

The Hardy-Schuster integral, therefore, is equivalent to the stunning triple integral

$$J = \int_0^\infty \int_x^\infty \int_x^\infty \cos(t^2 - u^2) dt\, du\, dx, \quad x \geq 0.$$

Hardy sounds less the mathematician and more the physicist or engineer when he goes on to write "If we integrate with respect to x, and ignore any difficulties in changing the order of integrations . . . the integral then falls apart . . . " and then, suddenly, he simply writes the final result. This all requires just six lines (four of which are prose), and he concludes in

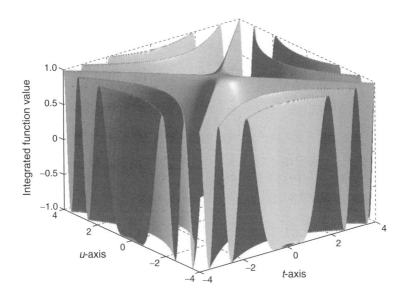

Figure 5.7.1. Integrand of the Hardy-Schuster integral

a flourish with the astonishing (to me, anyway) words "We thus obtain Schuster's result *almost without calculation* (my emphasis)." Figure 5.7.1 shows the integrand of the Hardy-Schuster integral.

Hardy's "mathematical instincts" were perhaps not quite happy with the devil-may-care attitude his "ignore any difficulties in changing the order of integration" might have implied. Writing that "it is plain that a strict analytical proof would be rather troublesome to write out," he went on to say "An alternative proof may be based on the theory of Fourier transforms." The rest of this section is a discussion of what Hardy did, and you'll see that the derivation is essentially Rayleigh's energy formula combined with a clever trick that you first saw back in chapter 4. The entire derivation is a tour de force.

We start with some preliminary remarks about *any* $f(x)$, not necessarily confined to $C(x)$ and $S(x)$. We'll specialize the analysis to them in just a bit, but not yet. Since the Fourier transform of any $f(x)$ is

$$F(\omega) = \int_{-\infty}^{\infty} f(x)e^{-i\omega x}\,dx = \int_{-\infty}^{\infty} f(x)\cos(\omega x)\,dx - i\int_{-\infty}^{\infty} f(x)\sin(\omega x)\,dx,$$

for the case where $f(x)$ is real *and zero* for $x < 0$, $F(\omega)$ *must* be complex since neither integrand on the right is either even or odd. But, suppose we make an *odd extension* of $f(x)$ for $x < 0$ (this is the "clever trick" I mentioned). Then the integral that is the real part of $F(\omega)$ vanishes (because its integrand is now odd). So, if we denote the odd extension of $f(x)$ by $\widehat{f}(x)$, then its Fourier transform is the purely imaginary $\widehat{F}(\omega)$ given by

$$\widehat{F}(\omega) = -i \int_{-\infty}^{\infty} \widehat{f}(x) \sin(\omega x)\,dx.$$

Since $\widehat{f}(x) = f(x)$ for $x > 0$ (and of course $\widehat{f}(x) = -f(x)$ for $x < 0$), we can write

$$\widehat{F}(\omega) = -i\left[\int_{-\infty}^{0} \widehat{f}(x)\sin(\omega x)\,dx + \int_{0}^{\infty} f(x)\sin(\omega x)\,dx\right].$$

In the first integral we change variables to $s = -x$, and then that integral becomes

$$\int_{\infty}^{0} \widehat{f}(-s)\sin(-\omega s)(-ds) = -\int_{0}^{\infty} \widehat{f}(-s)\sin(\omega s)\,ds = \int_{0}^{\infty} \widehat{f}(s)\sin(\omega s)\,ds,$$

where the last integral follows because $\widehat{f}(-s) = -\widehat{f}(s)$, because that's how we constructed $\widehat{f}$, that is, it is an *odd* function. Thus,

$$\widehat{F}(\omega) = -i\left[\int_{0}^{\infty} \widehat{f}(s)\sin(\omega s)\,ds + \int_{0}^{\infty} f(x)\sin(\omega x)\,dx\right]$$

or, since $\widehat{f} = f$ for *positive* arguments (notice that both integrals are indeed over positive intervals), and using the same variable for the dummy variable of integration, we arrive at

Box #1 $$\widehat{F}(\omega) = -i2 \int_{0}^{\infty} f(x)\sin(\omega x)\,dx.$$

Now, the energy of $\widehat{f}(x)$ is given by $\int_{-\infty}^{\infty} \widehat{f}^2(x)\,dx = 2\int_0^{\infty} f^2(x)\,dx$, as well as (à la Rayleigh) by $1/2\pi \int_{-\infty}^{\infty} |\widehat{F}(\omega)|^2\,d\omega$. Since $|\widehat{F}(\omega)|^2$ is even (because $\widehat{f}(x)$ is real) we can also write the energy as $1/\pi \int_0^{\infty} |\widehat{F}(\omega)|^2\,d\omega$, and so

Box #2

$$\int_0^{\infty} f^2(x)\,dx = \tfrac{1}{2\pi} \int_0^{\infty} |\widehat{F}(\omega)|^2\,d\omega .$$

Now we stop being general and restrict our *any* $f(x)$ to $C(x)$ and $S(x)$. We have our earlier definitions of them,

$$C(x) = \int_x^{\infty} \cos(t^2)\,dt, \quad S(x) = \int_x^{\infty} \sin(t^2)\,dt,$$

which we take to hold for all $x \geq 0$. Furthermore, we imagine we have made *odd* extensions of each (just as argued above for an arbitrary $f(x)$) for $x < 0$. Then, if we define two real functions $\phi(\omega)$ and $\psi(\omega)$ as

$$\phi(\omega) = \int_0^{\infty} C(x)\sin(\omega x)\,dx, \quad \psi(\omega) = \int_0^{\infty} S(x)\sin(\omega x)\,dx,$$

then as shown above (see Box #1) the Fourier transforms of $C(x)$ and $S(x)$ are

$$\widehat{F}_{C(x)}(\omega) = -i2\phi(\omega), \quad \widehat{F}_{S(x)}(\omega) = -i2\psi(\omega).$$

Then, from our result in Box #2,

$$\int_0^{\infty} C^2(x)\,dx = \frac{1}{2\pi} \int_0^{\infty} 4\phi^2(\omega)\,d\omega = \frac{2}{\pi} \int_0^{\infty} \phi^2(\omega)\,d\omega$$

and, similarly,

$$\int_0^{\infty} S^2(x)\,dx = \frac{2}{\pi} \int_0^{\infty} \psi^2(\omega)\,d\omega.$$

Thus, we have for the Hardy-Schuster integral,

$$J = \int_0^{\infty} \{C^2(x) + S^2(x)\}\,dx = \frac{2}{\pi} \int_0^{\infty} \{\phi^2(\omega) + \psi^2(\omega)\}\,d\omega.$$

We can find $\phi(\omega)$ and $\psi(\omega)$ via integration by parts. So, for $\phi(\omega) = \int_0^\infty C(x) \sin(\omega x) dx$ with $C(x) = \int_x^\infty \cos(t^2) dt$, recall that $\int_0^\infty u \, dv = (uv|_0^\infty - \int_0^\infty v \, du$ and set $u = C(x)$. Thus, $du/dx = -\cos(x^2)$. Also, with $dv = \sin(\omega x) dx$, $v = -\cos(\omega x)/\omega$. Thus,

$$\int_0^\infty C(x) \sin(\omega x) dx = \phi(\omega)$$

$$= \left\{ -\frac{\cos(\omega x)}{\omega} C(x) \Big|_0^\infty - \int_0^\infty -\frac{\cos(\omega x)}{\omega} \{-\cos(x^2) dx\}, \right.$$

or, as $C(\infty) = 0$,

$$\phi(\omega) = \frac{1}{\omega} C(0) - \frac{1}{\omega} \int_0^\infty \cos(\omega x) \cos(x^2) dx.$$

Since $C(0) = \int_0^\infty \cos(t^2) dt = \frac{1}{2}\sqrt{\pi/2}$, and since from integral tables we have the result

$$\int_0^\infty \cos(ax^2) \cos(2bx) dx = \frac{1}{2}\sqrt{\frac{\pi}{2a}} \left[\cos\left(\frac{b^2}{a}\right) + \sin\left(\frac{b^2}{a}\right) \right],$$

with $a = 1$ and $b = \frac{1}{2}\omega$ we have

$$\int_0^\infty \cos(\omega x) \cos(x^2) dx = \frac{1}{2}\sqrt{\frac{\pi}{2}} \left[\cos\left(\frac{1}{4}\omega^2\right) + \sin\left(\frac{1}{4}\omega^2\right) \right].$$

Thus,

$$\phi(\omega) = \frac{1}{2\omega}\sqrt{\frac{\pi}{2}} \left[1 - \cos\left(\frac{1}{4}\omega^2\right) - \sin\left(\frac{1}{4}\omega^2\right) \right].$$

For $\psi(\omega)$ the calculations are *almost* identical. I'll let you go through the details, and you should arrive at

$$\psi(\omega) = \frac{1}{2\omega}\sqrt{\frac{\pi}{2}} \left[1 - \cos\left(\frac{1}{4}\omega^2\right) + \sin\left(\frac{1}{4}\omega^2\right) \right].$$

So,

$$\phi^2(\omega) + \psi^2(\omega) = \frac{\pi}{8\omega^2}\left[\left\{1 - \cos\left(\frac{1}{4}\omega^2\right) - \sin\left(\frac{1}{4}\omega^2\right)\right\}^2\right.$$

$$\left. + \left\{1 - \cos\left(\frac{1}{4}\omega^2\right) + \sin\left(\frac{1}{4}\omega^2\right)\right\}^2\right]$$

$$= \frac{\pi}{8\omega^2}\left[2\left\{1 - \cos\left(\frac{1}{4}\omega^2\right)\right\}^2 + 2\sin^2\left(\frac{1}{4}\omega^2\right)\right]$$

$$= \frac{\pi}{4\omega^2}\left[1 - 2\cos\left(\frac{1}{4}\omega^2\right) + \cos^2\left(\frac{1}{4}\omega^2\right) + \sin^2\left(\frac{1}{4}\omega^2\right)\right]$$

$$= \frac{\pi}{2\omega^2}\left[1 - \cos\left(\frac{1}{4}\omega^2\right)\right].$$

Thus,

$$J = \frac{2}{\pi}\cdot\frac{\pi}{2}\int_0^\infty \frac{1 - \cos(\omega^2/4)}{\omega^2}\,d\omega = \int_0^\infty \frac{1 - \cos(\omega^2/4)}{\omega^2}\,d\omega.$$

This integral (whose integrand is perfectly well behaved at $\omega = 0$—see the discussion in note 8 for the behavior of a similar integrand)—can also be done by integration by parts. In the standard formula let $u = 1 - \cos(1/4\omega^2)$ and so $du = \frac{1}{2}\omega\sin(1/4\omega^2)\,d\omega$. Also, let $dv = d\omega/\omega^2$ and so $v = -1/\omega$. Thus,

$$J = \int_0^\infty \frac{1 - \cos(\omega^2/4)}{\omega^2}\,d\omega$$

$$= \left\{-\frac{1}{\omega}\left(1 - \cos\left(\frac{1}{4}\omega^2\right)\right)\right\}\Big|_0^\infty$$

$$- \int_0^\infty -\frac{1}{\omega}\cdot\frac{1}{2}\omega\sin\left(\frac{1}{4}\omega^2\right)\,d\omega$$

$$= \frac{1}{2}\int_0^\infty \sin\left(\frac{1}{4}\omega^2\right)\,d\omega.$$

Now, let $t = \frac{1}{2}\omega$ and so $dt = \frac{1}{2}d\omega$ and, at last, we have Hardy's result of

$$J = \frac{1}{2} \int_0^\infty \sin(t^2)2\,dt = \int_0^\infty \sin(t^2)\,dt = \frac{1}{2}\sqrt{\frac{\pi}{2}}.$$

What did Schuster think of all this? As far as I know he never commented in print on Hardy's solution, but I would not be surprised if he never bothered to fill in the gaps as I've done here. Schuster *knew* Hardy was a brilliant analyst (Hardy received a Royal Medal from the Royal Society of London in 1920, of which Schuster had been a Fellow since 1879 and Secretary from 1912 to 1919). It seems most likely that he was of course pleased that his conjecture had been confirmed by an expert, and simply let it go at that. Still, on the other hand, maybe Schuster *did* fill in the missing details—I think he certainly *could* have. He was, in fact, quite comfortable with Fourier's mathematics. There is a nice discussion of the Fourier transform and of Rayleigh's energy formula, for example, in Schuster's book *An Introduction to Optics* (all editions of which appeared before Hardy's paper, and which is dedicated, by the way, to Schuster's friend Lord Rayleigh). Perhaps I am being unfair to Hardy, but I have the impression from his paper that he meant it to be taken as a mere trifle tossed off while reading the cricket scores in *The Times* one morning over tea. "Let the physicists top *this*!" is the message I sense being sent (surely unconsciously, as by all accounts Hardy was fundamentally a kind person—but nevertheless he was *human*, too).

Chapter 6
Electronics and $\sqrt{-1}$

6.1 Why this chapter is in this book.

Up to now this book has treated mathematical topics for the sake of the *mathematics*, alone. However, since my writing was strongly motivated by an admiration for Euler the *engineering physicist* nearly as deep as my admiration for him as a mathematician, it seems appropriate to end with a chapter on some technological uses of the complex number mathematics he helped develop. Euler's large body of applied work shows that he took to heart, very seriously, the motto of the Berlin Academy of Sciences (of which he was a member from 1741 to 1766): "theoria cum praxi," which my tiny linguistic abilities in Latin translate as "theory *with* practice." No engineer could have said it better. (The motto is due, I should tell you, to a *mathematician*—Leibniz, who founded the Berlin Academy in 1700.)

In Euler's times there was no electricity other than lightning bolts in the sky, and so what I'll do here is show you some wonderful uses of complex numbers in *electronics* that I think Euler would have loved. I've made this choice because I am, after all, an electrical engineer (but I will *not* assume that *you* are—you do *not* need to know anything about electronics to read this chapter). Euler's own applied work in ship building, cannon ballistics, and hydraulics was pretty spectacular, too, but that has been written about many times before[1]—I want to do something just a bit different here.

6.2 Linear, time-invariant systems, convolution (again), transfer functions, and causality.

Electronic systems are actually constructed out of amazingly complex, tiny bits of matter created by equally complex manufacturing processes to precisely implement certain current/voltage relationships. These tiny bits of matter, mostly constructed from the common element silicon (every beach in the world has kilotons, at least, of silicon locked up in sand), behave like resistors, capacitors, transistors, diodes, and other devices with equally fancy names that are part of the workaday lingo of electrical engineers. Those engineers get paid a lot of money to know how electrons make their way through circuits constructed from millions of microscopic slivers of silicon, circuits that are often interconnected to a degree approaching the connectivity of the human brain. It's all very arcane stuff that often comes close to being magical, but we will need none of it here.

We will, instead, work here with what are called *technology independent block diagrams*. They are the diagrams you'll find on the blackboards in the offices of a project manager (I call them *management-level diagrams*). Sometimes they are simply called input-output diagrams. The simplest[2] such diagram is the single box shown in figure 6.2.1, with $x(t)$ as the input and $y(t)$ as the output. I'll tell you soon what that $h(t)$ inside the box means. In principle we can mathematically describe the behavior of any box by, of course, just writing an equation that tells us what $y(t)$ is as a function of $x(t)$. For example, if our box is an integrator we could write

$$y(t) = \int_{-\infty}^{t} x(s)\,ds.$$

For our purposes in this chapter, we'll limit ourselves to boxes that are what electrical engineers call *linear, time-invariant* (LTI) boxes. The *linear* part of LTI means that the *principle of superposition* applies, which in turn means (a right-pointing arrow is shorthand for *causes*):

if $x_1(t) \to y_1(t)$
and if $x_2(t) \to y_2(t)$,

Figure 6.2.1. The simplest block diagram

then $c_1 x_1(t) + c_2 x_2(t) \rightarrow c_1 y_1(t) + c_2 y_2(t)$
for c_1 and c_2 *any* constants (perhaps complex).

The *time-invariant* part of LTI means:

if $x(t) \rightarrow y(t)$,
then $x(t - t_0) \rightarrow y(t - t_0)$.

In words, if the input to an LTI box is shifted in time by t_0, then the output of the box will also be shifted in time by t_0.

Now, what's that $h(t)$ in figure 6.2.1? It is what is called the *impulse response* of the box. That is, *if* $x(t) = \delta(t)$ *then* $y(t) = h(t)$. The box output is $h(t)$ when the box input is the unit strength impulse $\delta(t)$. As I'll show you next, $h(t)$ represents *complete information* on the contents of the box. That is, if we know $h(t)$ then we can calculate the output $y(t)$ for *any* input $x(t)$. One reason the impulse response of a box "contains all knowledge" of the internal structure of the box is that the impulse has its energy (as we showed in chapter 5) distributed uniformly over *all* frequencies, from $-\infty$ to $+\infty$. *Everything* and *anything* inside the box will find some energy in the impulsive input at whatever frequency that "thing" needs to be stimulated. That is, there is *nothing* inside the box that won't contribute to the total output response. More dramatically, there is nothing inside a box that "keeps a secret"—*everything* inside the box "talks" about itself when stimulated by an input impulse.

One of the first things an undergraduate electrical engineering student learns to do is how to calculate $h(t)$ for an LTI box when the detailed internal structure of the box (e.g., a circuit schematic) is given. We won't get into that here, but rather will simply assume we already know $h(t)$. Even more direct is to just imagine we've *applied* $\delta(t)$ to the box and then *measured* what comes out—*by definition* that's $h(t)$. Of course, you might wonder how a $\delta(t)$ could actually be applied, since it's an infinite energy signal; how could one really generate a $\delta(t)$? That's a valid question, but it has an easy answer (which I'll tell you later in this section). On *paper*, of course, there's no problem in *mathematically* applying an impulse. For example, what's $h(t)$ for an integrator? Well, if $x(t) = \delta(t)$, then

$$y(t) = \int_{-\infty}^{t} x(s)\,ds = \int_{-\infty}^{t} \delta(s)\,ds = \begin{cases} 0, & t < 0, \\ 1, & t > 0, \end{cases}$$

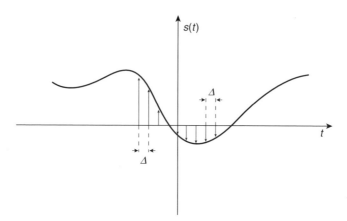

Figure 6.2.2. A train of impulses, uniformly spaced Δ in time

which says $h(t) = u(t)$. The impulse response of this integrator is the unit step function.

Now, to start, let me first prove to you that *if* we have $h(t)$, by whatever means (theory or experiment), then we can indeed calculate $y(t)$ for any $x(t)$. Imagine the input to our box in figure 6.2.1 is a train of impulses of varying strength (uniformly spaced in time) as determined by the arbitrary function $s(t)$, as shown in figure 6.2.2. That is, if the temporal spacing between sequential impulses is Δ we'll write $x(t)$ as

$$x(t) = \sum_{k=-\infty}^{\infty} s(k\Delta) \cdot \Delta \cdot \delta(t - k\Delta).$$

Be sure you understand what this means: $s(k\Delta) \cdot \Delta$ is the strength or *area* of the impulse at time $t = k\Delta$. (The *height* of each impulse is infinity, by the very concept of an impulse.) Since $x(t)$ is the input to an LTI box, by superposition and time invariance we can write the output as (because the *unit* strength impulse at time $t = k\Delta$ produces the output $h(t - k\Delta)$)

$$y(t) = \sum_{k=-\infty}^{\infty} s(k\Delta)\Delta h(t - k\Delta).$$

Let's now argue just like we did in Section 5.2, when we derived the Fourier transform *integral* from the Fourier series *summation*. That is, let's imagine that $\Delta \to 0$. Then,

    (i) $k\Delta$ "becomes" a continuous variable, which we'll write as τ;

    (ii) $\displaystyle\sum_{k=-\infty}^{\infty}$ "becomes" $\displaystyle\int_{-\infty}^{\infty}$;

    (iii) Δ "becomes" $d\tau$.

Our $x(t)$ and $y(t)$ expressions then become

$$x(t) = \int_{-\infty}^{\infty} s(\tau)\delta(t - \tau)d\tau,$$

$$y(t) = \int_{-\infty}^{\infty} s(\tau)h(t - \tau)d\tau.$$

From the sampling property of the impulse the equation for $x(t)$ reduces to just $x(t) = s(t)$. And so, if we substitute that result into the equation for $y(t)$, we arrive at our answer:

$$\boxed{y(t) = \int_{-\infty}^{\infty} x(\tau)h(t - \tau)d\tau = \int_{-\infty}^{\infty} x(t - \tau)h(\tau)d\tau} \quad ,$$

where the second integral follows by the obvious change of variable in the first integral. In other words, if you recall section 5.3, the output $y(t)$ of an LTI box with input $x(t)$ and impulse response $h(t)$ is the *time convolution* of $x(t)$ and $h(t)$, that is, $y(t) = x(t) * h(t)$. This result is a gold mine of information.

    For example, we can now answer the question of how to measure $h(t)$ experimentally without having to actually generate the *impossible to generate* input signal $\delta(t)$. Suppose our input is instead the *easy to generate* unit step function, that is, $x(t) = u(t)$. (The step, like the impulse, is an infinite energy signal but, unlike the impulse, we don't have to come up with all that infinite energy in a *single instant*; with the step, we literally can take forever.) The response of the box (logically enough called the *step response*) is

$$y(t) = \int_{-\infty}^{\infty} x(t-\tau)h(\tau)d\tau = \int_{-\infty}^{\infty} u(t-\tau)h(\tau)d\tau$$

$$= \int_{-\infty}^{t} h(\tau)d\tau.$$

Then, differentiating with respect to t, we immediately have

$$\frac{dy}{dt} = h(t).$$

That is, the impulse response of an LTI box is simply the derivative (easy to calculate) of the step response. What could be simpler?

As a second example of what the convolution integral can tell us, recall our result from section 5.3 when we took the Fourier transform of the time convolution integral,

$$\boxed{Y(\omega) = X(\omega)H(\omega)} \quad,$$

a result so important I put it in a box there and I've put it in a box here, too. It, in turn, immediately says that

$$\frac{1}{2\pi}|Y(\omega)|^2 = \frac{1}{2\pi}|X(\omega)|^2|H(\omega)|^2.$$

Since Rayleigh's energy formula tells us that $1/2\pi|X(\omega)|^2$ and $1/2\pi|Y(\omega)|^2$ are the energy spectral densities of the input and output signals, respectively, then we have

$$\text{energy of output} = \int_{-\infty}^{\infty} \frac{1}{2\pi}|Y(\omega)|^2 d\omega$$

$$= \int_{-\infty}^{\infty} (\text{ESD of input}) \cdot |H(\omega)|^2 d\omega.$$

The central role of $H(\omega)$ (the Fourier transform of the impulse response) in the study of LTI boxes can not be overstated. It is so important it has its own name: $H(\omega)$ is the *transfer function* of our LTI box, as it determines (as in the above integral) how the energy in the input signal is *transferred* to the output signal. Any frequency at which $H(\omega) = 0$ is a frequency at which there is *no energy* in the output (even if there is a lot of energy at that frequency in the input). You'll see how useful is this insight in building electronic systems before this chapter is done.

So far, I've said nothing in particular about the nature of $h(t)$. So, to end this section, let me introduce the concept of a *causal* $h(t)$. Mathematically, this simply means that $h(t) = 0$ for $t < 0$. There is a deep physical interpretation of this. The impulse response $h(t)$ of an LTI box is the output of the box *because* the input is $\delta(t)$, an input that occurs at the single instant $t = 0$. Any box that can be made from actual hardware that exists in the world as we know it *cannot* have an impulse response that exists *before* $t = 0$, hence the term *causal*. That is, the impulsive input is the *cause* of the $h(t)$ output signal. Any box that has an impulse response $h(t) \neq 0$ for $t < 0$ is called an *anticipatory* box because such a box "anticipates" the arrival of the impulse (at $t = 0$) and so begins to produce an output response *before* $t = 0$, that is, at $t < 0$. (That's why another name for a noncausal box that starts responding to an input that hasn't yet occurred is *time machine*!) To impose causality on $h(t)$ will obviously have an impact on the nature of its Fourier transform, the transfer function $H(\omega)$. What may not be so obvious, however, is just how great that impact is; it is profound. Let me show you why.

I'll start by explicitly writing the real and imaginary parts of the generally complex $H(\omega)$ as $R(\omega)$ and $X(\omega)$, respectively. That is,

$$H(\omega) = R(\omega) + iX(\omega).$$

In addition, I'll write $h(t)$, itself, as the sum of even and odd functions,

$$h(t) = h_e(t) + h_o(t).$$

The easiest way to prove you can *always* do this, for *any* $h(t)$, is to simply derive what $h_e(t)$ and $h_o(t)$ are (a proof by construction, the best sort of proof of all). By the very definitions of evenness and oddness we have

$$h_e(-t) = h_e(t),$$
$$h_o(-t) = -h_o(t).$$

Thus,

$$h(-t) = h_e(-t) + h_o(-t) = h_e(t) - h_o(t),$$

and so if we add and subtract $h(t)$ and $h(-t)$ we quickly arrive at

$$h_e(t) = \frac{1}{2}[h(t) + h(-t)],$$
$$h_o(t) = \frac{1}{2}[h(t) - h(-t)],$$

which shows that there *is* an $h_e(t)$ and an $h_o(t)$ for *any* given $h(t)$.

Now, since $h(t)$ is given as causal, that is, $h(t) = 0$ for $t < 0$, then

$$\left. \begin{aligned} h_e(t) &= \frac{1}{2}h(t) \\ h_o(t) &= \frac{1}{2}h(t) \end{aligned} \right\} \text{ if } t > 0$$

and

$$\left. \begin{aligned} h_e(t) &= \frac{1}{2}h(-t) \\ h_o(t) &= -\frac{1}{2}h(-t) \end{aligned} \right\} \text{ if } t < 0.$$

That is,

$$h_e(t) = h_o(t) \text{ if } t > 0,$$
$$h_e(t) = -h_o(t) \text{ if } t < 0,$$

which could of course be equally well written as

$$h_o(t) = h_e(t) \text{ if } t > 0,$$
$$h_o(t) = -h_e(t) \text{ if } t < 0.$$

And finally, we can write each of these *pairs* of statements very compactly, without having to give explicit conditions on t, as

$$
\begin{aligned}
h_e(t) &= h_o(t)\mathrm{sgn}(t), \\
h_o(t) &= h_e(t)\mathrm{sgn}(t)
\end{aligned}
$$

Since $h(t) = h_e(t) + h_o(t)$ we can write

$$H(\omega) = H_e(\omega) + H_o(\omega).$$

And since $h_e(t)$ is even (by definition), $H_e(\omega)$ is purely real, and since $h_o(t)$ is odd (by definition), $H_o(\omega)$ is purely imaginary (take a look back at section 5.2). Thus, it must be true that

$$H_e(\omega) = R(\omega),$$
$$H_o(\omega) = iX(\omega).$$

Recall now a Fourier transform pair that we derived in section 5.4,

$$\mathrm{sgn}(t) \longleftrightarrow \frac{2}{i\omega}.$$

Combining this with the *frequency* convolution theorem from section 5.3, and applying to the two equations written in the above box, we have

$$H_e(\omega) = R(\omega) = \frac{1}{2\pi}H_o(\omega) * \frac{2}{i\omega} = \frac{1}{2\pi}iX(\omega) * \frac{2}{i\omega}$$

and

$$H_o(\omega) = iX(\omega) = \frac{1}{2\pi}H_e(\omega) * \frac{2}{i\omega} = \frac{1}{2\pi}R(\omega) * \frac{2}{i\omega}.$$

Or,

$$R(\omega) = \frac{1}{\pi}X(\omega) * \frac{1}{\omega} = \frac{1}{\pi}\int_{-\infty}^{\infty}\frac{X(\tau)}{\omega - \tau}d\tau$$

and

$$X(\omega) = -\frac{1}{\pi}R(\omega) * \frac{1}{\omega} = -\frac{1}{\pi}\int_{-\infty}^{\infty}\frac{R(\tau)}{\omega - \tau}d\tau.$$

These two equations show that $R(\omega)$ and $X(\omega)$ each determine the other for a causal LTI box. The integrals that connect $R(\omega)$ and $X(\omega)$ are called *Hilbert transforms*,[3] a name introduced by our old friend G. H. Hardy. Hardy published the transform for the first time *in English* in 1909, but when he later learned that the German mathematician David Hilbert (remember him from section 3.1?) had known the transform integral since 1904 Hardy began to call it the Hilbert transform. But even Hilbert was not first—the transform appears in the 1873 doctoral dissertation of the Russian mathematician Yulian-Karl Vasilievich Sokhotsky (1842–1927). Notice that the Hilbert transform, unlike the Fourier transform, does not change domains. That is, the Hilbert transform takes an ω-function and transforms it into *another* ω-function (the Fourier transform, of course, transforms between *two* domains, those of t and ω).

The Hilbert transform integrals that relate the real and imaginary parts of the transfer function of a causal box might be called *local* constraints, because they show how the values of $R(\omega)$ and $X(\omega)$ are determined, for any *specific* ω. We can also derive what might be called *global* constraints on $R(\omega)$ and $X(\omega)$ for a causal $h(t)$ as follows. Applying Rayleigh's energy formula to the Fourier transform pairs

$$h_e(t) \longleftrightarrow R(\omega),$$

$$h_o(t) \longleftrightarrow iX(\omega),$$

we have (remember, both $R(\omega)$ and $X(\omega)$ are *real* functions of ω)

$$\int_{-\infty}^{\infty} h_e^2(t)\,dt = \frac{1}{2\pi}\int_{-\infty}^{\infty} R^2(\omega)\,d\omega,$$

$$\int_{-\infty}^{\infty} h_o^2(t)\,dt = \frac{1}{2\pi}\int_{-\infty}^{\infty} X^2(\omega)\,d\omega.$$

We showed earlier, for a causal $h(t)$, that $h_e(t) = h_o(t)\mathrm{sgn}(t)$, so $h_e^2(t) = h_o^2(t)$. Thus, the two time integrals (on the left) are equal, and therefore the two frequency integrals are equal, too,

$$\int_{-\infty}^{\infty} R^2(\omega)\,d\omega = \int_{-\infty}^{\infty} X^2(\omega)\,d\omega,$$

which shows how the integrated (global) behavior of $R(\omega)$ depends on the integrated (global) behavior of $X(\omega)$—and vice versa—*for a causal $h(t)$.*

It should now be clear that causality is a very strong constraint on the LTI boxes we can *actually build in real hardware,* as opposed to simply pushing symbols around on paper. Indeed, no matter how sophisticated a *super top secret* electronic gadget may be *on paper,* it *must* obey the "fundamental" law of cause and effect if we are to be able to actually build it in hardware.[4] If it can be shown that a proposed gadget violates cause and effect (i.e., that its impulse response is anticipatory), then that gadget *cannot* be built in actual hardware. Electrical engineers say that such a gadget is *unrealizable.* Put simply, there must be no output signal before there is an applied input signal. This may sound so trivially obvious that as to be hardly worth mentioning, but in fact some circuits that look quite benign on paper are not causal. Try as you might, they are simply impossible to build according to electrical engineering as it is presently understood, and to save yourself from an endless quest for what doesn't exist it is important to know how to tell if a theoretical paper design can actually be constructed. To demonstrate how this works, I'll now take you through the details of a specific example.

An important theoretical circuit in electronics is the *ideal unity-gain bandpass filter,* which allows energy located in a specified interval of frequencies to pass through it, while completely rejecting all input energy outside that interval. A plot of the *magnitude* of the transfer function of this idealized filter is shown in figure 6.2.3 (it is called a *unity-gain* filter because the magnitude of its transfer function $H(\omega)$ is one for those

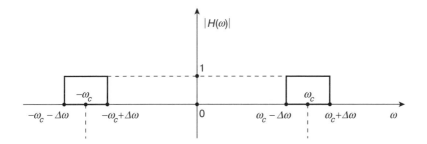

Figure 6.2.3. Transfer function (magnitude only) of the ideal unity-gain bandpass filter

frequencies at which energy transmission occurs). The plot of figure 6.2.3 is said to be idealized because $|H(\omega)| = 0$ when it isn't equal to one, that is, because of the so-called *vertical skirts* where $|H(\omega)|$ changes value (the term comes from the resemblance of the magnitude plot to a nineteenth-century hoop skirt). Actual filters always exhibit a less than vertical roll-off of the skirts.

The *bandwidth* of this ideal filter is $2\Delta\omega$, and the frequency interval over which the filter passes energy is called the *passband*, that is, the passband is $\omega_c - \Delta\omega < |\omega| < \omega_c + \Delta\omega$, where ω_c is the center frequency of the passband. Be clear about what this plot means. Suppose $\omega_c - \Delta\omega < \omega < \omega_c + \Delta\omega$ and we apply the signal $\cos(\omega t) = (e^{i\omega t} + e^{-i\omega t})/2$ as the input. Then half the energy of the input is at frequency $+\omega$ and half is at frequency $-\omega$, and $\cos(\omega t)$ "passes through" the filter, *unattenuated*, because the $e^{i\omega t}/2$ "gets through" since $|H(\omega)| = 1$ on the right of the ω-axis of figure 6.2.3 *and* the $e^{-i\omega t}/2$ "gets through" since $|H(\omega)| = 1$ on the left of the ω-axis of figure 6.2.3.

Knowledge of $|H(\omega)|$ is not enough to describe the ideal bandpass filter completely, of course, as it doesn't include *phase information*. That is, in general $H(\omega)$ is complex and so $H(\omega) = |H(\omega)|e^{i\theta(\omega)}$, where $\theta(\omega)$ is called the *phase function* of $H(\omega)$. To determine what should be used for the phase function, electrical engineers impose an additional constraint on the ideal filter called *zero phase distortion*. Phase distortion is said to occur in a filter if energies at different frequencies take different times to transit the filter, from input to output. Physically, *zero* phase distortion means that the input signal *shape* will be unaltered by passage through the ideal bandpass filter if *all* of the energy in the input signal is in the passband of the filter.

Consider, then, a particular frequency ω that we take as being in the passband of the filter. Further, suppose that all of the energy propagating through the filter experiences the same time delay t_0. Now, since the magnitude and shape of the input are unaltered by a unity-gain filter, the input signal $e^{i\omega t}$ experiences only a time delay of t_0. That is, the output signal will be $e^{i\omega(t-t_0)} = e^{i\omega t}e^{-i\omega t_0}$. But, by the very definition of the transfer function $H(\omega)$, the output signal is $H(\omega)e^{i\omega t}$. Thus, $H(\omega) = e^{-i\omega t_0}$, where ω is any frequency in the passband. ($H(\omega) = 0$ when ω is *not* in the passband, from the definition of the ideal bandpass filter.) So, an ideal zero phase distortion bandpass filter has a negative phase

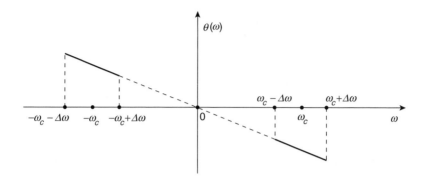

Figure 6.2.4. Phase function of the ideal bandpass filter

function that varies linearly with frequency (as shown in figure 6.2.4), that is, $\theta(\omega) = -\omega t_0$ for $\omega_c - \Delta\omega < |\omega| < \omega_c + \Delta\omega$.

We can show that this ideal filter is impossible to actually construct because its impulse response $h(t) \neq 0$ for $t < 0$. To do that we write the inverse Fourier transform integral

$$h(t) = \frac{1}{2\pi} \int_{-\infty}^{\infty} H(\omega) e^{i\omega t} \, d\omega$$

$$= \frac{1}{2\pi} \left[\int_{-\omega_c-\Delta\omega}^{-\omega_c+\Delta\omega} e^{-i\omega t_0} e^{i\omega t} \, d\omega + \int_{\omega_c-\Delta\omega}^{\omega_c+\Delta\omega} e^{-i\omega t_0} e^{i\omega t} \, d\omega \right],$$

which, after doing the integrals and applying Euler's formula, becomes

$$h(t) = \frac{1}{\pi t_0} \cdot \frac{\sin[\omega_c t_0 (1 + \Delta\omega/\omega_c)(t/t_0 - 1)] - \sin[\omega_c t_0 (1 - \Delta\omega/\omega_c)(t/t_0 - 1)]}{t/t_0 - 1},$$

which, after using a trig identity and doing just a bit more algebra (I'll let *you* fill in the details) becomes

$$h(t) = \frac{2}{\pi t_0} \cdot \frac{\cos[\omega_c t_0 (t/t_0 - 1)] \sin[\Delta\omega t_0 (t/t_0 - 1)]}{t/t_0 - 1}.$$

What does $h(t)$ "look like"? For plotting purposes, suppose that $t_0 = 1$, $\omega_c = 50$, and $\Delta\omega = 5$. Figure 6.2.5 shows $h(t)$ for these values (over the interval $-4 \leq t/t_0 \leq 3$) and, as you might expect, the maximum

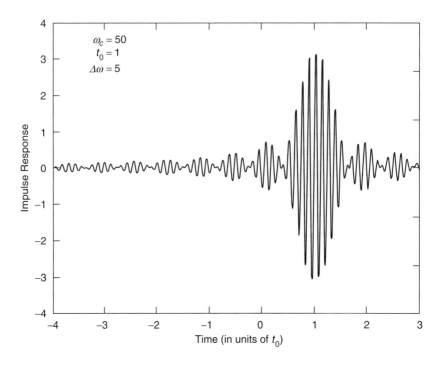

Figure 6.2.5. Impulse response of an ideal bandpass filter

response occurs at $t = t_0$; you can also see that $h(t)$ is quite "active" for $t < 0$.

So the ideal bandpass filter is impossible to build. But, nonetheless, it is so useful in helping electrical engineers when thinking about electronic systems that they use it anyway—at least they do on paper. You'll see how, later in this chapter. During a 1933 collaboration the American mathematician Norbert Wiener (recall the Wiener-Khinchin theorem from section 5.3) and the English mathematician Raymond Paley (1907–1933) discovered a necessary and sufficient condition for $|H(\omega)|$ to be the magnitude part of the transfer function of a causal filter: the so-called *Paley–Wiener integral* must be finite, that is,

$$\int\limits_{-\infty}^{\infty} \frac{|\ln|H(\omega)||}{1 + \omega^2}\, d\omega < \infty.$$

For the ideal bandpass filter, with $|H(\omega)| = 0$ for almost all ω, it is obvious that the Paley-Wiener integral is *not* finite, and so such a magnitude response is not possible in a causal filter. In the second volume of his autobiography, Wiener wrote of their joint work

> Paley attacked [the problem] with vigor, but what helped me and did not help Paley was that it is essentially a problem in electrical engineering [much of Wiener's work in mathematics was inspired by his interactions with colleagues in the MIT electrical engineering department]. It had been known for many years that there is a certain limitation on the sharpness with which an electric wave filter cuts a frequency band off, but the physicists and engineers had been quite unaware of the deep mathematical grounds for those limitations. In solving what was for Paley a beautiful and difficult chess problem, completely contained within itself, I showed at the same time that the limitations under which the electrical engineers were working were precisely those which prevent the future from influencing the past.[5]

The last sentence of this must have seemed mysterious to many readers of his autobiography (which is fascinating reading—Wiener was truly an odd duck), but he was of course referring to the anticipatory impulse response of a non causal circuit. The Paley-Wiener integral doesn't say anything about the *phase* of the transfer function; all it tells us is that *if* $|H(\omega)|$ passes the Paley-Wiener integral test *then* there exists some phase function that, when combined with $|H(\omega)|$, results in the transfer function of a causal filter.

One thing the Paley-Wiener integral test does tell us, however, is that along with the ideal bandpass filter it is also true that the ideal lowpass, the ideal highpass, and the ideal bandstop filters are impossible to build as well (the magnitude part of their transfer functions are shown in figure 6.2.6). But again, just as with the ideal bandpass filter, these three filters are so useful in aiding the thinking of electrical engineers that the engineers often pretend that such unrealizable filters actually do exist.

6.3 The modulation theorem, synchronous radio receivers, and how to make a speech scrambler.

Electrical engineers often find it useful to shift the energy of a signal both up and down in frequency. Fourier theory is invaluable for

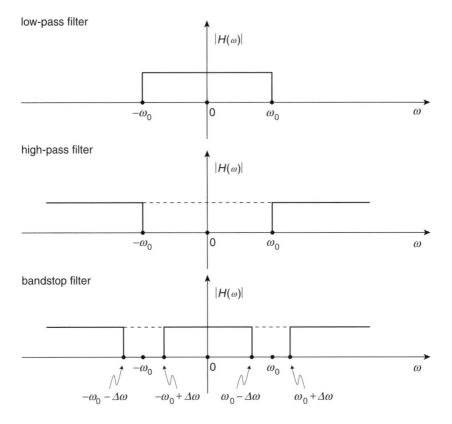

Figure 6.2.6. Magnitudes of the ideal lowpass, highpass, and bandstop filter transfer functions

understanding how that is accomplished using electronic circuitry. But first, to help motivate the mathematics in this section, let me give you an easy-to-appreciate example of *why* this is such an important task. I think every reader of this book would agree that radio is a marvelous electronic invention. It may not be commonly known, however, that frequency shifting is *essential* to radio. Here's why.

A common commercial use of radio (what is called amplitude-modulated or AM radio) allows the operation of many transmitters in the same geographical area. Why don't these multiple transmitters conflict with each other? That is, why can we select (*tune in*) from among all the multiple stations broadcasting to us the *one* to which we'll actually listen? To be simple about this, suppose Alice is talking into her microphone at Station A and Bob is talking into his microphone at Station B. Since the sounds produced by both are generated via the same physical process (vibrating human vocal chords), the energies of the two voice signals will be concentrated at essentially the same frequencies (typically, a few tens of hertz up to a few thousand hertz). That is, the frequency interval occupied by the electrical signals produced on the wires emerging from Alice's microphone is the same as the frequency interval occupied by the electrical signals produced on the wires emerging from Bob's microphone. This common interval of so-called *audio* frequencies determines what is called the *baseband spectrum*, that is, the voices of both Alice and Bob produce baseband, *bandlimited* (defined in section 5.3) energy.

To apply a baseband electrical signal of a microphone directly to an antenna will *not* result in the efficient radiation of energy into space; Maxwell's equations for the electromagnetic field tell us that for the efficient coupling of the antenna to space to occur the physical size of the antenna must be comparable to the wavelength of the radiation (you'll have to take my word on this bit of electrical physics). At the baseband frequency of 1 kHz, for example, a wavelength of electromagnetic radiation is *one million feet*, which is pretty big. So, to get a reasonably sized antenna, we need to reduce the wavelength, that is, to *increase* the frequency. What is done in commercial AM radio to accomplish that is to shift the baseband spectrum of the microphone signal up to somewhere between about 500 kHz to 1,500 kHz, the so-called *AM radio band*. (Each radio station receives a license from the Federal Communications Commission—the FCC—that gives it permission to do the upward frequency shift by a value that no other station in the same geographical area may use.) At 1,000 kHz, for example, the wavelength is a thousand times shorter than it is at 1 kHz—one thousand feet. If our station's antenna is constructed to be a quarter-wavelength in size, for example, then we are talking about an antenna 250 feet high (which you'll notice, the next time you drive by your local AM radio station's transmitter site, is just about what you'll actually see).

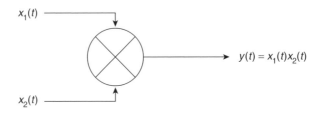

Figure 6.3.1. Block diagram symbol for a multiplier

So, let's suppose that at Station A we shift Alice's baseband signal up to 900 kHz, while at Station B we shift Bob's baseband signal up to 1,100 kHz. A radio receiver could then select which signal to listen to by using a *tunable* bandpass filter, that is, an adjustable[6] bandpass filter whose center frequency could be centered on either 900 kHz or 1,100 kHz (the *bandwidth* of the filter, for AM radio, is 10 kHz). Notice, carefully, that radio uses a frequency *upshift* for *two* reasons: (1) to move baseband energies up to radio frequencies to achieve efficient radiation of energy and (2) to *separate* the baseband spectra of the broadcasting stations by using a *different* upshift frequency, which will allow a bandpass filter to *select* a particular station. At the receiver we need a final frequency *downshift* to place the energy of the selected station signal back at the baseband frequencies to which our ears respond.

To accomplish these frequency shifts, both up and down, is, as you'll see, as simple as doing a multiplication—but you'll also soon see that doing direct multiplication in electronic circuitry is *not* so simple. We'll have to be just a bit clever at it, and Fourier theory will show us the way. To see that the mathematics of multiplication does indeed do the job of frequency shifting, let's first assume that we actually do have an electronic circuit with two inputs $x_1(t)$ and $x_2(t)$ and a single output $y(t)$ that is the product of the two inputs, as shown in figure 6.3.1. (Do you see that such a circuit *cannot* be an LTI box?[7]) Now, suppose that $x_1(t) = m(t)$, the baseband voice signal produced by a microphone, and that $x_2(t) = \cos(\omega_c t)$, where ω_c is the specific frequency that a radio station receives from the FCC as its assigned upshift frequency (the so-called *carrier frequency*). Our question is: where (in frequency) is the energy of $y(t) = m(t) \cos(\omega_c t)$?

To answer that question we of course need to find the Fourier transform of $y(t)$. So,

$$Y(\omega) = \int_{-\infty}^{\infty} y(t)e^{-i\omega t}\,dt$$

$$= \int_{-\infty}^{\infty} m(t)\cos(\omega_c t)e^{-i\omega t}\,dt$$

$$= \int_{-\infty}^{\infty} m(t)\frac{e^{i\omega_c t} + e^{-i\omega_c t}}{2}e^{-i\omega t}\,dt$$

$$= \frac{1}{2}[\int_{-\infty}^{\infty} m(t)e^{-i(\omega-\omega_c)t}\,dt + \int_{-\infty}^{\infty} m(t)e^{-i(\omega+\omega_c)t}\,dt]$$

$$= \frac{1}{2}[M(\omega - \omega_c) + M(\omega + \omega_c)].$$

Thus, the Fourier transform of the multiplier output is the transform of the baseband signal shifted both up *and* down by the frequency ω_c. This fundamental result is called the *modulation* or *heterodyne theorem*,[8] and is illustrated in figure 6.3.2, where the energy of $m(t)$ is confined to the frequency interval $-\omega_m \leq \omega \leq \omega_m$ (as mentioned before, ω_m is the maximum frequency in the baseband signal $m(t)$ at which there is energy and has a value of a few kHz).

The information in the baseband signal, $m(t)$, originally centered on $\omega = 0$, is now "riding piggyback" on $\omega = \omega_c$, and so you can appreciate why ω_c is called the *carrier* frequency. The multiplication done at the radio station transmitter facility is accomplished with a circuit called a *modulator* (about which I'll say more in section 6.4), and the multiplier output $y(t)$ is the signal applied to the antenna (after being boosted to a power level that, in commercial AM radio in America, can be as high as 50,000 watts). Note: my use of a triangle for the transform of $m(t)$ is just a metaphor for the actual transform, which could have a very *non*-triangular shape in reality (of course, since $m(t)$ is real, we know $|M(\omega)|$ is *always* even)—the important thing for us, here, is that $M(\omega)$ is *bandlimited.*

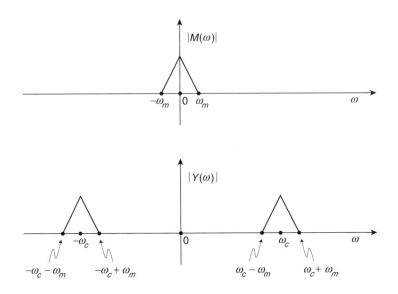

Figure 6.3.2. Fourier transform of a baseband signal, and its heterodyned version

Once the selection process is done at the receiver, the frequency *downshift* back to baseband can be accomplished with yet another multiplication by $\cos(\omega_c t)$, which results in moving the transform of $Y(\omega)$ in figure 6.3.2 both up *and* down by ω_c, that is, this newly shifted transform is that of a signal that has some of its energy centered around $2\omega_c$ and the rest around $\omega = 0$ (which is, of course, the original *baseband* signal). The baseband energy can then be selected by a lowpass filter that rejects the energy at and around $2\omega_c$ (in actual practice this filtering operation is *automatically* done by the fact that a mechanically "massive" loudspeaker simply cannot respond to the high frequency that is $2\omega_c$, a frequency that is greater than 1 MHz in commercial AM radio). This entire process at the receiver is called *detection* or *demodulation* by electrical engineers. Indeed, since it depends on having available the same $\cos(\omega_c t)$ signal used at the transmitter (any deviation in frequency from ω_c and/or the introduction of a phase shift, e.g., the use of $\cos(\omega_c t + \theta)$, with $\theta \neq 0$, will result in serious problems[9]), this form of radio receiver is called a *synchronous demodulator* (or *synchronous detector*). If $r(t)$ denotes the received signal (with a spectrum similar to that of the transmitted signal $y(t)$), then figure 6.3.3 shows the block diagram of a synchronous demodulation receiver. (The circle with the wavy curve inside it represents an

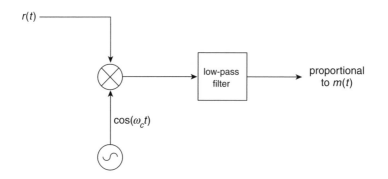

Figure 6.3.3. A synchronous demodulation radio receiver

oscillator circuit that produces a sinusoidal signal at frequency ω_c—it is not difficult to construct such a circuit.)

The receiver block diagram in figure 6.3.3 *looks* simple, but looks are deceiving. The requirement to *locally* generate a sinusoid at the receiver that very nearly matches in frequency and phase the sinusoid at the transmitter (which could be thousands of miles distant) is actually not an impossible task, but such synchronous receivers are expensive to build. That makes them unattractive for use in commercial AM radio. A modern AM radio receiver is so cheap, in fact, that nobody bothers to fix them—when they fail people just throw them away and buy a new one. For that reason alone, techniques other than synchronous demodulation for detecting the transmitted baseband signal are actually used in AM radio receivers.[10] *However,* if one is willing to pay for the additional circuit complexity of synchronous demodulation, then some near-magical new possibilities are opened up and I'll discuss an example of that in section 6.5.

Now, all of the above is well and good, but it of course assumes that we actually do have a multiplier circuit. Do we? Well, no—they're *really hard* to build to operate at radio frequencies. So, is all that I've been telling you just a big, outrageous shaggy dog story? Well, no. We just have to be clever about how we achieve multiplication (i.e., we have to do it without actually *multiplying*). To explain this mysterious statement, let's be just a bit less ambitious and suppose that all we have is a special case of a multiplier circuit, a circuit that produces the *square* of its single input (look again at Note 7). We can make a multiplier from our squarer if

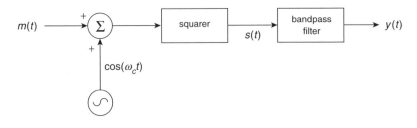

Figure 6.3.4. Multiplying by squaring and filtering

we assume that we also have a circuit that *adds* (shown in figure 6.3.4 as the circle with a summation sign inside it; you'll see why a bandpass filter is in the circuit, too, in just a moment). The input to the squarer is $m(t) + \cos(\omega_c t)$, and so the squarer output is

$$s(t) = [m(t) + \cos(\omega_c t)]^2 = m^2(t) + 2m(t)\cos(\omega_c t) + \cos^2(\omega_c t),$$

which includes the desired product $m(t)\cos(\omega_c t)$. The output $s(t)$ also includes, seemingly to our misfortune, two other terms. The fact is, however, to our great fortune, that it is possible to arrange matters so that the energy of the product term is completely separate and distinct (in frequency) from the energies of the two other terms. Therefore, if we apply $s(t)$ as the input to the appropriate bandpass filter, then that filter's output $y(t)$ will contain the energy only of the product term, that is, $y(t)$ will be proportional to $m(t)\cos(\omega_c t)$.

Consider now each of the three terms of $s(t)$, in turn. First, and the easiest to understand, is $2m(t)\cos(\omega_c t)$. By the heterodyne theorem the energy of this term is simply the energy of $m(t)$ shifted up and down the ω-axis, to be centered on the frequencies $\omega = \pm\omega_c$. Next, the $\cos^2(\omega_c t)$ term can be expanded with a trigonometric identity to give the equivalent expression (and its Fourier transform)

$$\frac{1}{2} + \frac{1}{2}\cos(2\omega_c t) \longleftrightarrow \pi\delta(\omega) + \frac{1}{2}\pi[\delta(\omega - 2\omega_c) + \delta(\omega + 2\omega_c)],$$

that is, all of the energy of $\cos^2(\omega_c t)$ is at the three specific frequencies of $\omega = 0$ and $\omega = \pm 2\omega_c$. (We found the Fourier transform of a constant, as well as the Fourier transform of a pure sinusoid, in section 5.4. In that section I analyzed $\sin(\omega_c t)$ instead of $\cos(\omega_c t)$, but the calculations are the same and the results are trivially different.) And finally,

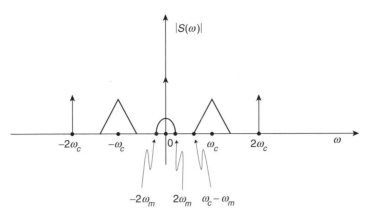

Figure 6.3.5. Where the energy is in the output $s(t)$ of the squarer of figure 6.3.4

from section 5.3 and the frequency convolution theorem, we know the Fourier transform of $m^2(t)$ is

$$m^2(t) \longleftrightarrow M(\omega) * M(\omega).$$

There we showed (with the aid of figure 5.3.1) that the Fourier transform of $m^2(t)$ is confined to the bandlimited interval $|\omega| \leq 2\omega_m$. All of these conclusions are shown in figure 6.3.5, where I've drawn all of the individual energies for the case of the heterodyned baseband energy (our desired product term) *not* overlapping the energies of the other two terms. It is clear from the figure that this will indeed be the situation *if* $2\omega_m < \omega_c - \omega_m$, i.e., *if* $\omega_c > 3\omega_m$. In AM radio this condition is more than satisfied, as you'll recall that I stated earlier that ω_c is greater than 500 kHz while ω_m is only a few kHz. Thus, if the output of the squarer, $s(t)$, is the input to the bandpass filter shown in figure 6.3.4 (with bandwidth $2\omega_m = 10$ kHz for AM radio) centered on ω_c, then the energy of the output of the filter is the energy associated with the product term $m(t)\cos(\omega_c t)$ (and *only* that energy). We have achieved multiplication by squaring-and-filtering with a circuit I think only Fourier theory can make it possible to understand.

Still, even though elegant, the squarer/filter circuit is not the easiest way to make a multiplier. The way multiplication is actually accomplished by an AM radio transmitter is incredibly *more* clever—I'm going to put off showing you the way it's actually done until the next section, just to let you think about this. Can *you* discover a method even simpler (and far cheaper) than figure 6.3.4? Don't peek ahead until you've thought about this, at least for a while. Again, Fourier theory will be the key to understanding what is happening.

To end this section, let me now show you how all of the ideas we've discussed in this chapter come together in the design of an electronic *speech scrambler*, a personal portable device that provides a modest level of privacy over public telephone circuits. This gadget (its block diagram is shown in figure 6.3.6, where HPF and LPF stand for *highpass filter* and *lowpass filter*, respectively) clamps onto the mouth *and* ear pieces (you'll see why *both*, in just a moment) of an ordinary telephone. It is sufficiently complex to keep the "innocent" from "accidently" listening on a conversation, but not so sophisticated that it would be beyond the legal intercept, crime-fighting capabilities of even the local police

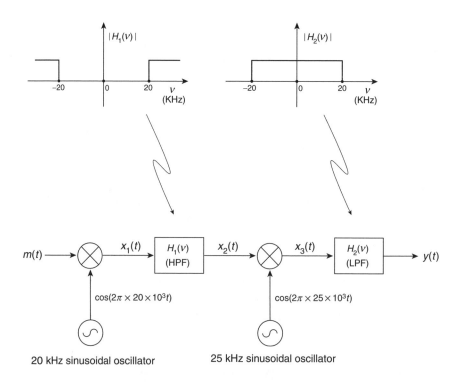

Figure 6.3.6. A speech scrambler

department. This scrambler circuit is quite old, dating back to just after the First World War. It was first used commercially on the radio-telephone link connecting Los Angeles and the offshore gambling casino and resort hotel on Santa Catalina Island, a place made famous, at least for a while, during my senior year of high school (1958) in a little town near Los Angeles, with the Top-10 pop-song "26 Miles" by the Four Preps:

> Twenty-six miles across the sea,
> Santa Catalina is a-waitin' for me.

(I'm singing this to myself as I write during the subzero New Hampshire winter of 2004—okay, that's enough nostalgia!)

Both filters in the figure are ideal; each has a skirt at 20 kHz that drops *vertically* to zero. The input signal to the scrambler, $m(t)$, is a voice baseband signal that is assumed to be bandlimited at 5 kHz (an assumption that can be assured by first lowpass filtering the input signal). The signals that exist in the circuit as we move from left to right, $x_1(t)$, $x_2(t)$, $x_3(t)$, and finally the (scrambled) output signal $y(t)$, have their respective Fourier transform magnitudes sketched in figure 6.3.7 (which, by Rayleigh's energy formula, tells us "where the energy is"); these sketches follow immediately from direct applications of the heterodyne theorem and the filter transfer functions. As the figure shows, the scrambler output has *inverted* the energy spectrum of the input, that is, input energy at frequency v kHz (remember, $\omega = 2\pi v$) appears in the output at frequency $(5 - v)$ kHz. This is sufficient mischief to make a conversation pretty much unintelligible to a casual eavesdropper.

We do have an obvious last question to answer, of course. We don't want *everybody* to be puzzled; how does the person on the *other end* of the conversation understand what is being sent? *That* person clearly needs a *de*scrambler. The fact that our scrambler works by inverting the input energy spectrum strongly hints at the amusing idea that the scrambler is its *own* descrambler (if we invert an inversion we should arrive back with what we started with). And, in fact, if you take the scrambled output spectrum $Y(v)$ from figure 6.3.7 (the received signal at the telephone earpiece at the other end of the link) and apply it to the scrambler input you'll find that you arrive at figure 6.3.8, which confirms our guess. So, each person simply uses the same scrambler circuitry, which doubles as its own descrambler, a feature that makes

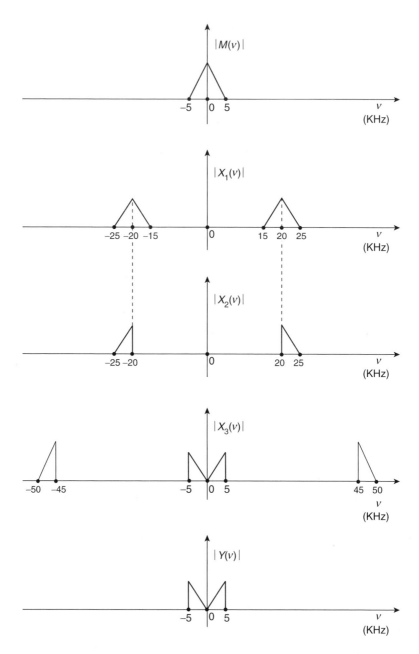

Figure 6.3.7. Scrambling an input

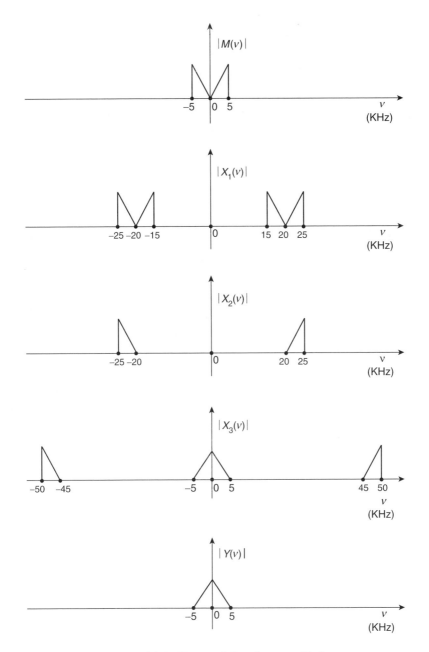

Figure 6.3.8. Unscrambling the scrambled

the circuit of figure 6.3.6 particularly attractive. Without Fourier theory, do you think it possible to "understand" how this gadget works? I don't think so.

6.4 The sampling theorem, and multiplying by sampling and filtering.

So, since reading the last section have you thought some more about how to multiply *without really multiplying*? To understand the clever way that is done to generate an AM radio transmitter's signal, I'll first derive a famous result in electrical engineering (dating from the 1930s) called the *sampling theorem*, which is usually attributed to either the American electrical engineer Claude Shannon (1916–2001) or the Russian electronics engineer V. A. Kotel'nikov (1908–2005), although the basic idea can be traced all the way back to an 1841 paper by a French mathematician (Cauchy).

In figure 6.4.1 I've illustrated the very simple *conceptual* idea of *mechanically sampling* the function $m(t)$; it shows a *rotating* switch that completes a rotation each T seconds; during each revolution the switch briefly connects to $m(t)$. If we denote the contact duration by τ, then we can write the *sampled* version of $m(t)$ that appears on the switch wire at the right as $m_s(t) = m(t)s(t)$, where $s(t)$ is also shown in the

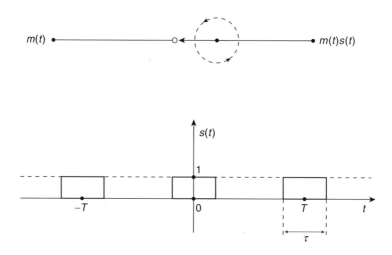

Figure 6.4.1. A mechanical sampler

figure (I am assuming that when the switch is not connected to $m(t)$ we have $m_s(t)=0$, a mathematical assumption that is easily satisfied in an actual electrical circuit with simple engineering details we needn't get into here).

Since $s(t)$ is a periodic function, we can write it as a Fourier series, and so, with $\omega_s = 2\pi v_s = 2\pi/T$ as the so-called *sampling frequency*, we have

$$m_s(t) = m(t) \sum_{k=-\infty}^{\infty} c_k e^{ik\omega_s t}, \omega_s = \frac{2\pi}{T}.$$

We *could*, if we want, now calculate the c_k, but as you'll soon see this is not necessary to get to the result we are after. You will notice, however, that I've drawn $s(t)$ as an even function in figure 6.4.4, and that means all of the c_k, whatever their values are, are real; that conclusion is not necessary for anything that follows, but it perhaps will make it easier to "see" what happens. Our central question, now, is simply this: where in frequency is the energy of $m_s(t)$?

As usual, this question is answered by calculating the Fourier transform and so, with $m_s(t) \longleftrightarrow M_s(\omega)$, we have

$$M_s(\omega) = \int_{-\infty}^{\infty} \left\{ m(t) \sum_{k=-\infty}^{\infty} c_k e^{ik\omega_s t} \right\} e^{-i\omega t} dt$$

$$= \sum_{k=-\infty}^{\infty} c_k \int_{-\infty}^{\infty} m(t) e^{i(\omega-k\omega_s)t} dt.$$

Recognizing the last integral is $M(\omega - k\omega_s)$, we have

$$M_s(\omega) = \sum_{k=-\infty}^{\infty} c_k M(\omega - k\omega_s)$$

.

This deceptively simple-appearing result says that the transform of $m_s(t)$ is just the transform of $m(t)$ *repeated*, *endlessly*, up and down the frequency axis at intervals of the sampling frequency ω_s.

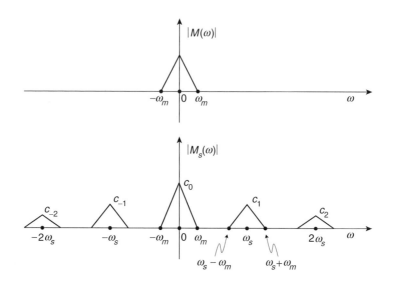

Figure 6.4.2. Where the energy is in $m(t)$ and in $m_s(t)$

Let's now suppose that $m(t)$ is a baseband, bandlimited signal, that is, that $|M(\omega)| = 0$ for $|\omega| > \omega_m$. Figure 6.4.2 shows what $|M_s(\omega)|$ looks like for such a signal, where the additional assumption has been made that the frequency-shifted copies of $M(\omega)$ do not overlap. That is clearly the case *if* $\omega_m < \omega_s - \omega_m$, that is, if $\omega_s > 2\omega_m$. If this condition is not satisfied (notice that the switch contact time τ plays no role), then the adjacent copies of $M(\omega)$ will overlap. Electrical engineers say the sampled signal is *aliased,* since energy at one frequency is now mixed up with energy at a different frequency, that is, if one thinks of frequency as the "name" of the energy at that frequency, then we have energy going under the different "name" of the overlapped energy. That is, going under an *alias.*

Now, in particular, you'll notice that even though $m_s(t)$ is a time-*sampled* version of $m(t)$, that is, "most" of the time the switch is *not* "looking at" $m(t)$, nevertheless $m_s(t)$ still contains *all* the information of $m(t)$. That is, perfect knowledge of $m_s(t)$ is equivalent to perfect knowledge of $m(t)$. We can see that this is so by observing that if we have $m_s(t)$, then we can perfectly *recover* $m(t)$ by simply lowpass filtering $m_s(t)$ and thereby select the transform copy of $M(\omega)$ centered on $\omega = 0$. We *cannot* do this, of course, if we have not sampled fast enough to prevent adjacent copies of $M(\omega)$ from overlapping. This result is what is called the sampling theorem—if a bandlimited signal $m(t)$ is sampled at a rate *more*[11] than twice the highest frequency at which energy is present in $m(t)$, then the sampling process has not lost any information. This mathematical result generally strikes most people as counterintuitive upon first encounter but it *is* true, both on paper and in real electronic systems.

Still, as neat as this result is, it isn't the one we are after here. We are more interested in the two copies of $M(\omega)$ that are centered on $\omega = \pm\omega_s$. *If* we select those two copies with a bandpass filter, and *if* we associate ω_s with what our AM radio transmitter calls ω_c, *then* as we showed in section 6.3 the bandpass filter output has the energy of $m(t)\cos(\omega_c t)$ (scaled by $|c_1|$). *At last* we have multiplied $m(t)$ and $\cos(\omega_c t)$ together in a really simple way—all we have to do is run $m(t)$ into a mechanically rotating switch spinning around at a rate greater than 500 kHz (half a million rotations per second) and bandpass filter the result.

Yeah, right. *You* try building *anything* mechanical that spins half a million times a second! Just don't stand too near it. In fact, instead of the easy-to-comprehend mechanical switch in figure 6.4.1, what is actually used in AM radio transmitters is an *electrical* switch in which *nothing moves*—it can be made from what is called a *radio-frequency transformer* (a sophisticated cousin to the humming, black cylindrical cans you often see attached to the tops of neighborhood power company poles bringing 60 Hz a-c electricity into clusters of homes) and a handful of *diodes* (either solid-state or "old-fashioned" vacuum tubes). This is the circuit I called a *modulator* in the previous section and, while not exactly cheap as dirt, it isn't really very expensive either. You can find electrical engineering discussions on modulator circuits in any technical book on radio.[12]

6.5 More neat tricks with Fourier transforms and filters.

Despite the fact that synchronous demodulation radio receivers are relatively expensive, they *are* used. Why? Because they have technical properties that can, in certain cases, be worth the cost. In this brief section I'll show you *one* such property. In the discussion of section 6.3 on AM radio, you probably came away with the idea that the different baseband signals from different radio stations *require* a different carrier frequency ω_c in order for a receiver to be able to select the baseband signal to which to listen without interference. It is true, in actual practice, that is indeed what is done in commercial AM radio. But, it is *not* impossible for the same carrier to "carry" *more* than one baseband signal. This strikes most people as contrary to some fundamental law of "how things should be," but I'll prove to you that it *can* be done with the best proof of all—I'll show you the actual circuitry (block diagrams, of course) of the synchronous transmitter and receiver that accomplish the astonishing feat of first placing *two* baseband signals on the same carrier and then separating them without interference. And we'll not have to write even a single equation; Fourier diagrams will be all we need to show us the way.

Let's call our two baseband signals and their transforms $m_1(t) \longleftrightarrow M_1(\omega)$ and $m_2(t) \longleftrightarrow M_2(\omega)$. To help keep straight what our transmitter and receiver are doing, I'll use our standard triangular shape for $M_1(\omega)$, but a different shape for $M_2(\omega)$, as shown in figure 6.5.1. Notice,

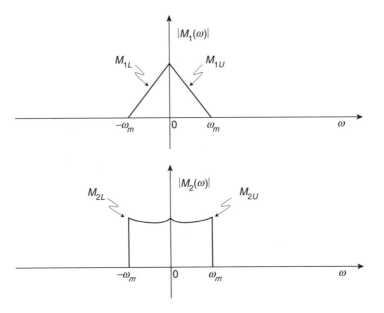

Figure 6.5.1. How the energy is distributed over frequency in two different baseband signals

however, that for both shapes I've drawn $|M_1(\omega)|$ and $|M_2(\omega)|$ as even, because $m_1(t)$ and $m_2(t)$ are each real. In both cases I've labeled the so-called upper and lower *sidebands* of the transforms; for example, M_{1U} is the *upper sideband* ($\omega > 0$) of $M_1(\omega)$ and M_{2L} is the *lower sideband* ($\omega < 0$) of $M_2(\omega)$. Then, the synchronous transmitter circuit of figure 6.5.2 generates a transmitted signal $y(t)$, with *both* baseband signals present, at the single carrier frequency ω_c. The ideal LPF (lowpass filter) and ideal HPF (highpass filter) each have their vertical skirt at $\omega = \omega_c$.

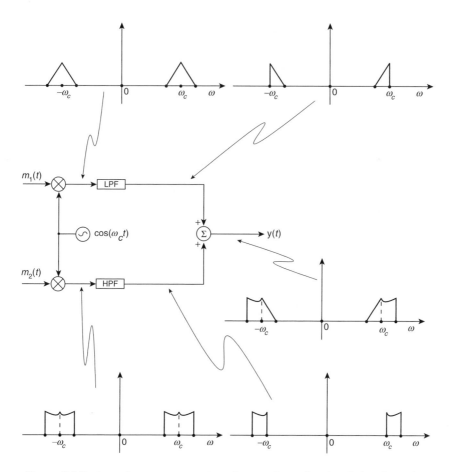

Figure 6.5.2. A synchronous transmitter that sends two baseband signals on the same carrier

At each point in the transmitter circuit I've indicated where the energy is, and you can see that while $|Y(\omega)|$ consists of an interesting amalgamation of the sidebands of $m_1(t)$ and $m_2(t)$, it is nevertheless true that $|Y(\omega)|$ is also even, as it *must* be since $y(t)$ is real. And finally, what we need to put our two scrambled baseband eggs back together again is shown in the synchronous demodulation radio receiver of figure 6.5.3. The input at the left is a (scaled) version of the transmitted $y(t)$ of figure 6.5.2. And again, all of the ideal filters in our receiver have their vertical skirts at $\omega = \omega_c$. As with the speech scrambler analysis in

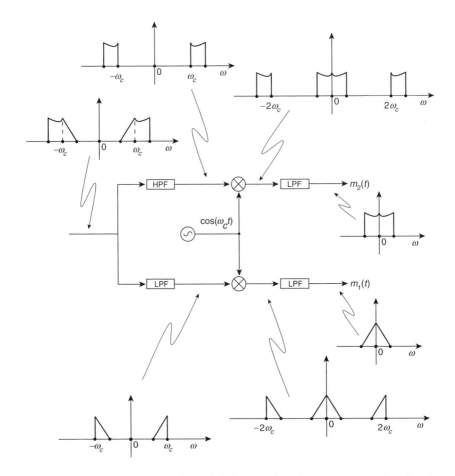

Figure 6.5.3. A synchronous demodulation receiver that recovers two baseband signals from the same carrier without interference

section 6.3, these sketches follow immediately from the heterodyne theorem (no equations required) and application of the behavior of ideal filters.

6.6 Single-sided transforms, the analytic signal, and single-sideband radio.

In this final section of the last chapter of the book, I want to show you how Fourier "thinking" explains a particularly beautiful electronic application of complex number mathematics. In the previous analyses of synchronous radio you surely noticed that, at various points in the circuits being discussed, we had just *half* of the transform of a signal (see figures 6.5.2 and 6.5.3 again). That was due to the presence of an ideal highpass or ideal lowpass filter that had its vertical skirt at just the right frequency to reject either the lower or the upper sideband of a real-valued signal. Because of the symmetry of the upper and lower sidebands of a real signal, *sideband rejection* results in no loss of information. There is a very practical reason for doing sideband rejection. If a real-valued, bandlimited baseband signal has energy at positive frequencies from zero to ω_m (an interval of ω_m), then after heterodyning up to the carrier frequency ω_c the signal has energy at positive frequencies from $\omega_c - \omega_m$ to $\omega_c + \omega_m$ (an interval of $2\omega_m$, that is, twice as large as at baseband). If, however, we transmit only *one* sideband, then all of the signal energy is confined (at positive frequencies) to an interval of, again, ω_m. That is, transmitting only one sideband *conserves frequency spectrum*, a major feature of what is called single-sideband (SSB) radio. To receive (i.e., *detect*) an SSB signal is, of course, a simple task (at least on paper): simply multiply the received signal by $\cos(\omega_c t)$ and then lowpass filter. (One does have the usual synchronous receiver complications mentioned earlier, that of generating *at the receiver* a near-perfect replica of the $\cos(\omega_c t)$ signal used at a distant transmitter.)

The most obvious way to generate an SSB signal is, as already observed, simply to use a highpass filter on the heterodyned baseband signal (to reject the *lower* sideband) or a lowpass filter on the heterodyned baseband signal (to reject the *upper* sideband). This was the initial method employed by the American electrical engineer John R. Carson (1887–1940) in the early years of the twentieth century. Carson, who worked for the American Telephone and Telegraph Company, was searching for a way to allow more messages to be simultaneously sent over AT&T's copper transmission lines (which have a much smaller bandwidth than modern fiber optic cables). After thinking about the symmetry of the transform of a real signal, and realizing that the upper and lower sidebands are redundant, he concluded that only *one* of them (*either* one) need be transmitted. Carson filed a patent application for the filtering approach to SSB signal generation in 1915 (granted in 1923), and when he died at age 54 his obituary notice in the *New York Times* specifically cited that invention.

Carson's simple SSB signal generation via sideband rejection filtering caught on quickly; by 1918 it was in use on a telephone circuit connecting Baltimore and Pittsburgh, and by 1927 there was commercial SSB *radio* linking New York City and London (employing electronics that could heterodyne a baseband signal of bandwidth 2.7 kHz up to anywhere in the interval 41 kHz to 71 kHz, this system transmitted the *lower* sideband). Still, while obvious, the filtering method has its problems. It works perfectly *if* one has either a highpass or a low-pass filter with a *vertical* skirt at the carrier frequency, which is of course an impossibility. A "real-life" filter would have to let a bit of rejected sideband "leak through," or would also have to reject a portion of the desired sideband in order to totally reject the undesired sideband.

Much more elegant than brute-force filtering would be the *direct* generation of an SSB signal from a given bandlimited baseband signal. And such a method does in fact exist. Called the *phase-shift method* (you'll see why soon), it was invented by Carson's colleague Ralph V. L. Hartley (1888–1970). Trained as a physicist, Hartley made many important contributions to electronics and information theory during a long career at the Bell Telephone Laboratories (created by AT&T as its research arm in 1925). Hartley filed for a patent in 1925 (granted in 1928), and it makes beautiful use of Fourier theory. To start the analysis of Hartley's SSB signal generator (I'll do the upper sideband), let's denote (as usual) our bandlimited baseband signal by $m(t)$ and write the Fourier transform pair $m(t) \longleftrightarrow M(\omega)$, where $M(\omega) = 0$ for $|\omega| > \omega_m$. Next, define a new pair that I'll write as $z_+(t) \longleftrightarrow Z_+(\omega)$, where $Z_+(\omega) = M(\omega)u(\omega)$, where $u(\omega)$ is the unit step function in the *frequency* domain. That is, $Z_+(\omega) = 0$ for $\omega < 0$, as shown in figure 6.6.1a. Clearly, $Z_+(\omega)$ is *not* a

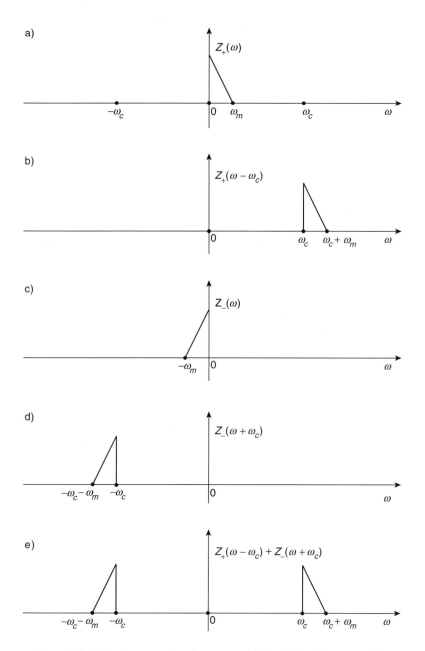

Figure 6.6.1. Piecing together the upper sideband signal generated by
Hartley's SSB modulator circuit

symmetrical transform (it is called a *single-sided* transform), and so $z_+(t)$ is *not* a real-valued time function. This will not be a problem for us, however, as we are not going to try to actually generate $z_+(t)$—which is good since that is an impossible task—$z_+(t)$ is merely an intermediate *mathematical* entity in our calculations.

Now, recall from section 5.4 the exotic Fourier transform pair

$$\frac{1}{2}\delta(t) + i\frac{1}{2\pi t} \longleftrightarrow u(\omega).$$

Since $Z_+(\omega) = M(\omega)u(\omega)$ is a *multiplication* in the ω-domain, we know from Section 5.3 that $z_+(t)$ is formed from a *convolution* in the *time* domain of $m(t)$ and the time function that pairs with $u(\omega)$, that is,

$$z_+(t) = m(t) * \left[\frac{1}{2}\delta(t) + i\frac{1}{2\pi t}\right] = \frac{1}{2}m(t) * \delta(t) + i\frac{1}{2\pi}m(t) * \frac{1}{t}$$

$$= \frac{1}{2}\int_{-\infty}^{\infty} m(\tau)\delta(t-\tau)d\tau + i\frac{1}{2\pi}\int_{-\infty}^{\infty}\frac{m(\tau)}{t-\tau}d\tau$$

or, using the sampling property of the impulse to evaluate the first integral, we have

$$z_+(t) = \frac{1}{2}\left[m(t) + i\frac{1}{\pi}\int_{-\infty}^{\infty}\frac{m(\tau)}{t-\tau}d\tau\right].$$

This quite odd-looking *complex*-valued time function is called an *analytic signal,* a name coined in 1946 by the Hungarian-born electrical engineer Dennis Gabor (1900–1979)—he received the 1971 Nobel Prize in physics for his work in holography—to describe any time function with a single-sided transform. If you look back at section 6.2, you'll see that the second integral (including the $1/\pi$ factor) is what we called the Hilbert transform of $m(t)$, which I'll write as $\overline{m}(t)$. Thus,

$$z_+(t) = \frac{1}{2}[m(t) + i\overline{m}(t)].$$

$z_+(t)$ is clearly a baseband signal—just look at $Z_+(\omega)$—and we can shift its energy *up* the frequency axis to $\omega = \omega_c$ by multiplying $z_+(t)$ by $e^{i\omega_c t}$. Mathematically we have

$$z_+(t)e^{i\omega_c t} = \frac{1}{2}[m(t) + i\overline{m}(t)][\cos(\omega_c t) + i\sin(\omega_c t)]$$

$$= \frac{1}{2}[m(t)\cos(\omega_c t) - \overline{m}(t)\sin(\omega_c t)]$$

$$+ i\frac{1}{2}[\overline{m}(t)\cos(\omega_c t) + m(t)\sin(\omega_c t)].$$

This complex-looking, *complex-valued* time signal has the *non-symmetrical* transform shown in figure 6.6.1b.

To get a *real*-valued signal that can actually be physically generated (and so transmitted as a radio signal by an antenna), we of course need a symmetrical transform. To accomplish that, let's next repeat all of the above but work with negative frequencies. That is, let's write $z_-(t) \longleftrightarrow Z_-(\omega)$, where $Z_-(\omega) = M(\omega)u(-\omega)$, as shown in figure 6.6.1c. Since we know from section 5.4 that

$$\frac{1}{2}\delta(t) - i\frac{1}{2\pi t} \longleftrightarrow u(-\omega),$$

we have

$$z_-(t) = m(t) * \left[\frac{1}{2}\delta(t) - i\frac{1}{2\pi t}\right] = \frac{1}{2}[m(t) - i\overline{m}(t)].$$

Thus, if we shift the energy of $z_-(t)$ *down* the frequency axis to $\omega = -\omega_c$ (see figure 6.6.1d) by multiplying $z_-(t)$ by $e^{-\omega_c t}$, then you should be able to confirm (with just a bit of easy algebra) that

$$z_-(t)e^{-i\omega_c t} = \frac{1}{2}[m(t)\cos(\omega_c t) - \overline{m}(t)\sin(\omega_c t)]$$

$$- i\frac{1}{2}[\overline{m}(t)\cos(\omega_c t) + m(t)\sin(\omega_c t)].$$

This reason we did this is now apparent—if we add the transforms of figures 6.6.1b and 6.6.1d we arrive at the symmetrical SSB transform

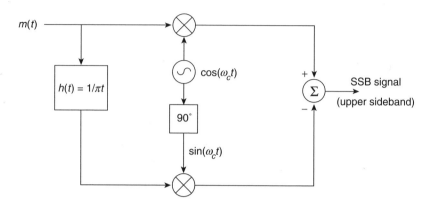

Figure 6.6.2. Hartley's SSB signal generator

shown in figure 6.6.1e. This symmetry means we have the transform of a *real*-valued time function, and indeed

$$z_+(t) + z_-(t) = m(t)\cos(\omega_c t) - \overline{m}(t)\sin(\omega_c t)$$

is an expression with zero imaginary part that tells us exactly how to construct Hartley's SSB generator *if* we can generate $\overline{m}(t)$. You'll see that figure 6.6.2 implements this expression (the box with "90°" inside it represents a circuit that shifts the phase of the $\cos(\omega_c t)$ signal to give the required $\sin(\omega_c t)$—operating at *just the single frequency* ω_c, this is an *easy* circuit to construct). The box with the *noncausal* impulse response $h(t) = 1/\pi t$ is, of course, the box that "makes" $\overline{m}(t)$—it is often called a *Hilbert transformer*—and of course it is *impossible* to make such a noncausal box.

Well, is this all just a big joke? After all, if we can't actually *build* a Hilbert transformer then all we have is a mathematical fairy tale. You of course know there *must* be an answer to this, as Hartley wouldn't have gotten a patent for a fairy tale. The answer, briefly, is to *approximate* $h(t)$, but of course that merely begs the question; how do you approximate an *unbounded* and *discontinous* (as $|t| \rightarrow 0$) impulse response? The precise answer is actually, I think, rather surprising, and as we develop the details you'll see why Hartley's method is called the "phase-shift" or *phasing* method.

For $x(t)$ any real-valued time function, define the analytic function

$$z(t) = x(t) + i\overline{x}(t).$$

Because $z(t)$ *is* analytic we can write it as

$$z(t) = 2x(t) * \left[\frac{1}{2}\delta(t) + i\frac{1}{2\pi t}\right],$$

from which the time convolution theorem says we can then write

$$Z(\omega) = 2X(\omega)u(\omega) = \begin{cases} 2X(\omega), & \omega > 0, \\ \\ 0, & \omega < 0. \end{cases}$$

This should be no surprise; since $z(t)$ was *constructed* to be analytic we *know* it has a single-sided transform. But notice that from the defining equation for $z(t)$ we can also write

$$Z(\omega) = X(\omega) + i\overline{X}(\omega)$$

where $\overline{x}(t) \longleftrightarrow \overline{X}(\omega)$, that is, $\overline{X}(\omega)$ is the Fourier transform of the Hilbert transform of $x(t)$. Combining these two expressions for $Z(\omega)$, we have

$$X(\omega) + i\overline{X}(\omega) = \begin{cases} 2X(\omega), & \omega > 0, \\ \\ 0, & \omega < 0. \end{cases}$$

from which it immediately follows that

$$\overline{X}(\omega) = \begin{cases} -iX(\omega), & \omega > 0, \\ \\ iX(\omega), & \omega < 0. \end{cases}$$

Now, the transfer function of the Hilbert transformer circuit is $H(\omega)$, where $\overline{X}(\omega) = X(\omega)H(\omega)$, and so

$$H(\omega) = \frac{\overline{X}(\omega)}{X(\omega)} = \begin{cases} -i, & \omega > 0, \\ \\ +i, & \omega < 0. \end{cases}$$

From this we see that $|H(\omega)| = 1$ for $-\infty < \omega < \infty$, that is, the Hilbert transformer does not affect the *amplitude* of its input, no matter what the frequency (and so it is often called an *all-pass* circuit), but it *does* affect the *phase* of the input. Indeed, all negative frequency inputs are given a $+90°$ phase shift (because the transfer function is $+i$ for $\omega < 0$) and all positive frequency inputs are given a $-90°$ phase shift (because the transfer function is $-i$ for $\omega > 0$).

Huh?

To see what the above means *physically*, focus your attention on some particular frequency component in $x(t)$; let's call that frequency ω_0. Then, to within some (irrelevant) amplitude scale factor, we can write that component of $x(t)$ as $\cos(\omega_0 t + \theta_0) = (e^{i(\omega_0 t + \theta_0)} + e^{-i(\omega_0 t + \theta_0)})/2$, where θ_0 is some arbitrary phase that I've included just to show you its particular value won't, in the end, matter. Now, imagine that we simply *delay* this component in time by one-fourth of a period (a time delay circuit is just as simple, conceptually, as a piece of wire, with the delay being directly proportional to the length of the wire). If T is the period, and if $\cos(\omega_0 t + \theta_0)$ is the input to the delay circuit, then the output is $\cos[\omega_0(t - T/4) + \theta_0]$, that is, since $\omega_0 T = 2\pi$ then $T/4 = \pi/2\omega_0$ and so the output of the delay circuit is

$$\cos\left[\omega_0(t - \frac{\pi}{2\omega_0}) + \theta_0\right] = \cos\left[\omega_0 t - \frac{\pi}{2} + \theta_0\right]$$
$$= \frac{e^{i(\omega_0 t - \pi/2 + \theta_0)} + e^{-i(\omega_0 t - \pi/2 + \theta_0)}}{2}$$
$$= \frac{e^{-i\pi/2}e^{i(\omega_0 t + \theta_0)} + e^{i\pi/2}e^{-i(\omega_0 t + \theta_0)}}{2}.$$

Since $e^{-i\pi/2} = -i$ and $e^{i\pi/2} = i$, we see that the positive frequency exponential is multiplied by $-i$ and the negative frequency exponential is multiplied by $+i$. When we take the Fourier transform of the input signal, then, the transform for $\omega > 0$ is multiplied by $-i$ and the transform for $\omega < 0$ is multiplied by $+i$, and so our simple time delay circuit *is* our Hilbert transformer. It's just that simple.

But if it's so simple, then why *can't* we build a Hilbert transformer? What makes something as simple as a one-fourth period time delay circuit noncausal? The problem is that the time delay circuit has to be a quarter-period delay not just for *one* frequency (like the 90° box in figure 6.6.2), but for *all* frequencies. Still, while we can't build such a circuit for all frequencies, we can now see how to build approximations to it in the frequency domain (something that is not at all obvious when looking at the unbounded, discontinuous impulse response of the Hilbert transformer in the time domain). We really only require the 90° phase-shifting to be over the *finite* frequency interval occupied by our *bandlimited* baseband signal $m(t)$, which has all of its energy in the interval $|\omega| \leq \omega_m$. Such circuits *can* be constructed and, not only that, they can be constructed from a *handful* of commonly available components (resistors and capacitors) that you can purchase from your local electronics store for the change in your pocket.[13]

Now, I should tell you that the mathematical presentation I just took you through is *not* historically correct. In the classic paper describing the New York City–London SSB radio link, for example, the mathematical level never exceeds the use of trigonometric identities.[14] The passage of decades did nothing to change that, either; for example, in December 1956 the Institute of Radio Engineers (IRE) devoted its entire *Proceedings* to the single topic of SSB radio, and *nowhere* in that entire issue do analytic signals appear. I think we can draw two conclusions from this: (1) the radio pioneers were *very* clever people who didn't need complex number mathematics to invent their circuits (but complex number mathematics certainly makes it a *lot* easier to understand those circuits), and (2) it took a very long time for even the more analytical electrical engineers to appreciate the usefulness of the analytic signal.

One "problem" with the Hilbert transform is that it is difficult to "see" what the transform does to a time function. (That discontinuity in the integrand doesn't help the intuition!) For example, is it obvious to you that the Hilbert transform of a constant is zero? Direct calculation of any particular Hilbert transform is almost always a tricky business; perhaps that "calculation sting" is what kept most of the early radio engineers away from analytic signals for such a long time. The development since the 1980s of software packages like MATLAB has taken nearly all of the sting out of such calculations for today's engineers, however. For example figure 6.6.3 shows the results of two such MATLAB-generated Hilbert transforms for two signals (a common, ordinary pulse, and a not so common two-cycle sinusoidal burst raised to the fourth power).[15]

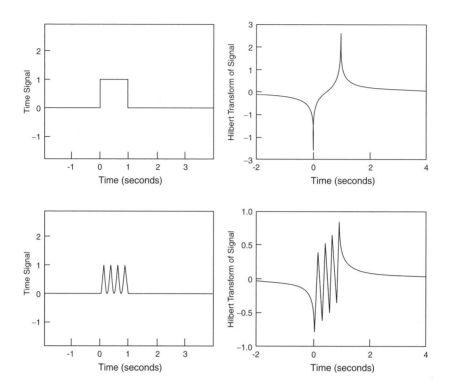

Figure 6.6.3. Computer-calculated Hilbert transforms

An aside: We have actually been working with analytic signals right from the start of this book. After all, $e^{i\omega_0 t}$ is a time function with energy only at the positive frequency $\omega = \omega_0$, that is, $e^{i\omega_0 t}$ has a single-sided transform. So, since $e^{i\omega_0 t} = \cos(\omega_0 t) + i\sin(\omega_0 t)$, $\sin(\omega_0 t)$ must be the Hilbert transform of $\cos(\omega_0 t)$. Now, what do you think is the Hilbert transform of $\sin(\omega_0 t)$ (it is not $\cos(\omega_0 t)$)? See if you can show that $\overline{\overline{x}}(t) = -x(t)$, that is, that the Hilbert transform of the Hilbert transform is the negative of the original time function (and so the Hilbert transform of $\sin(\omega_0 t)$ is $-\cos(\omega_0 t)$). The proof is easy if you remember the transfer function of a Hilbert transformer.[16] Figure 6.6.4 shows the result of taking the (computer-generated) Hilbert transform of the Hilbert transform of the sinusoidal burst signal in figure 6.6.3; and, sure enough, the result is the negative of the burst, with the slight deviations due to cumulative roundoff errors in all of the many calculations, as well as the truncated time span for the first Hilbert transform, that is, to faithfully reproduce the original

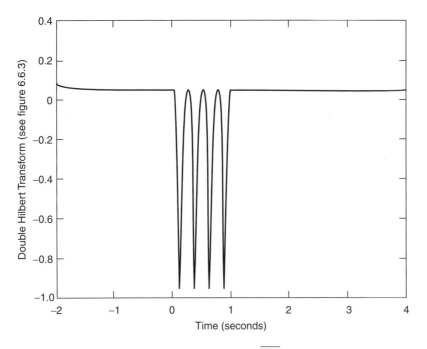

Figure 6.6.4. An illustration of $\overline{\overline{x(t)}} = -x(t)$

time function (actually, its negative), the second Hilbert transform should have knowledge of the first transform from minus infinity to plus infinity).

Still, while the 1956 issue of the *IRE Proceedings* had no analytic signals, it did have something else: the reporting of a new, *third* way to generate SSB signals that completely avoids the need to approximate the noncausal Hilbert transformer.[17] The author was Donald K. Weaver, Jr. (1924–1998), then a professor of electrical engineering at Montana State College (now Montana State University) in Bozeman. Weaver's stunning invention (it is included in every modern text on communication systems I've seen) shows that one can never be absolutely sure that even a well-studied topic has *really* been exhausted. Weaver's SSB signal generator (shown in figure 6.6.5) is today the modulator circuit of choice among radio engineers, and the explanation of how it works is made transparent by simply tracking "where the energy goes" with the aid of Fourier transforms, some simple complex number multiplications, and the modulation/heterodyne theorem.

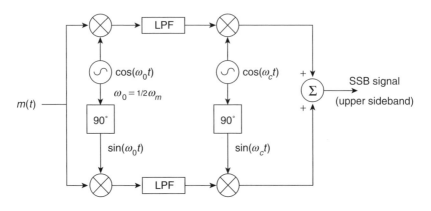

Figure 6.6.5. Weaver's SSB signal generator

Figure 6.6.6a shows the magnitude of the transform of our usual band-limited baseband signal $m(t)$, which of course is symmetrical since $m(t)$ is real. And again, as before, ω_m is the highest frequency at which $m(t)$ has energy. Now, what you probably noticed right away about Weaver's circuit is that there are *four* heterodyne operations taking place. To start, the energy of $m(t)$ is shifted up and down the frequency axis *twice*—once in the upper path of figure 6.6.5 by multiplication with $\cos(\omega_0 t)$, and again in the lower path by multiplication with $\sin(\omega_0 t)$. (Electrical engineers usually call these paths *channels*, with the upper channel called the *in-phase* or I-channel, and the lower channel called the *quadrature* or Q-channel, but I'll just call them the upper and lower paths.) The frequency ω_0 is equal to $\frac{1}{2}\omega_m$: ω_0 is called, in the lingo of electrical engineers, an *audio subcarrier frequency*. To understand the rest of figure 6.6.6, it is essential to keep in mind just what the heterodyne theorem says. In the upper path of Weaver's circuit, when we first multiply by $\cos(\frac{1}{2}\omega_m t)$, the transform of that multiplier output—I'll write $F\{x(t)\}$ as the Fourier transform of any time function $x(t)$—is

$$F\{m(t)\cos(\omega_0 t)\} = F\left\{ m(t)\frac{e^{i\omega_0 t} + e^{-i\omega_0 t}}{2} \right\}$$

$$= \frac{1}{2}F\{m(t)e^{i\frac{\omega_m t}{2}}\} + \frac{1}{2}F\{m(t)e^{-i\frac{\omega_m t}{2}}\}$$

or, in other words, the transform of $m(t)$ is shifted up the frequency axis by $\frac{1}{2}\omega_m$ and multiplied by $\frac{1}{2}$, *and* shifted down the frequency axis by $\frac{1}{2}\omega_m$

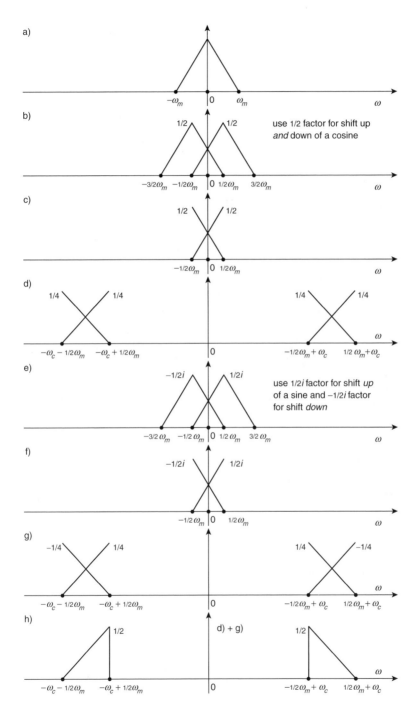

Figure 6.6.6. What Weaver's circuit does

and multiplied by $\frac{1}{2}$. It is almost *but not exactly* the same in the lower path, where we first multiply by $\sin(\frac{1}{2}\omega_m t)$. There the transform of the output of the first multiplier in that path is

$$F\{m(t)\sin(\omega_0 t)\} = F\left\{m(t)\frac{e^{i\omega_0 t} - e^{-i\omega_0 t}}{2i}\right\}$$

$$= \frac{1}{2i}F\{m(t)e^{i\frac{\omega_m t}{2}}\} - \frac{1}{2i}F\{m(t)e^{-i\frac{\omega_m t}{2}}\},$$

or, in other words, the transform of $m(t)$ is shifted up the frequency axis by $1/2\omega_m$ and multiplied by $1/2i$, *and* shifted down the frequency axis by $1/2\omega_m$ and multiplied by $-1/2i$.

These multiplication factors of $\frac{1}{2}$ for multiplication by a cosine, and $\pm 1/2i$ for multiplication by a sine, are crucial to understanding the operation of Weaver's circuit.

Figure 6.6.6b shows the up and downshifted baseband transform in the upper path after the first multiplication, where you'll notice that the two sidebands of $m(t)$ now overlap. This looks odd, I'll admit, but you'll soon see where this is all going. Beside each shifted transform I've written a "$\frac{1}{2}$" to indicate the multiplicative factor. Figure 6.6.6c shows the output of the lowpass filter (which I've assumed is an ideal filter with vertical skirts at $\omega = \pm\frac{1}{2}\omega_m$).[18] And figure 6.6.6d shows the second up and downshifted transforms in the upper path after the second multiplication (ω_c is, of course, a frequency very much greater than ω_m, and denotes the *radio* frequency at which occurs the efficient radiation of energy from an antenna). Again, there has been a multiplication by $\frac{1}{2}$ and so in figure 6.6.6d I've written a "$\frac{1}{4}$" by each shifted transform (because, dare I write this?, $\frac{1}{2} \cdot \frac{1}{2} = \frac{1}{4}$).

Starting with figure 6.6.6e, we repeat this whole business for the lower path, except that after the first multiplication we have a multiplicative factor of $1/2i$ for the upshifted transform and a multiplicative factor of $-1/2i$ for the downshifted transform. Figure 6.6.6f shows the output of the lower path lowpass filter, and figure 6.6.6g shows the second up and downshifted transforms in the lower path. Again, there has been a second multiplication by $\frac{1}{2i}$ for the upshifted transform and a second multiplication by $-\frac{1}{2i}$ for the downshifted transform, which should explain the "$\frac{1}{4}$" and the "$-\frac{1}{4}$" ($-\frac{1}{2i} \cdot -\frac{1}{2i} = -\frac{1}{4}$, $\frac{1}{2i} \cdot -\frac{1}{2i} = \frac{1}{4}$, and $\frac{1}{2i} \cdot \frac{1}{2i} = -\frac{1}{4}$).

Finally, as indicated in figure 6.6.5, the signals (i.e., transforms) in the upper and lower paths are added (see figures 6.6.6d and 6.6.6g) to give figure 6.6.6h, the transform of the Weaver circuit output. The $+\frac{1}{4}$ and $-\frac{1}{4}$ portions cancel each other to give zero, and the $+\frac{1}{4}$ and $+\frac{1}{4}$ *add.* The result is that only the *upper* sideband of the original $m(t)$ survives; the lower sideband has *self-canceled.* If, instead of a final summation we had *subtracted* the lower path from the upper path, then the output would be the *lower* sideband. Whether one adds or subtracts can be selected in actual practice with literally the flip of a switch.

Weaver's circuit is one of the prettiest applications of complex numbers in electrical engineering that I know of, one that I think Euler, himself—mathematician, pure *and* applied—would have loved. As I mentioned before, Weaver's discovery of his method for SSB signal generation caught electrical engineers by surprise. This reminds me of a wonderful passage from an article on Euler's generalization of the factorial function, from just the nonnegative integers to the *gamma function* which holds for *all real* numbers[19]:

> George Gamow, the distinguished physicist, quotes Laplace as saying that when the known areas of a subject expand, so also do its frontiers. Laplace evidently had in mind the picture of a circle expanding in an infinite plane. Gamow disputes this for physics and has in mind the picture of a circle expanding on a spherical surface. As the circle expands, its boundary first expands, but later contracts. This writer agrees with Gamow as far as mathematics is concerned. Yet the record is this: each generation has found something [new] of interest to say about the gamma function.

There will, I believe, *always* be something new to be learned in reading Euler. And that is probably as good a note as any on which to end this book.

Euler: The Man and the Mathematical Physicist

> God, when he created the world, arranged the course of all events so that every man should be every instant placed in circumstances to him most salutary. Happy the man who has wisdom to turn them to good account!
> —Leonhard Euler, whose own life is a testament to his words
> (from his *Letters to a Princess of Germany*, 3 vols., 1768–1772)

While there is a steady stream of biographies treating famous (or, even better from an entertainment point of view, *infamous*) persons in popular culture, there is still not even *one* book-length biography, in English, of Euler. There are a German-language biography (1929) and a French-language one (1927), as well as two more recent (1948, 1982) non-English works, but all are obsolete by virtue of the vast Eulerian scholarship that has occurred since they were written. Euler himself wrote so prodigiously that it would be a huge undertaking for a biographer to write, with true understanding, of what he actually did. Euler wrote, you see, more than any other mathematician in *history*. During his lifetime more than five *hundred*(!) books and articles by him were published (even more after his death), and his total work output accounts for a *third* of all that was published in Europe on mathematics, theoretical physics, and engineering mechanics from 1726 to 1800. In addition, there is another, equally enormous body of surviving personal letters (nearly 3,000 of them to and from hundreds of individuals), more thousands of pages of research notebooks, and voluminous diary entries that he continuously generated from his college days until his death. A dedicated biographer will have to read it all.

Still, scattered all about in the journal literature are numerous and (mostly) excellent essay-length biographical treatments on one of the greatest mathematicians who ever lived. Several have served as both my sources and my models. There is a lot of repetition among them, however, and to keep this essay from degenerating into an explosion of biographical citations I've simply listed my major references here in one note.[1] Unless I am specifically quoting from a particular author, I've not bothered with biographical citations. Perhaps by 2107 (the quadricentennial of Euler's birth) there will at last be an English-language, book-length biography of him that our grandchildren can curl up with next to a fireplace on a cold night.

Euler's life unfolded, in a natural way, in four distinct stages: his birth and youthful years in Switzerland, his first stay of fourteen years in Russia at the Imperial Russian Academy of Sciences in St. Petersburg, his departure from Russia for twenty-five years to join Frederick the Great's new Academy of Sciences in Berlin, and his return at Catherine the Great's invitation to the St. Petersburg Academy until his death. I'll treat each stage separately.

Euler's ancestors were present in Basel, a German-speaking area of Switzerland with a population of about 15,000, from 1594 onward. Most were artisans, but Euler's father, Paul (1670–1745), was a trained theologian (a 1693 graduate of the theological department of the University of Basel) and a Protestant minister. Paul's intellectual interests were broad, and while a student at the University he attended mathematics lectures given by Jacob (aka James) Bernoulli (1654–1705)—one of the founders of probability theory—at whose home he boarded. Another of the boarders Paul became friendly with was Jacob's brother Johann (aka John or Jean) Bernoulli (1667–1748), who would play an important role in the life of Paul's first son many years later.

Upon graduation at age twenty-three and after a short ministry at an orphanage, Paul was appointed pastor at a church next to the university. With that he felt secure enough to marry Margaret Brucker, herself the daughter of a minister, in April 1706. A year later to the month, on April 15, 1707, their first child and son Leonhard was born (Euler grew up with two younger sisters, but his only brother—Johann Heinrich who, following family tradition, became a painter—was born after Euler had left home by the age of twelve, if not sooner, for his first formal schooling). The year after his birth Euler's family moved a few miles from Basel, to the village of Riehen, where young Euler spent his youth.

It was not a life of luxury. His father's new parsonage had but two rooms; a study and *the* room in which the entire family lived. This simple life in the country with loving, educated parents certainly did nothing to make Euler either a snob or a fool. All through his life Euler impressed all by his calm disposition, his strong sense of practicality, and his deeply held religious views. He was strongly influenced by his parents, and his was at first a "home taught" education. His mother instructed him in classical subjects (blessed with what must have been a photographic memory, Euler could recite all 9,500 verses of Virgil's *Aeneid* by heart!), while his father introduced him to mathematics. In a short, unpublished autobiographical sketch he dictated to his eldest son in 1767, Euler recalled that his very first mathematical book—almost certainly given to him by his father—was an old (1553) edition of a text on algebra and arithmetic, and that he had faithfully worked through all 434 problems in it. There was, however, only so much that his parents could do, so, seeing that he was clearly an unusually talented boy, it was decided that young Euler needed a more formal educational setting. He was therefore sent back to Basel, to live with his now widowed grandmother (on his mother's side) while a student at Basel Latin school. This education was *supposed* to lead, at least in Paul's mind, to the son following in the father's theological footsteps.

Entering Basel Latin must have been quite a shock to the newly arrived country boy. Corporal punishment was employed on a regular basis and in abundance; no one would have said the rod was spared, as it wasn't unheard of for an unruly student to be beaten until the blood flowed. If the teachers weren't hitting the schoolboys, then the students did it themselves, with classroom fist-fights mixed in between parental assaults on the teachers! Curiously, the one subject that Euler would perhaps have thought would make such brutality tolerable—mathematics—was not part of the school's curriculum. Instead, Euler's parents found it necessary to employ a private mathematics tutor for their son, who did manage somehow to survive Basel Latin in one piece.

In the fall of 1720, at the age of thirteen, Euler entered the University of Basel. To be so young at the University was not at all unusual in those days, and it wasn't as if he had been admitted to the local equivalent of

Princeton, either. There were only a hundred students or so, along with nineteen underpaid and mostly (but not all, as you'll see) second-rate faculty. It should be no surprise that Euler stood out in such a crowd, and in June 1722 he graduated with honors with an undergraduate thesis in praise of temperance. He followed that with a master's degree in philosophy granted in June 1724, when a mere seventeen years old, with a thesis written in Latin comparing the philosophical positions of Descartes and Newton. While all of this would hardly seem the stuff of a proper education for one who would not too much later become the greatest mathematician of his day, the pivotal intellectual event in Euler's life was soon to take place.

Two years before Euler's birth, his father's mathematics teacher at Basel, Jacob Bernoulli, had died. He was succeeded as professor of mathematics at Basel by his younger brother Johann, Paul's old friend and fellow boarder in Jacob's home. (Johann's youngest son, Johann II [1710–1790], was similarly a friend of young Euler and a fellow master's degree recipient in 1724.) While many of the colleagues of the Bernoulli brothers at Basel may have been second-raters, Jacob and Johann certainly were not. Both were mathematicians with international reputations and, while young Euler of course never met Jacob, Johann was to have a profound influence on him.

Johann Bernoulli could have been at several far more prestigious places than the University of Basel, which was pretty much a backwater institution. He had, in fact, declined multiple attractive offers of professorial chairs in Holland because of the wishes of his wife's family. But just because he was at Basel didn't mean he had to like it, and he didn't—he was notorious for giving short-shrift to his elementary mathematics classes. He did, however, offer private instruction to the few he felt promising, and Euler joined that select group—the fact that Bernoulli knew Euler's father certainly didn't hurt—sometime around 1725. Probably because of his friendship with Johann II—and, through him, with the older Bernoulli sons, Nicholas (1695–1726) and Daniel (1700–1782)—Euler came to the attention of the Basel professor of mathematics. As Euler recalled in his 1767 autobiography,

> I soon found an opportunity to be introduced to a famous professor Johann Bernoulli ... True, he was very busy and so refused flatly to

give me private lessons; but he gave me much more valuable advice to start reading more difficult mathematical books on my own and to study them as diligently as I could; if I came across some obstacle or difficulty, I was given permission to visit him freely every Saturday afternoon and he kindly explained to me everything I could not understand ... and this, undoubtedly, is the best method to succeed in mathematical subjects.

Before coming under the influence of Bernoulli Euler had, in obedience to his father's wishes, devoted himself to the study of theology, Hebrew, and Greek, all in preparation for a life of ministry. Once Bernoulli came to realize the promise of his young student, however, Paul was eventually convinced that his son should follow in the footsteps of Johann, and not his. It says much to the love and compassion Paul must have had for his son that, even when faced with what had to be a great personal disappointment, he stood aside in favor of what his son so dearly wanted. Bernoulli's ever increasing admiration for Euler's blossoming genius can be measured by how he addressed his former student in their correspondence: from 1728 until Bernoulli's death, the salutations evolved through "The Very Learned and Ingenious Young Man" (1728), "The Highly Renowned and Learned Man" (1729), "The Highly Renowned and by Far Most Sagacious Mathematician" (1730), and "The Incomparable L. Euler, the Prince among Mathematicians" (1745). Johann Bernoulli was a man given to much professional competition and even jealousy, even with his own brother Jacob and with his own sons,[2] but not once does it seem that he challenged Euler's superiority as a mathematician. It was just so obvious that Euler was of a different class altogether (and Bernoulli was, himself, world class).

Once he had reached the ripe old age of nineteen (!) Euler was finished with his formal schooling, and he began to think about starting a career as an academic. He hoped to remain in Basel, near his family, and so when the professor of physics there died in September 1726 Euler applied for the job. Johann Bernoulli was delighted at the prospect of Euler joining him on the faculty, and encouraged him in his application. As part of that application, Euler prepared what is now considered to be equivalent to a doctoral dissertation, titled "Dissertatio physica de sono" ("A Physical Dissertation on Sound"). Short it was, a mere sixteen pages, but it was to become famous, and was cited by scholars for a century. It was essentially a research program in acoustics, which ended with discussions on six then hotly debated special problems of physics. As an example, one of them is now a classic included in just about all calculus-based college freshman physics texts: what would happen, if we ignore air friction and the earth's motion, to a stone dropped into a straight tunnel drilled to the center of the earth and onward to the other side of the planet?[3]

In this attempt at a professorship at Basel, Euler failed—he was simply too young—but yet another far more wonderful possibility existed for him, again because of his friendship with Johann Bernoulli (as well as with his sons Nicholas and Daniel). At almost the same time that he applied for the Basel position, Euler received an offer to join the new Imperial Russian Academy of Sciences in St. Petersburg. Founded by Peter the Great in 1724, the year before his death, the Academy was part of Peter's efforts to improve education and to encourage the development of western scientific thought in peasant Russia. The Russian Academy was created in the image of the existing scientific academies in Berlin and Paris, *not* in the image of the Royal Society in London, which, in Peter's view, was far too independent. The Russian Academy had direct governmental oversight. That was a feature that wouldn't bother Euler until after his arrival. Upon Peter's death his widow Empress Catherine I—not to be confused with Catherine II, better known as Catherine the Great—who shared his desire to improve education in Russia, became the Academy's benefactor.

Euler's invitation to go to Russia came about due to the cleverness of his Basel mentor Johann Bernoulli, who was himself the original choice of the Academy. Declining the invitation, he instead suggested that either of his two oldest sons would be good alternatives and—of course!—neither could go without the other. So, in 1725 Nicholas (appointed in mathematics) and Daniel (appointed in physiology) were off to Russia, thereby giving Euler two friends in St. Petersburg "on the inside." They promised their young friend that, at the first opportunity, they would champion him for an appointment. That opportunity soon came in a totally unexpected manner, when, in the summer of 1726, Nicholas suddenly died from appendicitis. Daniel assumed his late brother's mathematics appointment, and recommended that Euler be invited to fill his own now vacated spot in physiology. And so it happened that in the fall of 1726 Euler received his invitation to join the St. Petersburg Academy as a scholar not in mathematics but in physiology!

In November Euler accepted; it was, for a young man, not an unattractive proposition, carrying a small salary combined with free lodging, heat, and light, as well as travel expenses. In addition to Daniel Bernoulli, the Swiss mathematician Jakob Hermann (1678–1733), who was a second cousin of Euler's mother, would be a colleague at St. Petersburg. Euler's only condition of acceptance was to ask for a delay in his departure until the spring of 1727. His letter of acceptance cites weather concerns as the reason for the delay, but his real reasons were twofold. First, of course, was his desire to remain in Basel where the vacant physics position decision had not yet been made. And second, assuming that job failed to materialize (as it did not), Euler needed time to learn some anatomy and physiology so as not to arrive in Russia an ignoramus. Euler spent so much time in his early years studying the "wrong stuff"!

The academic year 1726–1727 was a "holding action" year for Euler, but, being Euler, he didn't just sit around with his fingers crossed hoping for a possible physics position in Basel and studying anatomy for the fall-back Russian job. He also wrote and submitted a paper to the Paris Academy of Sciences prize competition for 1727, in which the problem was to determine the best arrangement and size for the masts on ships, where "best" meant achieving the maximum propelling force from the wind without tipping the ship over. It is astonishing that Euler, still a teenager, nevertheless took second place, losing only to Pierre Bouguer, a professor well on his way to becoming a leading French nautical expert. Indeed, the Paris competition problem had been selected purposely to give Bouguer a huge head start on any competitors—he had been working on the problem for *years*. It must have been a disappointment to Euler to lose—later in his career Euler would win the Paris Academy competition a total of *twelve* times—but in fact Bouguer's win was a blessing in disguise. If Euler *had* won, maybe he would have gotten the Basel job, and would have taken a pass on Russia, where he would find compelling reasons to dedicate himself, totally and without distraction, to his academic work. In April 1727, as he turned twenty, Euler left for Russia and never set eyes on Basel again.

After an arduous seven-week journey by boat, wagon, and on foot, Euler arrived in St. Petersburg (then the capital of Russia) in late May, only to learn that Catherine had just died. The future of the Russian Academy was suddenly very much in jeopardy. The new tsar, Peter II, was a twelve-year-old boy, and the real power in Russia lay in the shadows. The nobility, who liked peasant Russia just as it was, ignorant and pliant, resented all of the foreign German, Swiss, and French intellectuals who had been recruited for the Academy and so withdrew financial support. The Academy appeared to be on the verge of physical collapse when the nobles moved the imperial court back to Moscow, and took the Academy's President with them to serve as the boy tsar's tutor. A number of the Academy's members despaired and, as soon as they could, returned home. But not Euler. His studies of anatomy and his second place finish in the Paris masting competition finally paid off for him— they brought him to the attention, of all things, the Russian Navy, which offered him a position as a *medical officer*. Even if the Academy sank, at least *Euler's* ship would still be afloat, perhaps even literally! With the medical appointment, along with his "doctoral" dissertation on sound, we might legitimately think of Euler as "Doctor" Euler!

Turmoil at the Academy continued until 1730, when, with the death of Peter II, Empress Anna Ivanovna's rise to power brought some stabilizing influence to the political situation. Euler's distant relative Jakob Hermann had resigned to return home to Switzerland, but Daniel Bernoulli had replaced Hermann as professor of mathematics at the Academy. Two years later Anna returned the capital of Russia to St. Petersburg. Euler's life flourished thereafter and, at age twenty-three, he was made professor of physics. When Bernoulli resigned in 1733 to accept a professorship back in Basel, Euler was selected to replace him as the premier mathematician of the St. Petersburg Academy. His personal life took a happy turn, as well, with his marriage to fellow Swiss Katharina Gsell (1707–1773), the daughter of a painter who taught at the school attached to the Academy. This event was celebrated by an Academy poet, who at one point gives us a hint as to how Euler's dedication to mathematics was viewed by others:

Who would have thought it,
That our Euler should be in love?

Day and night he thought constantly.
How he wanted more to calculate numbers,
. . .

Euler didn't think *always* of numbers: the first of Euler's thirteen
children was born late the next year.

Happily married, and the first among mathematicians in St.
Petersburg, Euler was a contented man. Both the Academy President and
its Secretary, the Prussian Christian Goldbach, were his close friends, and
his job security seemed assured. And, as long as he kept his nose firmly
planted in mathematics, his life would be equally sheltered from the
outside world of Russian political intrigue. Euler had, at last, "arrived,"
and the first period of his enormously productive career began to really
take-off. Indeed, it already had. In a letter dated December 1, 1729, for
example, Goldbach brought one of Fermat's conjectures (that $2^{2^n} + 1$ is
prime for all nonnegative integers n) to Euler's attention. By 1732 (and
probably earlier) he had shown the conjecture is false by factoring the
$n = 5$ case. During that same period, 1729–30, Euler discovered how
to generalize the factorial function for the nonnegative *integers* to the
gamma function integral, which holds for *all real* numbers. Before Euler,
writing $(-\frac{1}{2})!$ would have been without meaning, but after Euler the
world knew[4] that $(-\frac{1}{2})! = \sqrt{\pi}$.

By 1735 Euler had solved a problem that had stumped all
mathematicians—including both of the Basel professors of mathematics, Jacob and Johann Bernoulli—for almost a century. He calculated the
exact value of $\zeta(s) = \sum_{n=1}^{\infty} 1/n^s$—what we today call the *zeta function*—not
only for $s = 2$ (the original problem) but for *all* even integer s. This
wonderful calculation made Euler's reputation across all of Europe as
news of it spread through the mathematical world. Euler's old mentor back in Basel, Johann Bernoulli, was moved to say of his brother
Jacob, who had tried so hard and failed to do what Euler had done,
and had died not knowing the elegant solution Johann would live to
see, "if only my brother were still alive." In 1735 Euler defined what has
been called the most important number in mathematical analysis after
π and e—$\lim_{n \to \infty} \{\sum_{n=1}^{\infty} 1/n - \ln(n)\}$, called *Euler's constant* or *gamma*—
and calculated its value to *fifteen*(!) decimal places. In an age of
hand computation, that was itself an impressive feat. The year 1735

wasn't all glorious, however; he nearly died from a fever. Once recovered, though, he soon hit his stride again; for example, in 1737 he found a beautiful connection between the primes and the zeta function, which gave him the first new proof since Euclid of the infinity of the primes.[5]

It was in the 1730s, too, that Euler began his fundamental studies in extrema theory. In the "ordinary" calculus of Newton and Leibniz one learns how to find the values of the *variable* x such that the given function $f(x)$ at those values has a local minimum or maximum. In the calculus of variations, one moves up to the next level of sophistication: what *function* $f(x)$ gives a local extrema of $J\{f(x)\}$, where J (called a *functional*) is a function of the function $f(x)$? One of the pioneers in this sort of so-called *variational* problem was Johann Bernoulli, who in 1696 posed the famous *brachistochrone* problem: what is the shape of the wire (connecting two given points in a vertical plane) on which a point mass slides under the force of gravity, without friction, so that the vertical descent time from the high point to the low point is minimum? An even older question is the classic *isoperimetric problem*: what closed, non–self-intersecting curve of given length encloses the maximum area? Everyone "knew" the answer is a circle, but nobody could *prove* it! Those two problems,[6] and others like them, were all attacked by specialized techniques, different for each problem. There was no *general theory*.

Until Euler. In 1740 he finished the first draft of his book *Method of Finding Curves that Show Some Property of Maximum and Minimum* (it was published in 1744, after he had left St. Petersburg for Berlin). In it appears, for the first time, the *principle of least action*, about which I'll say more in just a bit. A line in an appendix to this work displays both the religious side of Euler and the deep attraction such problems had for him: "[S]ince the fabric of the universe is most perfect, and is the work of a most wise Creator, nothing whatsoever takes place in the universe in which some relation of maximum and minimum does not appear."

It wasn't all pure mathematics at St. Petersburg, however. In 1736 his two-volume book on mechanics appeared (*Mechanics, or the Science of Motion Set Forth Analytically*), in which he made extensive use of differential equations. This work was almost immediately recognized as a worthy successor to Newton's 1687 masterpiece, *Principia*; for example, Johann Bernoulli stated that the book showed Euler's "genius and acumen." Not everybody felt that way, however, notably the English gunnery expert Benjamin Robins (1707–1751), who thought the use of differential equations to be an admission of failure (to do experiment), and to represent an uncritical obedience to calculation. This is, of course, a distinctly odd position to the modern mind! Robins was no fool—he was the inventor of the ballistic pendulum, studied by every first-year college physics student to this day—but even in the 1730s Robins's negative view was that of a tiny minority. Euler and Robins would cross swords, of a sort, a few years later over a text authored by Robins, and again you'll see that Robins was singularly unappreciative of Euler.

In this short essay I can't even begin to do justice to what Euler did during his first stay in St. Petersburg, and my comments so far are a mere sampling of his technical accomplishments, out of *dozens* that could have been cited. But let me also mention here that he provided great immediate practical service to Russia with his astronomical observations at the St. Petersburg Observatory, work that played an important role in bringing the science of cartography (mapmaking) in Russia up from a primitive state to then modern standards. It was during that period that Euler's earlier, near-fatal brush with fever came back to haunt him—he began to lose vision in his right eye. Euler wrote to Goldbach in 1740 to say "Geography is fatal to me," believing that eyestrain from detailed attention to correcting landmaps was the cause of his difficulty. (Today it is believed that an eye abscess resulting from the earlier fever was the more likely cause.) By the time he wrote to Goldbach, Euler was nearly blind in the right eye. Later, a cataract in his left eye would leave him totally blind for the last twelve years of his life.

While in St. Petersburg Euler also worked on practical engineering problems involving naval ship design and propulsion, again using differential equations to study the motion of objects in a fluid. He brought all of that work together in his book *Naval Science*, mostly completed by 1738 while he was still at St. Petersburg, but not published until 1749, after he had left for Berlin.

That same year saw Euler's path cross, indirectly, that of a man he would be involved with in a most unpleasant encounter years later. That episode would be a war of words, and the "other side," Voltaire, was one of the great literary figures of those times and as much a master of the poison-pen as Euler was of mathematics. Voltaire was the pen name adopted in 1719 by the French writer/poet François-Marie Arouet

(1694–1778), who, during a forced exile in London from 1726 to 1729 (as an alternative to a prison sentence for the crime of exchanging insults with a man of higher social station than Voltaire's), became swept up by Newton's theories. Voltaire had attended Newton's funeral in 1727, talked of that impressive event for the rest of his life, and threw himself into writing what became a famous popularization of Newton's philosophy (*Éléments de la philosophie de Newton*), which appeared in 1738. It was a time during which the conflict between Newton's physics and Leibniz's metaphysics was a hot issue, and in *Éléments* Voltaire praised Newton while later (in 1759), in his famous satire *Candide,* he spoofed Leibniz (in the form of that work's character Dr. Pangloss). *Candide* is an attack on the view championed by Leibniz that we live in the "best of all possible worlds," and that all that happens is "for the best": the quote from Euler that opens this essay would have earned Voltaire's deepest scorn.

In the 1738 near-miss encounter with Euler the battleground was *scientific,* however, not literary, and while the prose of *Éléments* was elegant it is clear Voltaire did not have a deep understanding of Newton's mathematical and scientific concepts. As one writer put it, *Éléments* "made Newton's mathematics known to others if not to its author."[7] In the arena of analytical reasoning Voltaire, great writer of literary prose that he might be, was no match for Euler. The near-encounter with Euler was the result of the Paris Academy of Sciences prize competition, announced in 1736, to be awarded in 1738. The Academy's problem was for competitors to discuss the nature of fire. This was a time before there was any concept of a "chemical reaction," and philosophers still talked of the Aristotelian elements of air, earth, water, and fire as if they were fundamental entities. It was, not to be too tongue-in-cheek, also a hot topic. Being fascinated by science apparently convinced Voltaire that he could *do* science, even though he was completely without formal training. Voltaire was not a modest man. Euler's entry shared first place, but Voltaire did manage to snare an honorable mention, as did his lover Émilie du Châtelet (1706–1749), who, by all accounts, understood science and mathematics far better than did Voltaire. She prepared, for example, the first French translation of Newton's *Principia*, published after her death.

Émilie, with whom Voltaire had begun an affair in 1733 that would last until her early death shortly after childbirth (by a man neither Voltaire nor her husband), was an intelligent woman who employed experts in mathematics and physics as tutors. This is important in our story of Euler because one of her instructors (and yet *another* lover) was the French mathematician and astronomer Pierre Louis de Maupertuis (1698–1759) who in 1736 led an expedition to make measurements of the earth's shape and in 1738 published the book *La figure de la terre*, which supported the conclusion the planet is oblate and made Maupertuis famous as "the earth flattener." Another tutor was Samuel König (1712–1757), who had also studied for three years in Basel with Euler's old mentor Johann Bernoulli. While taking lessons from König, du Châtelet wrote a book titled *Institutions de physique* (published in 1740), treating the philosophical ideas of Descartes, Newton, and Leibniz, as well as the concepts of the natures of space, matter, force, and free will. König and du Châtelet fell out over her book, which König felt was simply a rehash of what *he* had taught his pupil. He essentially charged her, in private conversations with others, with stealing his work. Ten years later König would make a similar charge directed at Maupertuis, a charge that resulted in Voltaire, Maupertuis, and König clashing in a conflict that has been called one of the ugliest in the history of science. Euler would be swept up into it as well, and none of the four men would emerge unscathed.

The events that eventually led to that conflict started in mid-1740, when the new Prussian monarch Frederick II ("the Great") attempted to entice Euler away from St. Petersburg to join *his* newly energized Berlin Academy of Sciences.[8] Frederick neither knew nor appreciated mathematics but wanted Euler in his circle anyway, just because he knew *others* thought Euler was a genius. Euler was simply a prize to be bought as an ornament for his court (a type of faculty recruitment not unheard of in modern academia). Indeed, the entire Academy may have been just for show, at least at first: in a letter dated July 1737, to Voltaire, Frederick wrote that a "king needed to maintain an Academy of Sciences as a country squire needed a pack of dogs."

Euler initially declined the Berlin offer; several months later, when Empress Anna died leaving only an infant heir, which threw Russia once more into political turmoil and resulted in all the "foreigners" at the St. Petersburg Academy again being viewed with hostile suspicion, Euler (at his wife's insistence) reconsidered the king's invitation. He told Frederick what it would take to get him to come to Berlin, and in February 1741 the deal was struck. Euler's official reason to the St. Petersburg Academy for his wish to resign was that of health—he claimed he needed a less harsh climate, and that he was concerned for his eyesight. The Academy seemed to accept that, and Euler managed to leave Russia on good terms (which would work to his advantage in the future). His real reason for leaving was revealed shortly after he arrived in Berlin in late July 1741, when Frederick's mother, puzzled at why Euler seemed unwilling to answer any questions at length, bluntly asked him why he was so reserved, almost timid, in his speech. Euler's answer was equally blunt: "Madam, it is because I have just come from a country where every person who speaks is hanged."

When Euler came to Berlin Frederick's new Academy was still very much in a formative stage—there was not yet even a president. Frederick had offered the job the year before to another person, who had declined. Euler was therefore, at least in *his* mind, a candidate for the job, but so was Maupertuis, who had also been invited (at Voltaire's suggestion) to Berlin by the Francophile Frederick. It would only be after many years and disappointments that Euler would come to understand that a social snob like Frederick would never consent to a mere Swiss burgher being the head of his Academy, no matter how brilliant and accomplished he might be. Whenever possible—that is, when competence was not required—Frederick filled openings in government and military positions with nobility, and excluded commoners no matter how talented they might be. It did Euler's cause no good either that he also failed in the king's eye at being a witty conversationalist or the writer of *French* poetry. (So enamored with French culture was Frederick that in 1744 he ordered all the memoirs of the Berlin Academy to be published in *French*, not the usual Latin or even German.)

Simply being a mathematician hurt Euler, too. Frederick had written (January 1738), while still crown prince, to Voltaire—with whom he had corresponded already for two years—to tell the French writer what *his* plan of study would be: "to take up again philosophy, history, poetry, music. As for mathematics, I confess to you that I dislike it; it dries up the mind." Time did nothing to change the king's mind. Years later (January 1770) he wrote to Jean D'Alembert—the *French* mathematician Frederick wanted to be president of the Berlin Academy—to say "An algebraist, who lives locked up in his cabinet, sees nothing but numbers, and propositions, which produce no effect in the moral world. The progress of manners is of more worth to society than all the calculations of Newton."

This was the man to whom the naive Euler bowed his head. Of his Berlin appointment he wrote to a friend to say "I can do just what I wish [for his technical studies]. ... The King calls me his professor, and I think I am the happiest man in the world." Later, Euler would change his opinion. Things got off to an unsettled start when, nearly simultaneous with Euler's arrival in Berlin, Frederick was off to war with an invasion of neighboring Austria. His mind was not on either the Academy or Euler's possible role in it. The issue of the presidency would, in fact, remain unresolved for *five years*! The king's correspondence with Maupertuis shows that Euler was never, ever, in the running: more than a year before Euler's arrival, in a letter dated June 1740, Frederick wrote to the Frenchman to express his "desire of having you here, that you might put our Academy into the shape *you alone* are capable of giving it. Come then, come and insert into this wild crabtree the graft of the sciences, that it may bear fruit. You have shown the figure of the Earth to mankind; show also to a King how sweet it is to possess such a man as you" (my emphasis). When Maupertuis finally accepted in 1746, he was, in the words of Frederick himself, to be "the pope of the Academy." Euler received the consolation prize of being Maupertuis's chief deputy and director of the mathematics class of the Academy.

Euler's Berlin years were a time of stunning brilliance. The list of his accomplishments is simply enormous (he prepared 380 works, of which 275 were published!), but to select just a few, let me mention analyses of proposed government-supported lotteries, annuities, and pensions, studies that led to Euler's writings in probability theory; translation from English to German of Benjamin Robins's (mentioned earlier) 1742 book *New Principles of Gunnery*, a work of tremendous interest to the warrior-king Frederick (Robins was greatly irritated with Euler because he added supplementary material *five times* longer than the original work!); authorship of the book *Introductio in analysin infinitorum* (in which he clearly states what I have called "Euler's formula" all through this book), a text one prominent historian[9] of mathematics has ranked as important as Euclid's *Elements*; studies in the technology of constructing optical lenses, toothed gears, and hydraulic turbines; and finally (for this list), assorted studies in differential geometry, hydrodynamics, and lunar/planetary motion. Euler's extraordinary intellectual and physical powers were at their peak in his Berlin period.

While denied the presidency he so desired, Euler's administrative responsibilities at the Academy were nevertheless extensive. He served as the de facto president during Maupertuis's absences, selected Academy personnel, oversaw the Academy's observatory and botanical gardens, and provided oversight of numerous financial matters (most important of which was the publication of calendars, maps, and almanacs, the sale of which generated the *entire* income of the Academy). On this last matter, in particular, Euler learned early on that while Frederick might be a mathematical novice, when it came to money the king could count. In January 1743 Euler wrote to the king to suggest more money could be raised by selling almanacs in the newly conquered territory in Austria. In reply, Frederick wrote "I believe that, being accustomed to the abstractions of magnitude in algebra, you have sinned against the ordinary rules of calculation. Otherwise you would not have imagined such a large revenue from the sale of almanacs." Two decades later the matter of almanac revenue would drive the final wedge between Euler and the king.

Frederick's private view of Euler, briefly hinted at in the above response, was more openly expressed in his correspondence with others. In an October 1746 letter to his brother, for example, the king called Euler a necessary participant in the Academy because of his prodigious abilities, but, said that persons such as Euler were really nothing more than "Doric columns in architecture. They belong to the understructure, they support the entire structure." That is, good enough to hold the roof up, but that was it. What Voltaire and Maupertuis had, that Euler didn't and Frederick valued most, was the ability to generate light-hearted, clever conversation and correspondence (often at the expense of others). The fact that Euler couldn't compose a minuet or a flowery poem was a fatal lacking, in the king's view. Despite all this, Euler's life under Frederick seems to have been a full one, as well as one of increasing financial well-being. Since 1750, for example, his now widowed mother had lived with Euler, and in 1753 he had the resources to purchase an estate on the outskirts of Berlin that she managed for him.

Then, in 1751, we can see the beginning of the end of Euler's hopes for a lifelong career in Berlin. A few years earlier, just after assuming the presidency of the Academy, Maupertuis put forth what he claimed to be a new scientific principle, called *least action*.[10] The fact that Euler had enunciated essentially the same ideas in 1744 seems to have escaped him, and Maupertuis claimed the principle of least action in his 1750 book *Essai de cosmologie*. As he wrote there, "Here then is this principle, so wise, so worthy of the Supreme Being: Whenever any change takes place in Nature, the amount of action [a term most ambiguously defined by Maupertuis] expended in this change is always the smallest possible." Despite Maupertuis first laying claim to the presidency that Euler so wanted, and then to a technical concept that Euler had mathematically refined far beyond Maupertuis's mostly theological statement and which Euler surely felt was really *his*, Euler remained supportive of Maupertuis. Then Samuel König entered.[11]

König, who since 1749 had been the librarian to the royal court at The Hague, had been proposed by Maupertuis for election to the Berlin Academy, which was done in 1749. Nonetheless, König then accused Maupertuis of having stolen the least action concept from an October 1707 letter by Leibniz to the Swiss mathematician Jakob Hermann (Euler's distant relative who had been at St. Petersburg with him from 1727 to 1730), a copy of which König claimed to have seen. Accusing the president of the Berlin Academy of plagiarism was a serious charge, and he was of course asked to substantiate the charge. König wasn't able to produce the copy, and a search of the surviving letters of Leibniz to Hermann failed to produce the original. An Academy committee, headed by Euler, was formed to investigate this awkward mess, which concluded that it was König who was the fraud. (Modern historians generally believe König was in the right on this matter, and that there was indeed such a letter from Leibniz, but it still has not been found to this day.) But that wasn't the end of the matter. Voltaire, who had earlier fallen out with Maupertuis over both a squabble on filling a vacancy at the Academy and Maupertuis's refusal to provide a false alibi to help Voltaire escape blame in a stock swindle(!), felt he had reason to take revenge on his previous friend. He claimed that Maupertuis had earlier been in a lunatic asylum and, in his opinion, was *still* crazy! Maybe König, suggested Voltaire, wasn't a fraud after all.

When Frederick publicly sided with Maupertuis, Voltaire was stung by the royal rebuff and decided to really retaliate. "I have no scepter," he wrote, "but I have a pen." The result was the 1752 satire *Diatribe du Dokteur Akakia,* in which a thinly disguised Maupertuis was plainly portrayed as an idiot: he is finally "reduced" to the principle of least action, that is, *death,* by a bullet going at the square of its speed! *Diatribe* made Maupertuis the laughing-stock of Europe, and the ridicule was devastating to him. In 1753 Maupertuis returned to France and then, only at Frederick's demand that he return because the Academy was in chaos with his absence, he came back the next year—only to leave again in 1756 for good. As a supporter of Maupertuis, the episode did Euler no good, either. In a sequel to *Diatribe,* Voltaire inserted a snide reference to Euler by name. At one place Maupertuis and König are imagined to sign a peace treaty, which includes the following passage:

> our lieutenant general L. Euler hereby through us openly declares
> I. that he has never learnt philosophy and honestly repents that by
> us he has been misled into the opinion that one could understand
> it without learning it, and that in future he will rest content with
> the fame of being the mathematician who in a given time has filled
> more sheets of paper with calculations than any other.

While it is said the king laughed until he cried at reading Voltaire's cruel spoof (thus showing he loved what passed for satiric wit more than friendship), Voltaire's book nonetheless was a public insult to the head of Frederick's Academy. The king had a bonfire made of copies of *Diatribe* and Voltaire, too, found it expedient to return to France. He, like Maupertuis, never returned to Berlin. The assassin had brought himself down along with his victim.

To Euler's despair, even with the downfall of Maupertuis the king continued to overlook Euler as the logical person to be the next Academy president. Frederick clearly preferred the disgraced Frenchman to the one-eyed Swiss mathematician, even though it was Euler who was now keeping the king's Academy from total disintegration. There was simply no spiritual connection between the king and the half-blind man he mocked (behind Euler's back) as a "limited cyclops." So, again, the Academy went for *years* without a president, with Euler again serving de facto in that role. Then, in 1763, Frederick offered the presidency to the *French* mathematician Jean D'Alembert, who declined. If offering the job to another wasn't enough of an insult to Euler, the year after

D'Alembert's refusal to come to Berlin the king named *himself* president! The hurt to Euler must have been enormous. And yet he remained in Berlin for two more years. What finally made Euler's decision to leave was yet one more insult from Frederick.

At the end of 1763 the King believed the Academy's income from the sale of its almanacs could be increased by changing the administrative structure of the Academy. That is, Euler would no longer be the man in charge, but would be just one voice on a *committee*. Euler wrote in protest, with the king replying in a sharp, unpleasant manner. Frederick's decision stood—Euler was, in no uncertain terms, in a certain sense "demoted." Euler had at last had enough with this sort of treatment, and looked for a way out. He didn't have to look far. His way out of Berlin had been laid, in fact, years earlier. In early July 1763 Euler had received a letter from Grigorij Teplov, Assessor of the St. Petersburg Academic Chancellery, sent by the authority of Russian Empress Catherine the Great, offering Euler the position of Director of the Mathematical Division of the St. Petersburg Academy. In addition, he offered Euler the post of Conference Secretary of the Academy, and positions for all of his sons. Euler quickly wrote back to Teplov to say

> I am infinitely sensitive to the advantageous offers you have made by order of Her Imperial Majesty and I would be particularly happy if I were in a position to profit from it immediately...if...Mr.D'Alembert or another Frenchman had accepted the President's position of the [Berlin] Academy, *nothing could have stopped me from my immediate resignation and I could not have been refused under any pretext. It is understood that everyone would have blamed me for submitting to such a President. ...* However, not only did Mr. D'Alembert refuse this offer, but he subsequently did the wrong thing by highly recommending me[12] to the King, and if I wished to give my resignation I would be met with the most obstinate refusal. This would place me in a decidedly difficult, if not impossible, situation for any subsequent steps. (my emphasis)

Euler's comments seem to indicate that at the time he wrote he still thought he had a chance to be named president of the Academy. By 1766 those hopes were finally dead, and Euler revisited the St. Petersburg offer. Proving himself a tough negotiator, he got everything he asked for including, *at last*, the directorship of the St. Petersburg Academy. It took four letters of request to Frederick to obtain permission to leave Berlin, but in May 1766 the king finally relented and let Euler go, and in June Euler and his family left for Russia. Of their unhappy parting, the king wrote at the end of July to D'Alembert to say "Mr. Euler, who is in love even to madness with the great and little bear, has travelled northward to observe them more at his ease." To replace Euler as director of the mathematical class in Berlin, D'Alembert recommended to the king that the position be offered to the Italian-born Frenchman Joseph Lagrange (1736–1813), who accepted. In his July letter to D'Alembert, the king thanked D'Alembert for his aid in replacing Euler, and also got in one last insulting shot (which he somehow imagined to be funny) at Euler, writing "To your care and recommendation am I indebted for having replaced a half-blind mathematician by a mathematician with both eyes, which will especially please the anatomical members of my academy." Such a wit was Frederick.

The final stage of Euler's life, his last seventeen years in St. Petersburg, was the mirror image of the Berlin years. In St. Petersburg he was a celebrity, and there was no greater admirer of him than the Empress herself. His personal life, however, was not so uniformly rosy. Soon after his arrival he lost nearly all the vision in his remaining eye, and a failed cataract operation[13] in 1771 left him almost totally blind. That same year saw a fire destroy his home; he escaped serious injury, perhaps death, only with the aid of a heroic rescue. And in late 1773 his wife died; three years later he married again, to his first wife's half-sister. The powerful Euler intellect was not to be stopped by these events, however, and his scientific output continued to be enormous. About half of his total lifetime output was generated after his return to St. Petersburg. He started off with a bang by publishing what today we would call a bestseller, his famous *Letters to a Princess of Germany*.[14] This work found its origins in lessons, in the form of letters, given by Euler to a fifteen-year-old second cousin of Frederick's. Those letters covered a wide range of topics, including general science, philosophy, and physics. *Letters* was a huge success, with many editions in French, English, German, Russian, Dutch, Swedish, Italian, Spanish, and Danish. His more advanced work in Russia included other books and papers on algebra, geometrical optics, calculus, and the probability mathematics of insurance.

Perhaps a true sign of fame, of 'having arrived,' is when people start making-up stories about you. There is a famous example of this, famous, at least, in the mathematical world, in the case of Euler. To quote a well-known historian of mathematics,

> The story goes that when the French philosopher Denis Diderot paid a visit to the Russian Court, he conversed very freely and gave the younger members of the Court circle a good deal of lively atheism. There upon Diderot was informed that a learned mathematician was in possession of an algebraical demonstration of the existence of God, and would give it to him before all the Court, if he desired to hear it. Diderot consented. Then Euler advanced toward Diderot, and said gravely, and in a tone of perfect conviction: "Monsieur, $a + b^n/n = x$, donc Dieu existe: répondez!" Diderot, to whom algebra was Hebrew, was embarrassed and disconcerted, while peals of laughter rose on all sides. He asked permission to return to France, which was granted.[15]

This story is absurd on the face of it—Denis Diderot (1713–1784) was *not* a mathematical illiterate, and it is unimaginable that a man like Euler would have participated in such a stupid stunt. Modern historians have demonstrated quite convincingly that this tale is a fairy tale, probably started by Frederick (who greatly disliked Diderot) or one of his sycophants.[16]

Euler had a long, almost unbelievably productive life, but it all came to an end on September 18, 1783. As his biographical entry in the *Dictionary of Scientific Biography* describes his final hours,

> Euler spent the first half of the day as usual. He gave a mathematics lesson to one of his grandchildren, did some calculations with chalk on two boards on the motions of balloons; then discussed with [two colleagues] the recently discovered planet Uranus. About five o'clock in the afternoon he suffered a brain hemorrhage and uttered only "I am dying," before he lost consciousness. He died about eleven o'clock in the evening.

Analysis incarnate, as Euler was known, would calculate no more, and he was buried with great fanfare. He lies today in the Alexander Nevsky Lavra cemetery in St. Petersburg,[17] beneath an enormous headstone erected in 1837. His tomb is near some of the greatest Russian musical talents, including Mussorgsky, Rimsky-Korsakov, and Tchaikovsky.

The end came suddenly for Euler, but, as a deeply religious man, I suspect that even if he had had some advance warning he would have been at peace. Just after writing the words in the quote that opens this essay, Euler went on to write "[The] idea of the Supreme Being, as exalted as it is consolatory, ought to replenish our hearts with virtue the most sublime, and effectually prepare us for the enjoyment of life eternal." Euler clearly believed there is something beyond the grave, and it *is* comforting to imagine him now, vision restored with pen in hand, finishing new calculations that have at last revealed to him the value of $\zeta(3)$. But, no matter. Euler will never die. The brilliance of his mind, the clarity of his thought, lives everywhere in mathematics.[18]

$\mathcal{N}$otes

A number of citations to my book *An Imaginary Tale: The Story of* $\sqrt{-1}$ (Princeton University Press 1998) are made in the following notes. To keep those citations from repeating that title needlessly, the format is simply *AIT* followed by page numbers.

Preface

1. *Boston Globe*, May 16, 2002, p. A16.

2. My association of eroticism with technical creativity is not a gratuitous one. See, for example, Arthur I. Miller, "Erotica, Aesthetics, and Schrödinger's Wave Equation," in *It Must Be Beautiful: Great Equations of Modern Science* (edited by Graham Farmelo), (Granta Books, 2002). The Irish mathematician and theoretical physicist John L. Synge (1897–1995) briefly alluded to this issue, too, in his marvelously funny and erudite fantasy novel *Kandelman's Krim* (Jonathan Cape, 1957, p. 115). There he has one character (the Orc) remark in conversation with others, during the novel's extended exchange on mathematics, "I am recovering from our recent discussion about passion and sex. . . . You will admit that the ascent from that level to the square root of minus one calls for some adjustment." Synge's book ought to be read by anyone interested in mathematics, and most certainly by all who plan to teach mathematics.

3. The use of math in films to manipulate an audience's perception of a character is not new. In the 1951 movie *No Highway in the Sky*, for example, the hero (James Stewart) is supposed to be a bit of an odd duck, an aeronautical engineer who discovers that metal fatigue (and the resulting cracks) can be fatal to jet airplanes. To establish that he is not quite "ordinary," we learn early on that he has been spending time thinking about something weird called the "Goldbach conjecture." (You can almost hear a 1950s theater audience sucking in its collective breath, and the scattered, horrified exclamations of "Good Lord!" in the dark.) Now, there actually *is* a Goldbach conjecture, named after the Prussian mathematician Christian Goldbach (1690–1764). The conjecture dates from a 1742 letter he wrote to Euler, and it is easy to state: every even integer greater than 2 can be written as the sum of two primes. Computer calculations have shown it is true for all even integers up to 10^{14},

but its general truth is still an open question. In 2000 the British publisher Faber and Faber announced a prize of $1,000,000 for a solution to Goldbach's conjecture, if received by them by March 15, 2002 and published by March 15, 2004 in a reputable math journal. This was a publicity stunt to attract attention to their publication of a novel by Apostolos Doxiadis, *Uncle Petros and Goldbach's Conjecture.* There was a double irony to the contest: first, one had to be a legal resident of either the United Kingdom or the United States, and so Goldbach himself (if resurrected) could not have even entered, and second, the publisher was inviting people to enter a contest—almost surely doomed to failure (and so it was, since no proof was ever published)—based on a novel about an old mathematician who feels his life is a failure because he spent it all in a failed attempt to prove Goldbach's conjecture!

4. The definitive work on Dirac, and of his views on mathematical beauty, is Helge Kragh, *Dirac: A Scientific Biography* (Cambridge University Press, 1990).

5. Pollock's method is more accurately called "drip painting," because he simply let paint drip off the end of a stick or through a hole in the bottom of a suspended can as the stick or the can moved over a canvas laid flat on the ground. In the case of the can, a gravity-driven mechanical system did all of the "creative" work. Rockwell made subtle fun of this mechanistic way to "paint" in a cover he did for the *Saturday Evening Post* in 1962—see the color plate of *The Connoisseur* facing p. 357 in the biography by Laura Claridge, *Norman Rockwell: A Life* (Random House, 2001). For more on the "mathematics" of Pollock's paintings, see Richard P. Taylor, "Order in Pollack's Chaos," *Scientific American*, December 2002, pp. 117–21.

Introduction

1. K. Devlin, "The Most Beautiful Equation," *Wabash Magazine*, Winter/Spring 2002.

2. *American Mathematical Monthly*, January 1925, pp. 5–6. One of Peirce's sons was the philosopher and logician Charles Sanders Peirce (1839–1914) who led a greatly troubled personal and professional life. Some years after his death a handwritten note was found in one of his books in which he said of his father "He . . . had a superstitious reference for 'the square root of minus one.' " And of his brother James Mills Peirce (1834–1906), like their father a professor of mathematics at Harvard, C. S. Peirce declared *him* to be a "superstitious worshipper of . . . $\sqrt{-1}$." Clearly, for the Peirce men, the square root of minus one was also the root of some personal conflict. See *American Mathematical Monthly*, December 1927, pp. 525–27.

3. Peirce's blackboard statement is equivalent to $e^{i\pi} + 1 = 0$. Just raise both sides to the ith power to get $e^{i\pi/2} = i$ and then square (which gives $e^{i\pi} = i^2$). Since $i^2 = -1$, the equivalence is now clear.

4. Le Lionnais's essay can be found in *Great Currents of Mathematical Thought*, vol. 2 (Dover, 1971) pp. 121–58.

5. For more on Dirac and his impulse function, see section 5.1. For an "explanation" of why *all* mathematics eventually finds a "use," see the amusing story by Alex Kasman (a professor of mathematics at the College of Charleston, in South Carolina) "Unreasonable Effectiveness," *Math Horizons*, April 2003. Another counter-example to Hardy's "uselessness" criterion is found today in prime number theory. Hardy himself was fascinated by prime numbers and certainly thought the study of them to be one of *pure* mathematical beauty. Hardy would no doubt be horrified, if alive today, to learn they are now at the center of many modern cryptographic systems used by practically everyone (e.g., the sending of encoded messages by governments, and the financial transactions conducted via the Internet by individuals buying things with credit cards).

6. The difference between a sufficient condition and a necessary one is a matter of strength. A sufficient condition is *at least* as strong as a necessary one, that is, it may demand more than is necessary. For example, it is certainly *sufficient* that a map maker have ten million colors on hand—there is no map that will give him any trouble then. That is, ten million colors are *more than* necessary. But only four colors are absolutely necessary, which means there are no maps that four colors can't handle, but there *are* maps that three colors can *not* handle. It should be clear that a necessary condition may not be enough to ensure whatever is at question. A *necessary and sufficient* condition achieves the perfect balance between requiring "too much" and "not enough." Somewhat oddly, however, if one draws maps on surfaces *more* complicated than a plane there are very pretty traditional (that is, they don't use a computer) proofs of the map coloring theorem. For example, seven colors are sufficient *and* necessary to color all possible maps on the surface of a torus (a donut). There is even a nice *sufficiency-only* proof for five colors for all possible maps on a plane. There is still no "beautiful" proof for four colors being sufficient *and* necessary for all possible planar maps. The complete story of the four color theorem, as I write, can be found in Robin Wilson, *Four Colors Suffice: How the Map Problem Was Solved* (Princeton University Press, 2003). More technical is the book by Thomas L. Saaty and Paul C. Kainen, *The Four-Color Problem: Assaults and Conquest* (McGraw-Hill, 1977). They tell (p. 8) the following wonderful story about Hermann Minkowski (1864–1909), the youthful Einstein's math teacher in Zurich: "The great mathematician Hermann Minkowski once told his students that the four-color conjecture had not been settled because only third-rate mathematicians had concerned themselves with it. "I believe I can prove it," he declared. After a long period, he admitted, "Heaven is angered by my arrogance; my proof is also defective."

7. K. Appel and W. Haken, "The Four Color Proof Suffices," *Mathematical Intelligencer* 8, no. 1, 1986, pp. 10–20.

8. For a long essay on the computer proof of the four-color problem (and on computers and truth, in general), see Donald MacKenzie, *Mechanizing Proof: Computing, Risk, and Trust* (MIT Press, 2001). For even more on computers and the four-color problem, and also the more recent use of computers to "resolve" a famous ancient

conjecture (Kepler's maximum density sphere-packing problem), see George G. Szpiro, *Kepler's Conjecture: How Some of the Greatest Minds in History Helped Solve One of the Oldest Math Problems in the World* (John Wiley, 2003), particularly chapter 13, "But Is It Really a Proof?".

9. Euler thought the equation $x^4 + y^4 + z^4 = w^4$ had no integer solutions. He probably arrived at that conclusion from the observations that $x^2 + y^2 = w^2$ *does* have integer solutions, while $x^3 + y^3 = w^3$ does *not* (this is a special case of Fermat's last theorem, of course, discussed in chapter 1), but $x^3 + y^3 + z^3 = w^3$ *does* (e.g., $3^3 + 4^3 + 5^3 = 6^3$). That is, *in general*, Euler conjectured at least n nth powers are required to sum to an nth power. To have integer solutions for $n = 5$, for example, Euler thought one needed to have *at least five* integers on the left, that is, $x^5 + y^5 + z^5 + u^5 + v^5 = w^5$. In 1966 this was shown, *by direct computer search*, to be false, with the discovery of the counterexample $27^5 + 84^5 + 110^5 + 133^5 = 144^5$.

10. The divergence is fantastically slow; for example, the first ten *billion* terms sum to only just 23.6. An even more surprising related result is that, if one sums the reciprocals of *just* the primes, then the sum still diverges. This produces a new way, due to Euler (see *AIT*, pp.150–52), different from Euclid's ancient proof, for showing the infinitude of the primes. In 1919 the Norwegian mathematician Viggo Brun (1885–1978) showed that the sum of the reciprocals of just the *twin* primes— consecutive primes differing by 2—is finite (which unfortunately does not settle the question of whether or not the twin primes are infinite in number, one of the most famous open questions in mathematics). By the way, this sum, called *Brun's constant,* is

$$B = \left(\frac{1}{3} + \frac{1}{5}\right) + \left(\frac{1}{5} + \frac{1}{7}\right) + \left(\frac{1}{11} + \frac{1}{13}\right) + \cdots = 1.90216\cdots,$$

and it was when calculating B in 1994 that Thomas Nicely, a professor of mathematics at Lynchburg College in Virginia, discovered the infamous division algorithm flaw in the math coprocessor of Intel's Pentium® chip (Intel had actually discovered the error itself, earlier, but had decided it was "not important"). In calculating $4,195,835/3,145,727$, for example, the result was $1.33373906\ldots$ instead of the correct $1.33382044\ldots$. This affair provided yet another reason for the uneasy feeling many have for proofs based on voluminous computer calculations.

11. You can find it, for example, in *AIT*, pp. 146–47.

12. See Marilyn vos Savant, *The World's Most Famous Math Problem* (St. Martin's Press, 1993), pp. 60–61. The review cited in the text is by Nigel Boston and Andrew Granville, *American Mathematical Monthly,* May 1995, pp. 470–73. Somewhat more restrained is the review by Lloyd Milligan and Kenneth Yarnall, *Mathematical Intelligencer* 16, no. 3, 1994, pp. 66–69. But even there vos Savant's book is characterized as being both superficial and full of distortions. Those reviewers also bluntly (and, in my opinion, correctly) charge the high-IQ vos Savant with simply not knowing what she is talking about. It is very difficult, in the face of such arrogant ignorance, to remain temperate. So, let me instead just quote a comment, directed to vos Savant, by the Australian mathematician Alf van der Poorten, in his book *Notes on Fermat's Last*

Theorem (Wiley-Interscience, 1996), p. 27: "Up a gum tree," which van der Poorten says is what his teachers would say in response to hearing "irrelevant nonsense."

13. Perhaps the most elementary example possible of a realm of real integers in which unique factorization fails is that of the positive even integers, that is, the realm is the infinite set 2,4,6,8,10,.... The product of any two integers in this realm is also in the realm (since even times even is even). Some numbers in this realm can be factored (e.g., $4 = 2 \cdot 2$ and $12 = 6 \cdot 2$), but the rest cannot because there are no smaller integers, *All in the realm*, whose product is the given integer, for instance, $2, 6, 10, 18$, and 30. The integers that can't be factored are the *primes* of this realm. *Unique* factorization into a product of primes fails in this realm, as shown by the counterexample $60 = 6 \cdot 10 = 2 \cdot 30$. This single illustration ($180 = 18 \cdot 10 = 6 \cdot 30$ is another) demonstrates that *unique* factorization must be formally established for each new realm. Arguing for unique factorization in the realm of the ordinary integers by simply invoking "intuition" or "obviousness" is a priori a false argument. In chapter 1 I will tell you of a famous episode in nineteenth-century mathematics precipitated by a mathematician who made a similar, disastrous mistake when trying to prove Fermat's "last theorem" with complex numbers.

14. Euler's product expansion of the sine and his original calculation of the sum of the reciprocals of the positive integers squared can both be found in *AIT*, pp. 155–56 and 148–49, respectively. In chapter 4 I'll rederive his famous sum in a different way, using complex exponential Fourier series.

15. What Euler was actually studying is the *zeta function*, $\zeta(s) = \sum_{n=1}^{\infty} 1/n^s$. As discussed in the text, $\zeta(1)$ diverges, and Euler showed $\zeta(2)$ is equal to $\pi^2/6$. He also successfully evaluated $\zeta(s)$ for *all even* integer values of s. The values of $\zeta(s)$ for s *odd* remain, *without exception*, unknown. The values of $\zeta(s)$ for all even s are irrational, while for odd s the irrationality (or not) of $\zeta(s)$ is known only for one specific value of s; it was big news in the mathematical world when the French mathematician Roger Apéry (1916–1994) was able to show (in 1979) that $\zeta(3)$ is (whatever its value may be) irrational. Then, in 2000, the French mathematician Tanguy Rivoal showed that, for an *infinity* of odd (but unspecified) s, $\zeta(s)$ is irrational. In 2001 he made this much more specific by showing that there is at least one odd integer s in the interval 5 to 21 such that $\zeta(s)$ is irrational. That same year the Russian mathematician Wadim Zudilin reduced the range on s to 5 to 11.

16. The formula that drove Watson to such heights of poetic passion was

$$\int_0^{\infty} e^{-3\pi x^2} \frac{\sinh(\pi x)}{\sinh(3\pi x)} dx$$

$$= \frac{1}{e^{2\pi/3}\sqrt{3}} \sum_{n=0}^{\infty} \frac{e^{-2n(n+1)\pi}}{(1 + e^{-\pi})^2(1 + e^{-3\pi})^2 \cdots (1 + e^{-(2n+1)\pi})^2}$$

which *is*, I must admit, pretty darn awe-inspiring!

17. These lectures were collected in book form: *The Beauty of Doing Mathematics: Three Public Dialogues* (Springer-Verlag, 1985).

18. One of Charles Darwin's friends once wrote of the great nineteenth-century naturalist that "He had . . . no natural turn for mathematics, and he gave up his mathematical reading before he had mastered the first part of Algebra, having a special quarrel with Surds and the Binomial Theorem." Quoted from volume 1 of *The Life and Letters of Charles Darwin* (Basic Books, 1958), p. 146. This quotation lends some credibility for the common attribution to Darwin of the definition of a mathematician as "a blind man in a dark room looking for a black cat which isn't there."

Chapter 1: Complex Numbers

1. From the 1943 book *Mathematics for the General Reader* by British mathematician Edward Charles Titchmarsh (1899–1963), who taught at Oxford for decades. Not *everybody* thinks $\sqrt{-1}$ to be a "simple concept." In Robert Musil's 1906 novella *Young Törless*, for example, two schoolboys who have just left a classroom after a mathematics lecture conclude their discussion about the mystery of $\sqrt{-1}$ with one saying

> Why shouldn't it be impossible to explain? I'm inclined to think it's quite likely that in this case the inventors of mathematics have tripped over their own feet. Why, after all, shouldn't something that lies beyond the limits of our intellect have played a little joke on the intellect? But I'm not going to rack my brains about it: these things never get anyone anywhere.

A very interesting psychological analysis of this literary work can be found in Harry Goldgar, "The Square Root of Minus One: Freud and Robert Musil's *Törless*," *Comparative Literature* (Spring 1965), pp. 117–32. As Goldgar writes of the central character's confusion,

> A visit to the young mathematics master solves nothing; the man merely advises Törless that he will understand later, and meanwhile might read Kant, of whom, of course, the boy cannot make head nor tail. And so the "imaginary numbers," the "square root of minus one," come to represent to Törless, and to us, the irrational forces within him; they become, from now until the end of the book, a controlling symbol for the unconscious.

2. The "puzzle" of $\sqrt{-1}$ has probably not been completely vanquished, even today. There is, for example, an amusing story told by Isaac Asimov about an argument he had in college, with a professor of sociology, who declared "Mathematicians are mystics because they believe in numbers that have no reality." When Asimov asked what numbers the professor was referring to, he replied "The square root of minus one. It has no existence. Mathematicians call it imaginary. But they believe it has some kind of existence in a mystical way." See Asimov's essay "The Imaginary That Isn't," in *Adding a Dimension: Seventeen Essays on the History of science* (Doubleday

1964), pp. 60–70. Historically, as I discuss in *AIT*, the real puzzle for the early thinkers on $\sqrt{-1}$ was why certain *cubic* equations with obviously *real* roots had they come out in terms of the mysterious $\sqrt{-1}$. When I use the quadratic equation $x^2 + 1 = 0$ as my starting point in this book, therefore, I am actually taking an approach more modern than it is historical.

3. I am assuming you know how to multiply matrices. If that isn't the case, here's a quick summary of what you need to know: if **A** and **B** are two 2×2 matrices, then

$$\mathbf{AB} = \left[\begin{array}{cc} a_{11} & a_{12} \\ a_{21} & a_{22} \end{array} \right] \left[\begin{array}{cc} b_{11} & b_{12} \\ b_{21} & b_{22} \end{array} \right]$$

$$= \left[\begin{array}{cc} (a_{11}b_{11} + a_{12}b_{21}) & (a_{11}b_{12} + a_{12}b_{22}) \\ (a_{21}b_{11} + a_{22}b_{21}) & (a_{21}b_{12} + a_{22}b_{22}) \end{array} \right].$$

That is, the (j, k) entry in the product matrix is the product of the jth row vector of **A** and the kth column vector of **B**: $[a_{j1} \ a_{j2}] \left[\begin{array}{c} b_{1k} \\ b_{2k} \end{array} \right] = a_{j1}b_{1k} + a_{j2}b_{2k}$. **A** and **B** are said to *commute* if $\mathbf{AB} = \mathbf{BA}$, a property we take for granted in the realm of the ordinary real numbers. It is *not* generally true for matrices, however; for example, if $\mathbf{A} = \left[\begin{array}{cc} 3 & 1 \\ 2 & 6 \end{array} \right]$ and $\mathbf{B} = \left[\begin{array}{cc} -1 & 1 \\ 2 & 0 \end{array} \right]$, then $\mathbf{AB} = \left[\begin{array}{cc} -1 & 3 \\ 10 & 2 \end{array} \right]$ while $\mathbf{BA} = \left[\begin{array}{cc} -1 & 5 \\ 6 & 2 \end{array} \right]$. And finally, when a matrix is multiplied by an ordinary number c, one merely multiples each entry of the matrix by that number, that is,

$$c\mathbf{A} = c \left[\begin{array}{cc} a_{11} & a_{12} \\ a_{21} & a_{22} \end{array} \right] = \left[\begin{array}{cc} ca_{11} & ca_{12} \\ ca_{21} & ca_{22} \end{array} \right].$$

4. In control theory one often encounters the strange-looking time-dependent object $e^{\mathbf{A}t}$, which is called the *state-transition matrix* (the matrix **A** is formed from the equations that describe how a control system moves from any one of its states to any other of its states). Now, that *is* an interesting concept, raising e to a *matrix* power! What could that possibly mean? Simply this: since $e^{\lambda t} = 1 + \lambda t + (\lambda t)^2/2! + (\lambda t)^3/3! + \cdots$, replacing λ with **A** and 1 with **I** gives the matrix

$$e^{\mathbf{A}t} = \mathbf{I} + \mathbf{A}t + \mathbf{A}^2 \frac{t^2}{2!} + \mathbf{A}^3 \frac{t^3}{3!} + \cdots.$$

So now you see why control theorists are interested in calculating arbitrarily high powers of matrices. The idea of forming $e^{\mathbf{A}t}$ in this way is due to the American mathematician William Henry Metzler (1863–1943), in an 1892 paper, in which he inserted **A** into the power series expansion of *any* transcendental function desired, e.g.,

$$\sin(\mathbf{A}) = \sum_{n=0}^{\infty} (-1)^n \frac{\mathbf{A}^{2n+1}}{(2n+1)!}.$$

The only constraint is that **A** have only constants for its entries, that is, $\mathbf{A} \neq \mathbf{A}(t)$.

5. The story of Ramanujan is one that is at once uplifting and tragic. The great English mathematician G. H. Hardy, who brought Ramanujan to Cambridge for a while to learn mathematics "properly," once wrote that his discovery of the Indian genius was the only romantic episode in his (Hardy's) life. You can find a good presentation of Ramunujan's life and his work in Robert Kanigel, *The Man Who Knew Infinity*, (Scribner's, 1991).

6. This, and all the other computer-generated plots in this book, were created with programs written in MATLAB. As a rather unsophisticated programmer my codes are also unsophisticated (mostly brute force computations), but the power of MATLAB more than makes up for my average programming skills. The two plots of Ramanujan's sum, for example, requiring together over 334,000 floating-point operations, were generated in less than one second on an 800 MHz PC.

7. Dennis C. Russell, "Another Eulerian-Type Proof," *Mathematics Magazine*, December 1991, p. 349.

8. The reason for the possibility of much higher frequencies in electronics, as compared to mechanical things, is that it is *very* tiny charge carriers (e.g., electrons) with extremely small masses, that are doing the oscillating in electronic circuits. In mechanical systems, it is generally relatively massive chunks of metal that are moving to and fro. For example, to achieve the required thrust for launch to orbit, the high-pressure centrifugal fuel turbopump (6,500 pounds per square inch) for the the space shuttle's three main engines has to deliver liquid hydrogen fuel at such a prodigous rate—800 gallons *per second*—that it has internal parts spinning at nearly 600 revolutions per second. A high-performance automobile engine (the same size as the turbopump), when redlined on the dashboard tachometer (i.e., about to explode), is running at "only" about 100 revolutions per second.

9. The term cycle per second has been officially replaced with *Hertz*, abbreviated Hz, in honor of the German physicist Heinrich Hertz (1857–1894). In 1887 Hertz experimentally detected electromagnetic waves at the microwave frequency of somewhere between 50×10^6 to 500×10^6 cycles per second (oops, I mean between 50 and 500 MHz) that had been predicted fifteen years earlier by Maxwell's theory of electromagnetism. This "name game" does have its odd aspects. Electrical engineers still use radians per second for *angular* frequency—why hasn't *that* been renamed? In fact, now and then a new name *has* been proposed; the *Steinmetz*, in honor of the German-born American electrical engineer Charles Proteus Steinmetz (1865–1923). There is a pleasing symmetry with having Sz as the unit for radians per second to go with Hz for cycle per second, as well as some irony (notice Steinmetz's initials)—the cycle per second should be Sz and radians per second should be what Hz stands for!

10. See *AIT*, pp. 155–56 for how to derive Wallis's formula for π from Euler's infinite product for the sine. For a sketch on how to use Wallis's integral to derive Wallis's formula, see E. Hairer and G. Wanner, *Analysis by Its History* (Springer, 1996) (corrected second edition), pp. 233–34. Using the same approach used in the text, you can read how to integrate $\sin^{2n}(\theta)$ and $\cos^{2n}(\theta)$ over *any* interval, not just 0 to 2π, in Joseph Wiener, "Integrals of $\cos^{2n}(x)$ and $\sin^{2n}(x)$," *American Mathematical*

Monthly, January 2000, pp. 60–61. Wallis's integral is then simply a special case of Wiener's more general results.

11. Ralph Palmer Agnew, *Differential Equations* (second edition) (McGraw-Hill, 1960), p. 370.

12. The inequality gets its name from the French mathematician Augustin-Louis Cauchy (1789–1857) and the German mathematician Hermann Schwarz (1843–1921), who each discovered different forms of the inequality. The Russian mathematician Viktor Yakovlevich Bunyakovsky (1804–1889), however, is the one who actually first published (in 1859) the form of the inequality as derived in this book. Bunyakovsky had earlier worked with Cauchy in Paris, where he learned Cauchy's methods of complex analysis (discussed in *AIT*, pp. 187–221).

13. Given a line segment of any length, a second line segment of length equal to the square root of the first can easily be constructed by compass and straightedge: see *AIT*, pp. 31–32, 243. I assume that it is indeed obvious to you that the other operations are also so doable. To construct $\sqrt{17}$ we can actually be even more direct than the general approach. Given a line segment of unit length, erect a perpendicular at one end. On that perpendicular use the compass to mark off a line segment of length four. The right triangle formed by the two segments, of lengths one and four, and its hypotenuse, has just what we need—the hypotenuse length is $\sqrt{17}$. For the other square roots, however, the general construction is required.

14. An *elegant* construction of the 17-gon is given in H.M.S. Coxeter, *Introduction to Geometry* (2nd edition) (Wiley, 1969), p. 27. It is due to the English mathematician Herbert Richmond (1863–1948), who published it in 1893.

15. $F_5 = 2^{32} + 1 = 4,294,967,297$, which Euler showed factors into primes as $(641) \cdot (6,700,417)$. F_5 can be factored in a flash today with the aid of modern computers, but it was far too big for Fermat. Even in Euler's day it was a nontrivial calculation. (You can read how it was done in William Dunham, *Journey Through Genius: The Great Theorems of Mathematics* (John Wiley, 1990), pp. 229–34). But, even if F_5 is not a Fermat prime, what about the possibility of F_p being prime for $p > 5$? If that ever happens, the next Fermat prime after F_4 will be enormous, as it is known (as of May 2003) that F_p is *not* prime for every p from 5 to 32. There presently is no known sixth Fermat prime. In 1844 the German mathematician F. G. Eisenstein (1823–1852) conjectured that there are an infinity of Fermat primes, and that is still an open question. His track record on prime conjectures is not good, however, as he also conjectured that *all* the numbers in the infinite sequence $2^2 + 1, 2^{2^2} + 1, 2^{2^{2^2}} + 1, \ldots$ are primes. But just the fourth one is F_{16}, which, as stated above, is composite.

16. The next constructable *odd*-gon after Gauss's 17-gon is, of course, the F_0F_2-gon = 51-gon. If there are just five Fermat primes, then the total number of constructable odd-gons is easy to calculate; it is the total number of ways to form products (without repetition) from the Fermat primes using any one prime or any two primes or any three primes and so on. This number is, in terms of binomial coefficients,

$$\binom{5}{1} + \binom{5}{2} + \binom{5}{3} + \binom{5}{4} + \binom{5}{5} = 5 + 10 + 10 + 5 + 1 = 31.$$

17. For a step by step guide through Gauss's construction of the regular 17-gon, using the same approach described in the text for the 5-gon, see Arthur Gittleman, *History of Mathematics* (Charles E. Merrill, 1975), pp. 250–52.

18. Michael Trott, "cos($2\pi/257$) à la Gauss," *Mathematica in Education and Research* 4, no. 2, 1995, pp. 31–36. You can read the details of what is involved in constructing the 257-gon, at a level a bit deeper than here, in Christian Gottlieb, "The Simple and Straightforward Construction of the Regular 257-gon," *Mathematical Intelligencer*, Winter 1999, pp. 31–36. The author's title is a bit of tongue-in-cheek humor, as he ends his essay with the line "I wish the reader who wants to pursue the construction in all details good luck."

19. *Arithmetica* is a collection of 130 problems, published in thirteen volumes. It was apparently completely lost forever when the legendary Library of Alexandria was burned, but then many centuries later Arabic translations of six of the volumes were discovered. Fermat's copy of the surviving portions of *Arithmetica* was the 1621 translation from Greek to Latin by Claude Bachet. The problem that prompted Fermat's most famous marginal note required that a given square number be written as the sum of two squares.

20. It is amusing to note that in a 1989 episode of *Star Trek: The Next Generation* ("The Royale"), set in the twenty-fourth century, we learn that Jean-Luc Picard finds relaxation from his duties as a starship captain by trying to resolve Fermat's last theorem. As he explains to his second in command, Commander Riker, "I find it stimulating. It puts things in perspective. In our arrogance, we feel we are so advanced yet we still can't unravel a simple knot tied by a part-time French mathematician [Fermat's day job was as a jurist] working alone without a computer." Alas for Hollywood, reality made this fine sentiment "historically" wrong when, as we all now know, Wiles beat Picard to the punch by several centuries. Its romantic origin surely accounts for the popular fame of Fermat's last theorem (plus being a generalization of the Pythagorean theorem, which most people can remember at least hearing about in the dim past of high school). The mathematical machinery invented to resolve the Last Theorem is beautiful and powerful, but the actual result is of little (if any) continuing importance in mathematics. The great Gauss refused to work on the problem because he found it intrinsically uninteresting. Indeed, other quite similar conjectures have been around in mathematics longer, and even *solved*, without all the excitement that surrounds Wiles's accomplishment. Consider, for example, the Catalan conjecture. Named after the Belgian mathematician Eugène Catalan (1814–1894), it asserts that the *unique* solution in integers to $x^m - y^n = 1$ is $3^2 - 2^3 = 1$. That is, 8 and 9 are the only two consecutive numbers that are powers. Catalan made his claim in 1844, but the origin of the problem is far more ancient. The medieval French mathematician and astronomer Rabbi Levi ben Gerson (1288–1344) proved (around 1320) that if $3^m - 2^n = 1$, then $m = 2$ and $n = 3$. In 1738 Euler established the reverse case; if $x^3 - y^2 = 1$, then $x = 2$ and $y = 3$. Other specialized results followed in the years after Euler, but a general proof eluded the world's greatest mathematicians. Then, suddenly, the Romanian

mathematician Preda Mihăilescu (at the University of Paderborn, Germany) put it all to rest in April 2002 with an elegant proof. His was work of the highest caliber, an absolutely first-rate intellectual accomplishment that has received much praise and admiration from mathematicians around the world. The Catalan conjecture is now a *theorem*. And yet I very much doubt we will ever hear a reference to it in a television show (other than perhaps a *NOVA* episode, or something akin to that).

21. You can find the outline of a proof that unique factorization fails with the polynomials Lamé studied in Ian Stewart and David Tall, *Algebraic Number Theory and Fermat's Last Theorem* (A. K. Peters, 2002), pp. 122–24.

22. Indeed, unique prime factorization is valid in $a + ib\sqrt{D}$ only for the following values of D: 1, 2, 3, 7, 11, 19, 43, 67, and 163. For why D is specified as having no square factors, see Stewart and Tall, pp. 61–62.

23. If I were a mathematician I would say that the numbers in **S** are "members of the imaginary quadratic field $\mathbf{Q}(\sqrt{-6})$ and they form a ring."

24. Even in complex number systems where unique factorization holds, there are still surprises. For example, in the ordinary integers 5 is a prime. But in the Gaussian integers it is not. This is easy to see once you notice that 5 can be factored into the product of two other Gaussian integers: $5 = (1 + i2)(1 - i2)$. If you're *really* observant you might think there's *another* factorization, $5 = (2 + i)(2 - i)$. It is not hard to show, just as done in the text, that $1 + i2$, $1 - i2$, $2 + i$, and $2 - i$ are all primes in the Gaussian integers (with the norm $N(a + ib) = a^2 + b^2$), and so it appears as though we have two *different* prime factorizations of 5. That probably makes you wonder why the Gaussian integers are said to enjoy *unique* prime factorization. This "puzzle" occurs because I didn't tell you *everything*: prime factorization is unique only up to order (just as with the ordinary integers) *and* to multiplicative factors called *units*. In the ordinary integers the units are ± 1, but in the Gaussian integers the units are ± 1 *and* $\pm i$, and you'll notice that $i(2 - i) = 1 + i2$. I only tell you this so you'll know there *is* an answer to the "puzzle." If you want to read more on this, see Stewart and Tall, pp. 76–79.

25. Leibniz's rule for differentiating an integral is

$$\frac{d}{dt} \int_{g(t)}^{h(t)} f(x, t)\, dx = \int_{g(t)}^{h(t)} \frac{\partial f}{\partial t}\, dx + f\{h(t), t\} \frac{dh}{dt} - f\{g(t), t\} \frac{dg}{dt}.$$

If the limits on the integral are *not* functions of the differentiation variable (t), then the last two terms are zero and so the derivative of the integral is the integral of the derivative. But that, in general, is not the case and we need all three terms. You can find a freshman calculus derivation of Leibniz's rule in my book *The Science of Radio* (2nd edition) (Springer-Verlag, 2001), pp. 415–18.

26. We can *mathematically* show that $g(\infty) = 0$ as follows:

$$|g(y)| = \left| \int_{0}^{\infty} e^{-uy} \frac{\sin(u)}{u}\, du \right| \leq \int_{0}^{\infty} \left| e^{-uy} \frac{\sin(u)}{u} \right| du = \int_{0}^{\infty} |e^{-uy}| \left| \frac{\sin(u)}{u} \right| du.$$

Since $e^{-uy} \geq 0$ for *all* real u and y, and $| \sin(u)/u | \leq 1$,

$$| g(y) | \leq \int_0^\infty e^{-uy} \, du = \left(\frac{e^{-uy}}{-y} \Big|_0^\infty \right) = \frac{1}{y}.$$

Thus, $\lim_{y \to \infty} | g(y) | = 0$ and our result immediately follows.

27. See, for example, Bernard Friedman, *Lectures on Applications-Oriented Mathematics*, Holden-Day, 1969, pp. 17–20.

Chapter 2: Vector Trips

1. See *AIT*, pp. 92–94.

2. As far as I know this problem has not appeared elsewhere, and so I have felt free to name it. I call it the *generalized harmonic walk* since, for $\theta = 0$, $\mathbf{p} = 1 + \frac{1}{2} + \frac{1}{3} + \frac{1}{4} + \cdots$, which, of course, is the harmonic series. In *AIT* I discuss a similar sort of walk, with the steps after each spin being of lengths $1, \frac{1}{2}, \frac{1}{4}, \frac{1}{8}, \cdots$ (see *AIT*, pp. 107–9). That walk proves to be much more tractable than the walk considered here; I didn't give that earlier walk a name in *AIT*, but since I believe it to be my creation, as well, I'll now christen it the *generalized geometric walk*, for the obvious reason.

3. That symmetry was *not* assumed in generating the bottom right plot of figure 2.1.1. Rather, the plot was created by letting θ vary over the entire $0°$ to $360°$ interval, and the symmetry was then *looked for* as a partial check on the correctness of the MATLAB coding.

4. The discussion in this section was motivated by a challenge problem that appeared decades ago by T. H. Matthews, in *American Mathematical Monthly*, October 1944, p. 475, and its solution (by Gordon Pall) the following year (same journal, December 1945, pp. 584–85). Matthews and Pall were both at McGill University, where Pall was a young mathematician who later went on to a distinguished career. I know nothing of Matthews, except that at that time the registrar at McGill was one T. H. Matthews.

5. This result often puzzles students, who wonder why the effects of flying with and against the wind don't just cancel. The answer is that the *time* spent flying with the wind (at the faster speed) is *less* than the time spent flying against the wind (at the slower speed). A much more sophisticated treatment of problems of this type can be found in M. S. Klamkin and D. J. Newman, "Flight in an Irrotational Wind Field," *SIAM Review* April 1962, pp. 155–56.

6. That paper was, in fact, the primary motivation for this section: R. Bruce Crofoot, "Running with Rover," *Mathematics Magazine*, October 2002, pp. 311–16. Also helpful reading was Junpei Sekino, "The Band Around a Convex Set," *College Mathematics Journal*, March 2001, pp. 110–14.

7. MATLAB has a number of very powerful differential equation solvers. If I were really interested in getting answers of precision, I would use one of of them, and not

my quick-and-dirty code (called *Euler's method* in differential equations texts, dating from 1768). My interest here, however, is simply to demonstrate the *sort* of solutions one expects.

Chapter 3: The Irrationality of π^2

1. An elegant discussion of continued fraction expansions, in particular, a derivation of Lambert's $\tan(x)$ expansion, and a proof of the irrationality of $\tan(x)$ for all nonzero rational x, can be found in E. Hairer and G. Wanner, *Analysis by Its History* (Springer, 1996) (corrected second edition), pp. 68–79.

2. Given a polynomial equation of some finite degree $n \geq 1$, with integer coefficients c_0 through c_n,

$$c_n x^n + c_{n-1} x^{n-1} + \cdots + c_2 x^2 + c_1 x + c_1 x + c_0 = 0,$$

its solutions (the n values of x that satisfy the equation) are called *algebraic numbers*. For example, since $i = \sqrt{-1}$ solves $x^2 + 1 = 0$ ($n = 2$, here) then i is an algebraic number. The algebraic numbers include all the rationals and many (but not all) of the irrationals. The irrationals that are not algebraic are called the *transcendentals*. In 1873, for example, the French mathematician Charles Hermite (1822–1901) showed that e is transcendental (Euler had shown in 1737 that e is irrational).

3. Ivan Niven, *Irrational Numbers*, (American Mathematical Society, 1954).

4. The fastest path to fame and glory for a young mathematician is to solve a problem on Hilbert's list. Some still remain unsolved, and the pot of "reputation gold" remains a powerful lure. See Jeremy J. Gray, *The Hilbert Challenge*, (Oxford, 2000), and Benjamin H. Yandell, *The Honors Class: Hilbert's Problems and Their Solvers* (A. K. Peters, 2002).

5. There is an amusing side to this work. In 1919 Siegel was in a lecture audience when Hilbert stated his belief that $2^{\sqrt{2}}$ is "transcendental or at least an irrational number," but he added that he didn't think anyone in the audience would live long enough to see a proof.

6. The quote is from pp. 52–53 of John L. Synge's novel, *Kandelman's Krim* (Jonathan Cape, 1957). See note 2 for the preface.

7. E. C. Titchmarsh, *Mathematics for the General Reader* (Dover, 1981), p. 196. The near mystical aura that surrounds π can make even skilled mathematicians and hard-nosed physicists a little weak in the knees. My favorite example of that sort of thing is a passage from Richard Feynman's well-known essay "What is Science," *Physics Teacher*, September 1969, pp. 313–20. There he recalls his teenage puzzlement over the appearance of π in the formula for the resonant frequency of certain electrical circuits. "Where is the circle?" (that we associate π with) in such circuits, he says he asked himself. He ends his romantic nostalgia by asserting (p. 315) "in my heart I still don't quite know where that circle is, where that pi comes from." That is simply

Feynman being Feynman. He certainly *had* learned long before 1969 where the π "comes from" and "where the circle is," and so have you *if* you read section 1.4 carefully. As discussed there, the relationship between frequency measured in radians per second (ω) and in hertz (ν) is $\omega = 2\pi\nu$ (*there's* that rascally π!), and the "circle" that the young Feynman searched for is just the *circular* path traveled by the tips of two counter-rotating complex exponential vectors as they combine to form the *real-valued* ac signals $\sin(\omega t)$ and $\cos(\omega t)$.

8. But perhaps this will not always be so. In a somewhat goofy 1967 episode of *Star Trek* titled "Wolf in the Fold," Mr. Spock finds great practical use for the irrationality of π, to drive an electromagnetic evil presence out of the ship's computer. He does this by asking the computer for the last digit of π, and of course there isn't a "last digit" and that conundrum (for some less than obvious reason) gets rid of the evil presence. An even more mysterious puzzle, relating the irrationalities of π and e, is the question posed in the following little jingle:

> *pi goes on and on and on …*
> *And e is just as cursed.*
> *I wonder: which is larger*
> *When their digits are reversed?*

Less goofy is a short story about pi that appeared in a math journal many years ago. Written by João Filipe Queiró, "The Strange Case of Mr. Jean D.", *Mathematical Intelligencer*, 5, no. 3, 1983, pp. 78–80, tells of a mathematics teacher who has a nightmare dream in which the horror is the discovery, by computer, that π's decimal expansion starts to repeat after the five millionth digit. That is, π is rational! In the end, of course, the professor wakes up.

9. A completely different proof of the irrationality of π^2, also at the undergraduate level (but which doesn't make use of Euler's formula), can be found in a terrific book by George F. Simmons, *Calculus Gems*. (McGraw-Hill, 1992), pp. 283–84. As presented there the proof looks much shorter than the proof here, but that's because a *lot* of steps are skipped over, just as in Siegel's book. See also D. Desbrow, "On the Irrationality of π^2," *American Mathematical Monthly*, December 1990, pp. 903–6.

10. See note 25 for chapter 1.

Chapter 4: Fourier Series

1. My sources for much of the historical discussion in the first three sections of this chapter are the following: Edward B. Van Vleck, "The Influence of Fourier Series upon the Development of Mathematics," *Science*, January 23, 1914, pp. 113–24; H. S. Carslaw, *Introduction to the Theory of Fourier's Series and Integrals* (Macmillan, 1930), pp. 1–19; Carl B. Boyer, "Historical Stages in the Definition of Curves," *National Mathematics Magazine*, March 1945, pp. 294–310; Rudolph E. Langer, "Fourier's Series: The Genesis and Evolution of a Theory," *American Mathematical Monthly*,

August–September 1947 (supplement), pp. 1–86; Israel Kleiner, "Evolution of the Function Concept: A Brief Survey," *College Mathematics Journal*, September 1989, pp. 282–300.

2. The nature of the differentiability of Riemann's function was determined only in relatively recent times, and it turned out that Riemann was *wrong*; his function *does* have points at which the derivative exists, although it *is* true that his function is *almost* everywhere nondifferentiable. See Joseph Gerver, "The Differentiability of the Riemann Function at Certain Rational Multiples of π," *American Journal of Mathematics*, January 1970, pp. 33–55.

3. Newton actually stated his second law of motion in the following more general way: "force is equal to the rate-of-change of momentum," that is, $F = d(mv)/dt = m\,dv/dt + v\,dm/dt$, where $v = dx/dt$ is the speed of mass m (and so dv/dt is the acceleration). If $dm/dt = 0$ then this reduces to the well-known "force equals mass times acceleration." For the launchphase of a rocket, for example, we must use the general expression because the rocket's mass *is* changing with time; as fuel is burned and ejected in the exhaust, the rocket's mass decreases with increasing time. For our vibrating string problem, however, we can usually use the reduced expression (unless the string's mass somehow changes, for example, a cotton string, vibrating in humid air, that becomes soggy with absorbed water and so experiences a mass increase with increasing time).

4. All we can really say about $\sin(\infty)$ and $\cos(\infty)$ is that they *might* be zero, but they could just as well have *any* value from -1 to $+1$. That is, $\lim\limits_{t\to\pm\infty}\sin(t)$ and $\lim\limits_{t\to\pm\infty}\cos(t)$ do not exist. An old but still fascinating historical (Fourier's work appears more than once) *and* mathematical discussion on this can be found in J.W.L. Glaisher, "On sin ∞ and cos ∞," *Messenger of Mathematics*, 1871, pp. 232–44. Glaisher (1848–1928) was a prolific (over 400 papers) and talented writer on many things mathematical, mostly written for the trained mathematician, but also often surprisingly within the reach of the interested, not-so-trained as well. He taught at Cambridge all his professional life, and was president of the London Mathematical Society from 1884 to 1886.

5. I am *not* going to derive the heat equation in this book. I did derive the wave equation because of its historical position of being the first equation of theoretical physics to be solved in the form of an infinite trigonometric series. You can find derivations of the heat equation in literally every textbook on partial differential equations in print. An old work (but still among the best), with *many* worked examples, is R. V. Churchill, *Fourier Series and Boundary Value Problems*, in many editions, first published in 1941.

6. Thomson's telegraph cable diffusion analysis predates the discovery of the fundamental equations of electricity and magnetism, the so-called *Maxwell equations of the electromagnetic field* (named after Thomson's good friend and fellow Scott James Clerk Maxwell), but under certain restrictions the diffusion analysis gives satisfactory results. See my book *Oliver Heaviside*, (Johns Hopkins University Press, 2002), in particular, chapter 3, "The First Theory of the Electric Telegraph."

7. Thomson's Fourier series and the age-of-the-Earth papers appear in *Mathematical and Physical Papers* (Cambridge University Press, 1882 and 1890). "On Fourier's Expansions of Functions in Trigonometrical Series" (1841) is on pp. 1–6 of volume 1, and "On the Secular Cooling of the Earth" (1862) is on pp. 295–311 of volume 3. You can find a modern historical discussion of Thomson's mathematics in his age-of-the-Earth analysis in my paper "Kelvin's Cooling Sphere: Heat Transfer Theory in the 19th Century Debate over the Age-of-the-Earth," in *History of Heat Transfer* (American Society of Mechanical Engineers, 1988), pp. 65–85.

8. Fourier's unpublished 1807 paper vanished until rediscovered in the late 1880s, and is now available in the original French with invaluable historical commentary in English; see I. Grattan-Guinness, *Joseph Fourier, 1768–1830* (MIT Press, 1972).

9. *The Analytical Theory of Heat* appeared in English in 1878, with marginal notes of great historical interest by the translator, Alexander Freeman, a Fellow of St. Johns College, Cambridge. Nearly two centuries after its first appearance, Fourier's book still astonishes, and it should be read by any serious student of mathematics and physics. It was reprinted by Dover Publications in 1955.

10. Fourier does indeed use this modern approach to calculate the series coefficients, but only as his *last* technique. At first he used an extremely complicated limiting procedure, based on *differentiation*. For example, to solve for the coefficients in the trigonometric equation

$$1 = c_1 \cos(y) + c_3 \cos(3y) + c_5 \cos(5y) + \cdots$$

he first differentiated an *even* number (n) of times to obtain the infinite sequence of equations

$$0 = c_1 \cos(y) + 3^2 c_3 \cos(3y) + 5^2 c_5 \cos(5y) + \cdots,$$

$$0 = c_1 \cos(y) + 3^4 c_3 \cos(3y) + 5^4 c_5 \cos(5y) + \cdots,$$

$$\cdots,$$

$$0 = c_1 \cos(y) + 3^n c_3 \cos(3y) + 5^n c_5 \cos(5y) + \cdots,$$

$$\cdots$$

Then, setting $y = 0$, he had an infinite number of linear algebraic equations with an infinity of coefficients. To solve for them he used the first m equations, with the assumption that he need pay attention only to the first m coefficients. Then having solved for the first m coefficients, he determined their limiting values as $m \to \infty$. This whole business is nothing less than bizarre. As Professor Van Vleck (see Note 1) wrote, "Fourier uses his mathematics with the delightful freedom and naïveté of the physicist or astronomer who trusts in a mathematical providence." Later he used Euler's method without attribution, and was apparently told of Euler's priority: sometime in 1808 or 1809 Fourier wrote to an unnamed correspondent (probably Lagrange) to explain "I am sorry not to have known the mathematician who first made use of this method [Euler's method for finding the coefficients of a trigono-metric series] because I would have cited him." See John Herivel, *Joseph Fourier: The Man and the Physicist* (Oxford University Press 1975), pp. 318–19.

11. *AIT*, pp. 155–57.

12. An extensive history of the mathematics in this section is in the paper by Edwin Hewitt and Robert E. Hewitt, "The Gibbs-Wilbraham Phenomenon: An Episode in Fourier Analysis," *Archive for History of Exact Sciences* 21, 1979, pp. 129–60. A very brief note giving an outline of the history, only, had appeared earlier in Fred Ustina's "Henry Wilbraham and Gibbs Phenomenon in 1848," *Historia Mathematica* 1, 1974, pp. 83–84. The Hewitts' paper, however, is a detailed mathematical discussion that explores events well into the twentieth century. But neither paper tells us *anything* about Wilbraham, himself.

13. *Nature*, October 6, 1898, pp. 544–545.

14. *Nature*, October 13, 1898, pp. 569–570.

15. *Nature*, December 29, 1898, p. 200 (for the Michelson and Gibbs letters), and pp. 200–201 for Love's letter.

16. *Nature*, April 27, 1899, p. 606.

17. *Nature*, May 18, 1899, p. 52 (for Poincaré's letter) and June 1, 1899, pp. 100–101 (for Love's final letter).

18. "A New Harmonic Analyzer," *American Journal of Science*, January 1898, pp. 1–14. Michelson's coauthor was the University of Chicago physicist Samuel Wesley Stratton (1861–1931), who went on to be the founding director of the National Bureau of Standards, as well as president of MIT (1923–1930).

19. "The Tide Gauge, Tidal Harmonic Analyzer, and Tide Predictor," in volume 6 of *Kelvin Mathematical and Physical Papers*, (Cambridge University Press, 1911), pp. 272–305 (this paper was originally published in 1882).

20. See the illustration of the tidal analyzer in Crosbie Smith and M. Norton Wise, *Energy and Empire: A Biographical Study of Lord Kelvin* (Cambridge University Press, 1989), p. 371.

21. A photograph of the Michelson/Stratton harmonic analyzer is the frontispiece illustration in the book by J. F. James, *A Student's Guide to Fourier Transforms: With Applications in Physics and Engineering* (Cambridge University Press, 1995).

22. Henry Wilbraham, "On a Certain Periodic Function," *Cambridge and Dublin Mathematical Journal* 3, 1848, pp. 198–201. The text from Fourier's Treatise that Wilbraham cites can be found in the Dover reprint (see Note 9) on p. 144.

23. Henry Wilbraham, "On the Possible Methods of Dividing the Net Profits of a Mutual Life Assurance Company Amongst the Members." The date on this paper is actually October 1856. The full title of the publishing journal is *Journal of the Institute of Actuaries and Assurance Magazine.*

24. As a great fan of the Indiana Jones persona, I wish I could tell you that my discoveries came as the result of worldwide globe-trotting with numerous narrow escapes from death (along, of course, with a few light-hearted romantic flings with beautiful, mysterious women who are strangely attracted to slightly portly,

gray-bearded electrical engineering professors who wear thick glasses). Alas, all of my detective work was done while slouched in my back-supporting chairs in front of my computer terminals at home and the office. The one great clue to Henry was the one he himself provided when he listed, after his name on the 1848 paper, "B.A., Trinity College, Cambridge." An e-mail to the Trinity Library Reference Desk gave me his birthdate, the names of his parents, and his school admission and graduation dates. (My thanks to Jonathan Smith/Trinity.) From there I was off (electronically) to Princeton University, which owns a copy of *The Royal Society Catalogue of Scientific Papers*, and where a search for all of Henry's published papers was done for me. (My thanks to Mitchell C. Brown at Princeton's Fine Hall Math/Physics Library.) Then, on the Web one day, I stumbled across a searchable site devoted to the historical description of brass memorials in all the churches of England and Wales. And there I found entries for Henry's parents *and* Henry; his memorial inscription included the year of his death. (My thanks to the English historian William Lack for maintaining this site.) And with that date in hand, I was able to obtain a copy of Henry's will, which opens with the confirmation that he was, indeed, a "late Fellow of Trinity," as well as a copy of his death certificate. (My thanks to the Chesire and Chester County Archivist—Wilbraham lies buried in the graveyard of the Church of St. Mary the Virgin in Weaverham, Chesire County—for promptly sending me a copy of the will, and to the General Register Office for very prompt service on my request for a copy of the death certificate.)

25. Harrow School, located in London, has a long and distinguished history. Founded in 1752 under Royal Charter from Elizabeth I, it boasts a long line of famous students, including Lord Byron and Winston Churchill. Many of the younger readers of this book are almost surely familiar with Harrow School without knowing it—it is the movie setting of Harry Potter's school, Hogwarts School of Witchcraft and Wizardry. Indeed, the real location of Professor Flitwick's Magical Charms Class is Harrow School's Fourth Forum Room, upon whose walls and desks (tradition has it) every Harrow student up until 1847 carved his name. If so, *Henry Wilbraham* is somewhere in that room, and if a reader should happen to visit Harrow and find Henry's name, please let me know (sending me a photo would be even better)!

26. In 1914 the German mathematician and historian Heinrich Burkhardt (1861–1914) published a huge encyclopedia article on the history of trigonometric series and integrals up to 1850, and so Wilbraham's 1848 paper just made the cut. But, again, nobody paid any attention; for example, three years later the Scottish mathematician Horatio Carslaw (1870–1954) published a paper in the *American Journal of Mathematics* on the Gibbs phenomenon that includes the comment, on *Gibbs's* work, "it is most remarkable that its [the Gibbs phenomenon] occurrence in Fourier Series remained undiscovered till so recent a date." Poor Wilbraham—*still* unrecognized as late as 1917. Then, at last, in 1925, both Carslaw and the American mathematician Charles Moore (1882–1967) (who had done his doctoral dissertation under Bôcher) simultaneously published historical notes in the widely

read mathematics journal *Bulletin of the American Mathematical Society* proclaiming Wilbraham's priority. Not that it has done his memory much good—nearly all electrical engineering *and* mathematics textbooks *still* ignore Wilbraham.

27. You can find a presentation of how Gauss solved for $G(m)$ using non-Fourier reasoning in Trygve Nagell, *Introduction to Number Theory* (John Wiley, 1951), pp. 177–80.

28. For a nice tutorial on the many different possibilities for these sums, see Bruce R. Berndt and Ronald J. Evans, "The Determination of Gauss Sums," *Bulletin of the American Mathematical Society*, September 1981, pp. 107–29.

29. See *AIT*, pp. 175–80. The Fresnel integrals are named after the French mathematical physicist Augustin Jean Fresnel (1788–1827), who encountered them in 1818 when conducting research into the nature of light. We'll see these integrals again in section 5.7.

30. See my book *When Least Is Best* (Princeton University Press, 2004), pp. 251–57.

31. You can find this formula *stated* in many calculus textbooks. If you want to see two different *derivations* of it, see *When Least Is Best* (note 30), pp. 352–58.

Chapter 5: Fourier Integrals

1. See, for example, Kevin Davey, "Is Mathematical Rigor Necessary in Physics?" *British Journal for the Philosophy of Science*, September 2003, pp. 439–63. Davey's answer of "probably not" would surely have gotten Dirac's approval. In thinking about Dirac's successful use of impulses and integrals, I'm reminded of a famous remark made by the University of Virginia mathematician E. J. McShane (1904–1989), during his Presidential Address at the 1963 annual meeting of the American Mathematical Society: "There are in this world optimists who feel that any symbol that starts off with an integral sign must necessarily denote something that will have every property that they would like an integral to possess. This of course is quite annoying to us rigorous mathematicians; what is even more annoying is that by doing so they often come up with the right answer." You can find McShane's complete address, "Integrals Devised for Special Purposes," in the *Bulletin of the American Mathematical Society* 69, 1963, pp. 597–627. Like Dirac, McShane had an undergraduate degree in engineering.

2. Paul Dirac, "The Physical Interpretation of Quantum Mechanics," *Proceedings of the Royal Society of London* A113, January 1, 1927, pp. 621–41.

3. *Paul Dirac: The Man and His Work* (Cambridge University Press, 1998), p. 3.

4. John Stalker, *Complex Analysis: Fundamentals of the Classical Theory of Functions*, (Birkhäuser, 1998), p. 120. Professor Stalker was commenting on the validity of interchanging the order of two infinite summations.

5. There is one qualification concerning this property of $R_f(\tau)$ that I should tell you about. If $f(t)$ is *periodic* with period T, then $R_f(\tau)$ will also be periodic with period T, and so $R_f(\tau)$ achieves its maximum value not only at $\tau = 0$, but at $\tau = \pm kT$ where

k is any integer. This is generally stated in the following way: If there is a constant $T > 0$ such that $R_f(0) = R_f(T)$, then $R_f(\tau)$, is periodic with period T.

This is easy to prove (with the aid of the Cauchy-Schwarz inequality that we derived in Section 1.5); if $f(t)$ and $g(t)$ are any two real-valued functions, then

$$\left\{ \int_{-\infty}^{\infty} f(t)g(t)\,dt \right\}^2 \leq \left\{ \int_{-\infty}^{\infty} f^2(t)\,dt \right\} \left\{ \int_{-\infty}^{\infty} g^2(t)\,dt \right\}.$$

If we define $g(t) = f(t - \tau + T) - f(t - \tau)$, then this inequality becomes

$$\left\{ \int_{-\infty}^{\infty} f(t)[f(t - \tau + T) - f(t - \tau)]\,dt \right\}^2$$

$$\leq R_f(0) \int_{-\infty}^{\infty} [f(t - \tau + T) - f(t - \tau)][f(t - \tau + T) - f(t - \tau)]\,dt.$$

Now, the left-hand side is $[R_f(\tau - T) - R_f(\tau)]^2$. The integral on the right is

$$\int_{-\infty}^{\infty} f(t - \tau + T)f(t - \tau + T)\,dt - \int_{-\infty}^{\infty} f(t - \tau)f(t - \tau + T)\,dt$$

$$- \int_{-\infty}^{\infty} f(t - \tau + T)f(t - \tau)\,dt + \int_{-\infty}^{\infty} f(t - \tau)f(t - \tau)\,dt$$

which is $R_f(0) - R_f(T) - R_f(T) + R_f(0) = 2R_f(0) - 2R_f(T)$. Thus,

$$[R_f(\tau - T) - R_f(\tau)]^2 \leq 2R_f(0)[R_f(0) - R_f(T)] = 0,$$

where the equality to zero follows because we are *given* that $R_f(0) = R_f(T)$. Now, the left-hand side (something *squared*) can never be negative, and so the inequality is actually an equality. That is, $[R_f(\tau - T) - R_f(\tau)]^2 = 0$, which says $R_f(\tau - T) = R_f(\tau)$ for all τ. But this is the very *definition* of periodicity with period T, and we are done.

6. A fascinating discussion on Einstein's priority in the history of the Wiener-Khinchin theorem can be found in a 1985 paper by A. M. Yaglom, "Einstein's 1914 Paper on the Theory of Irregularly Fluctuating Series of Observations," originally published in Russian and reprinted in English in *IEEE ASSP Magazine,* October 1987, pp. 7–11.

7. Notice that the ESD of df/dt is $1/2\pi \mid i\omega F(\omega) \mid^2 = \omega^2(1/2\pi) \mid F(\omega) \mid^2$, which is ω^2 times the ESD of $f(t)$. That tells us that the ESD of df/dt is *greater* than the ESD of $f(t)$ for $|\omega| > 1$, that is, time differentiation *enhances* the energy of high-frequency signals, that is, "noise." For this reason, alone, engineers do *not* use electronic differentiators to build analog computer circuits to solve differential equations (which

is what most people think would be the "natural" thing to do). The reason is now obvious; if any noise signals are present in the circuitry—and noise is practically impossible to completely eliminate—differentiators will make the noise worse. For that reason differential equations are solved by analog electronic circuitry based on *integrators* (which *suppress* high-frequency noise).

8. Notice that the integrand of the integral representation for $|t|$ is well-behaved for all ω, including $\omega = 0$. Recall the power series expansion for $\cos(x)$—then

$$\cos(x) = 1 - \frac{x^2}{2!} + \frac{x^4}{4!} - \cdots,$$

and so

$$\cos(\omega t) = 1 - \frac{(\omega t)^2}{2!} + \frac{(\omega t)^4}{4!} - \cdots,$$

and so

$$1 - \cos(\omega t) = \frac{(\omega t)^2}{2!} - \frac{(\omega t)^4}{4!} + \cdots,$$

and so

$$\frac{1 - \cos(\omega t)}{\omega^2} = \frac{t^2}{2!} - \frac{\omega^2 t^4}{4!} + \cdots,$$

and so

$$\lim_{\omega \to 0} \frac{1 - \cos(\omega t)}{\omega^2} = \frac{t^2}{2!}.$$

And as $|\omega| \to \infty$ it is obvious that the integrand goes to zero as $\frac{1}{\omega^2}$ since the magnitude of the numerator is never greater than 2 for *any* ω and/or t. Thus, the integrand "doesn't do anything weird" for *any* value of ω, including $\omega = 0$.

9. We could, of course, substitute in *any* value for t and get a "Poisson summation formula." The use of $t = 0$ gives the classic result. In his derivation Poisson did indeed use Fourier series mathematics, but in a far different way from the approach I've shown you in the text. His result was, in fact, *more general* than than the result in the text. See *Mathematics of the 19th Century* (volume 3), edited by A. N. Kolmogorov and A. P. Yushkevich (translated from the Russian by Roger Cooke) (Birkhäuser, 1998), pp. 293–294.

10. See *AIT*, pp. 177–178 for the derivation of $\int_{-\infty}^{\infty} e^{-x^2} dx = \sqrt{\pi}$. Then, making the change of variables $x = t\sqrt{\alpha}$ gives $\int_{-\infty}^{\infty} e^{-\alpha t^2} dx = \sqrt{\pi/\alpha}$.

11. The Gaussian pulse is not the only function with the property of being its own Fourier transform. A whole *class* of functions based on what are called *Hermite polynomials* also has this property. See Athanasious Papoulis, *The Fourier Integral and*

Its Applications (McGraw-Hill, 1962) (in particular, see problems 6 and 7 on p. 76 and their solutions on p. 78).

12. A quite interesting little book on these functions is Richard Bellman's *A Brief Introduction to Theta Functions* (Holt, Rinehardt and Winston), 1961.

13. Arthur Schuster, "On the Total Reflexion of Light," *Proceedings of the Royal Society of London*, 107A, 1925, pp. 15–30.

14. E. C. Titchmarsh, "Godfrey Harold Hardy, 1877–1947," *Obituary Notices of Fellows of the Royal Society of London* 6, 1949, pp. 447–470.

15. All of Hardy's papers, written either alone or in collaboration with others, have been reprinted in seven volumes totaling 5,000 pages: *Collected Papers of G. H. Hardy* (Oxford University Press, 1979).

16. This paper, "A Definite Integral Which Occurs in Physical Optics," was published in vol. 24 (1925) of the *Proceedings of the London Mathematical Society*, but it is more readily found in volume 4 of the *Collected Papers* (Note 15), pp. 522–23.

Chapter 6: Electronics and $\sqrt{-1}$

1. See, for example, Michael Eckert, "Euler and the Fountains of Sanssouci," *Archive for History of Exact Sciences* 56, 2002, pp. 451–68; C. Truesdell, "Euler's Contribution to the Theory of Ships and Mechanics. An Essay Review," *Centaurus* 26, 1983, pp. 323–35; A. J. Aiton, "The Contributions of Newton, Bernoulli and Euler to the Theory of the Tides," *Annals of Science* 11, 1956, pp. 206–23; J. A. Van den Broek, "Euler's Classic Paper 'On the Strength of Columns,'" *American Journal of Physics*, July–August 1947, pp. 309–18.

2. I actually had a boss, once, who drew figure 6.2.1 on his office blackboard for me, except that he wrote $x(t)$ as the word *problem* , $y(t)$ as the word *answer*, and $h(t)$ as the word *solution*. He then pointed at *solution* and said to his young, newest employee, "Go get it, tiger!" (You have to be able to say stuff like that, without blushing, to be a project manager.) I was fresh out of graduate school, with no experience to tell if he was joking or not. To this day I still wonder which it was.

3. They are sometimes also called the *Kramers-Kronig relations*, after the Dutch physicist Hendrik Kramers (1894–1952) and the American physicist Ralph Kronig (1904–1995), who encountered the Hilbert transform in the 1920s when studying the spectra of x-rays scattered by the atomic lattice structures of crystals. The expressions derived in the text are not the only way they are written. An alternative form is based on the observation that, since $h(t)$ is real, we know (as explained in Section 5.1), that $R(-\omega) = R(\omega)$ and $X(-\omega) = -X(\omega)$. Thus, the first Hilbert transform integral expression in the text can be written as

$$\pi R(\omega) = \int_{-\infty}^{\infty} \frac{X(\tau)}{\omega - \tau} d\tau = \int_{-\infty}^{0} \frac{X(\tau)}{\omega - \tau} d\tau + \int_{0}^{\infty} \frac{X(\tau)}{\omega - \tau} d\tau.$$

In the first integral on the right make the change-of-variable $s = -\tau$. Then,

$$\pi R(\omega) = \int_{\infty}^{0} \frac{X(-s)}{\omega + s}(-ds) + \int_{0}^{\infty} \frac{X(\tau)}{\omega - \tau} d\tau$$

$$= \int_{0}^{\infty} \frac{X(-s)}{\omega + s} ds + \int_{0}^{\infty} \frac{X(\tau)}{\omega - \tau} d\tau = -\int_{0}^{\infty} \frac{X(s)}{\omega + s} ds + \int_{0}^{\infty} \frac{X(\tau)}{\omega - \tau} d\tau$$

$$= \int_{0}^{\infty} X(\tau) \left[\frac{1}{\omega - \tau} - \frac{1}{\omega + \tau} \right] d\tau = \int_{0}^{\infty} X(\tau) \frac{2\tau}{\omega^2 - \tau^2} d\tau.$$

Thus,

$$R(\omega) = \frac{2}{\pi} \int_{0}^{\infty} \frac{\tau X(\tau)}{\omega^2 - \tau^2} d\tau.$$

If *you* repeat this argument for the second Hilbert transform integral expression in the text then *you* should find that

$$X(\omega) = -\frac{2\omega}{\pi} \int_{0}^{\infty} \frac{R(\tau)}{\omega^2 - \tau^2} d\tau.$$

4. At least we *think* causality is a fundamental law. It is *not* a consequence of any of the other known fundamental laws of physics, and so maybe causality is *not* an absolute requirement. We just think it is, based on how the world *appears* to behave. That belief would be shown to be wrong when (if) the first time machine is ever built—see my book *Time Machines: Time Travel in Physics, Metaphysics, and Science Fiction* (Springer-Verlag, 1999).

5. Norbert Wiener, *I Am a Mathematician: The Later Life of a Prodigy* (MIT Press, 1956), pp. 168–69. A derivation of the Paley-Wiener integral is far beyond the level of this book, but if you're interested you can find an outline of how to do it in Athanasious Papoulis, *The Fourier Integral and Its Applications* (McGraw-Hill, 1962), pp. 215–17.

6. This explanation, while *theoretically* correct, is not quite the way the electronics in an AM radio receiver actually works. For practical *engineering* reasons there are some variations (e.g., the bandpass selection filter is *not* tunable). If you want to read more on the details of how AM radio really works, see my book *The Science of Radio* (2nd edition) (Springer-Verlag, 2001).

7. Suppose we connect the two multiplier inputs together, which reduces the multiplier to the special case of a single-input *squarer*. The output of the circuit of figure 6.3.1 is then $y(t) = x^2(t)$, where $x_1(t) = x_2(t) = x(t)$. Doubling the input to an LTI box would double the output, but the output of a squarer obviously *quadruples*.

8. The *mathematics* of this result was known long before AM radio was invented, of course, but the name of the theorem is due to the American electrical engineer Reginald Fessenden (1866–1932), who patented the multiplication idea in 1901 for use in a radio circuit. The word *heterodyne* comes from the Greek *heteros* (for external) and *dynamic* (for force). Fessenden thought of the $\cos(\omega_c t)$ input as the "external force" being generated by the radio receiver circuitry itself (indeed, radio engineers call that part of an AM radio receiver the *local oscillator circuit*).

9. Here's why. The synchronous receiver of figure 6.3.3 shows a final detected signal at the output of the lowpass filter that is proportional to $m(t)$, which is what we want. However, if the local oscillator signal has a phase shift relative to the distant transmitter oscillator, then the multiplier output is $r(t) \cos(\omega_c t + \theta)$, or, since the received signal $r(t)$ is proportional to $m(t) \cos(\omega_c t)$, the multiplier output is proportional to $m(t) \cos(\omega_c t) \cos(\omega_c t + \theta) = \frac{1}{2} m(t)[\cos(\theta) + \cos(2\omega_c t)]$. That is, the output of the lowpass filter is proportional not to $m(t)$ but to $m(t) \cos(\theta)$. So, a phase error appears in the synchronous demodulation receiver as an *amplitude attenuation factor*, which might not be considered a fatal problem (unless $\theta = 90°$, of course, because then there is *no* lowpass filter output). For $\theta \neq 90°$ you might argue that one could counter the attenuation by simply "turning up the volume." The problem with that is that θ is almost certainly not a constant in time, but rather $\theta = \theta(t)$. One could listen to such a receiver only by constantly fiddling with the volume knob. Nobody would buy such a receiver! A similar problem occurs if there is a frequency mismatch (which I'll leave for *you* to show). From a historical viewpoint, it is interesting to note that as early as 1930 at least one patent had already been granted for a carrier synchronization circuit remarkably similar to a method that is often used today (for the engineering-minded reader, it was a *negative feedback, phase-locked loop* circuit).

10. In an actual AM radio receiver the baseband signal is recovered by an incredibly simple (and *cheap*) circuit called an *envelope detector*, which avoids having to do this last multiplication by $\cos(\omega_c t)$. Again, see my book *The Science of Radio* 6, pp. 59–60.

11. I should explain the *more*. This means $\omega_s > 2\omega_m$ is required to be able to recover $m(t)$ from $m_s(t)$ by lowpass filtering, not $\omega_s \geq 2\omega_m$. This may seem a trifling point from the way I have drawn figure 6.4.2, since $\omega_s = 2\omega_m$ allows the individual spectrum copies to *just touch*, and not to overlap. This might not seem to be a problem (other than the theoretical impossibility of building a real lowpass filter with a vertical skirt that can select the one copy of $M(\omega)$ centered on $\omega = 0$), but what if $m(t)$ has *impulses* at $\omega = \pm\omega_m$? That would occur if $m(t)$ has a sinusoidal component at $\omega = \omega_m$. Then $\omega_s = 2\omega_m$ would have the impulses in adjacent copies of $M(\omega)$ falling on top of each other. To avoid that, we must insist on $\omega_s > 2\omega_m$. Another way to see this, thinking strictly in the time domain, is to suppose that $m(t)$ is purely a sinusoid at frequency $\omega = \omega_m$. If we sample this $m(t)$ at $\omega_s = 2\omega_m$, then it is possible that each and every "look" at $m(t)$ falls right when $m(t)$ is going through zero, that is, every value of $m_s(t) = 0$! It would clearly not be possible to reconstruct our nonzero $m(t)$ from the "always zero" $m_s(t)$. Again, we avoid this theoretical possibility by insisting on $\omega_s > 2\omega_m$.

12. See *The Science of Radio* (Note 6), pp. 248–49.

13. See, for example, Sidney Darlington, "Realization of a Constant Phase Difference," *Bell System Technical Journal*, January 1950, pp. 94–104, and Donald K. Weaver, Jr., "Design of RC Wide-Band 90-Degree Phase-Difference Network," *Proceedings of the IRE*, April 1954, pp. 671–76. Weaver's paper includes a design example (using just six resistors and six capacitors) that, for any input with a frequency from 300 Hz to 3 kHz, generates two outputs with a phase difference that deviates from 90° by no more than 1.1° (he mentions in his paper that he had constructed other more complicated circuits that had total deviations from 90° that never exceeded 0.2° as the frequency varied over the same frequency interval). You can find a computer simulation of Weaver's design example in *The Science of Radio* (Note 6), p. 275.

14. Raymond Heising, "Production of Single Sideband for Trans-Atlantic Radio Telephony," *Proceedings of the IRE*, June 1925, pp. 291–312.

15. Figure 6.6.3 was created with the use of the MATLAB function hilbert($\mathbf{x}$), which takes the complex-valued vector $\mathbf{x}$—into which the original time function has been inserted as the real part—and places the Hilbert transform of the real part into the imaginary part of $\mathbf{x}$. That is, MATLAB creates the analytic signal. From the final version of $\mathbf{x}$, plotting figures such as figure 6.6.3 is easy. Each of the plots in figure 6.6.3 has a time step of 0.001 second and, on a 3 GHz machine, took a total of 0.11 seconds to generate (after nearly two million floating point operations).

16. If $y(t) = \overline{x}(t)$, then we know that

$$Y(\omega) = \begin{cases} -iX(\omega), \omega > 0, \\ iX(\omega), \omega < 0. \end{cases}$$

Since $\overline{y}(t) = \overline{\overline{x}}(t)$, then

$$\overline{Y}(\omega) = \begin{cases} -iY(\omega), \omega > 0 \\ iY(\omega), \omega < 0 \end{cases} = \begin{cases} -X(\omega), \omega > 0, \\ -X(\omega), \omega < 0. \end{cases}$$

That is, $\overline{Y}(\omega) = -X(\omega)$ for *all* ω. In other words, the Fourier transform of $\overline{\overline{x}}(t)$ is $-X(\omega)$, which means $\overline{\overline{x}}(t) = -x(t)$, and we are done.

17. Donald K. Weaver, Jr., "A Third Method of Generation and Detection of Single-Sideband Signals," *Proceedings of the IRE*, December 1956, pp. 1703–5. (This is the same Weaver as in Note 13.)

18. You'll often find electrical engineering textbooks that, in their discussion of the Weaver circuit, stipulate that $m(t)$ should have a nonzero width gap around $\omega = 0$ in which there is no energy (often, however, they don't say *why*). Weaver himself stipulated this. If that's the case, then the lowpass filters do not have to have vertical skirts, which is a big plus for Weaver's circuitry (over and above the elimination of the need for a Hilbert transformer). In Weaver's 1956 paper he presented the actual electrical circuitry for a specific SSB radio signal generator ($\omega_c = 2\pi \cdot 10^6$ Hz) with a zero energy gap located at -300 Hz $< \omega < 300$ Hz (absence of energy at those low

frequencies has no appreciable affect on the intelligibility of speech or the quality of music), and $\omega_m = 2\pi \cdot 3,300$ Hz. His ω_0 was set equal to the middle of the *nonzero* energy interval, i.e., $\omega_0 = 2\pi((300 + 3,300)/2) = 2\pi \cdot 1,800$ Hz.

19. Philip J. Davis, "Leonhard Euler's Integral: A Historical Profile of the Gamma Function," *American Mathematical Monthly*, December 1959, pp. 849–69.

Euler: The Man and the Mathematical Physicist

1. David Brewster, "A Life of Euler," in *Letters of Euler on Different Subjects in Natural Philosophy Addressed to a German Princess*, J.&J. Harper, 1833; J. J. Burckhardt, "Leonhard Euler, 1707–1783," *Mathematics Magazine*, November 1983, pp. 262–73; A. P. Yushkevich's entry on Euler in the *Dictionary of Scientific Biography*, vol. 4, pp. 467–84; C. Truesdell, "Leonhard Euler, Supreme Geometer (1707–1783)," in *Irrationalism in the Eighteenth Century* (edited by Harold E. Pagliaro), Press of Case Western Reserve University, 1972, pp. 51–95; Ronald Calinger, "Leonhard Euler: The Swiss Years," *Methodology and Science*, vol. 16-2, 1983, pp. 69–89; Ronald Calinger, "Leonhard Euler: The First St. Petersburg Years (1727–1741)," *Historia Mathematica*, May 1996, pp. 121–66.

2. For some discussion on this aspect of Bernoulli's personality, see my book *When Least is Best* (Princeton University Press, 2004), pp. 211, 244–45.

3. One historian (Calinger, 1983—see Note 1) says Euler "proposed the first correct solution" to this question, and that "Euler rejected both existing opinions that the stone would either make a dead stop at the center [of the earth] or travel beyond it. Instead, he argued that it would rebound at the center and return by the same path to the surface." This is puzzling since that is *not* the correct answer! A reference is given to a discussion of how this problem perplexed even the great Newton, but that discussion reveals that the "puzzle" was essentially over the nature of the gravitational force *inside* the earth. If R is the radius of the earth, and r is the distance of the stone from the center of the earth, then of course gravity famously varies as the *inverse square* of r if $r \geq R$. By 1685 Newton knew that the variation of gravity with r (for a uniformly dense spherical earth) is *directly* as r if $r < R$, and from that it is easy to show that the stone will execute what is called *simple harmonic motion*, as it oscillates sinusoidally from one side of the earth, through the center, *all the way to the other side*, and then all the way back to its starting point. Then it does the trip all over again, endlessly. It is easy to calculate that for a uniformly dense sphere of radius $R = 4,000$ miles and a surface acceleration of gravity of 32 ft/second^2 (that is, the earth) one complete round trip takes 85 minutes, and that as the stone transits the earth's center it is moving at 26,000 ft/second *towards the other side* of the planet (there is no "center rebound"). For a complete, modern discussion on this issue, see Andrew J. Simoson, "Falling Down a Hole Through the Earth," *Mathematics Magazine*, June 2004, pp. 171–89.

4. See *AIT*, pp. 176–178, and Philip S. Davis, "Leonhard Euler's Integral: A Historical Profile of the Gamma Function," *American Mathematical Monthly*, December 1959, pp. 849–69.

5. See *AIT*, pp. 150–152.

6. Both the isoperimetric and brachistochrone problems are discussed at length, and solved in my *When Least is Best* (Note 2); in particular, on pp. 200–78 there is an introduction to the calculus of variations.

7. David Eugene Smith, "Voltaire and Mathematics," *American Mathematical Monthly*, August–September 1921, pp. 303–5. An example of the emptiness of Voltaire's mathematics in *Éléments* is found in his "explanation" of Snel's law of the refraction of light—he tells his readers that there is a relationship between a light ray's angle of incidence and angle of refraction at the boundary of two distinct regions (e.g., air and glass, or air and water) and that this relationship involves something called a *sine*, but he fails to tell his readers what a sine *is*, because that's too technical! Voltaire didn't actually think much of his readers, in fact—in a letter to a friend before he began to write *Éléments*, he said his goal was "to reduce this giant [Newton's *Principia*] to the measure of the dwarfs who are my contemporaries."

8. My primary sources of historical information on the Berlin phase of Euler's life include Ronald S. Calinger, "Frederick the Great and the Berlin Academy of Sciences," *Annals of Science* 24, 1968, pp. 239-49; Mary Terrall, "The Culture of Science in Frederick the Great's Berlin," *History of Science*, December 1990, pp. 333–64; Florian Cajori, "Frederick the Great on Mathematics and Mathematicians," *American Mathematical Monthly*, March 1927, pp. 122–30; and Ronald S. Calinger, "The Newtonian-Wolffian Controversy (1740–1759)," *Journal of the History of Ideas*, July–September 1969, pp. 319–30.

9. C. B. Boyer, "The Foremost Textbook of Modern Times," *American Mathematical Monthly*, April 1951, pp. 223–226.

10. See *When Least Is Best*, pp.133–34.

11. See Calinger, 1969 (Note 8).

12. The relationship between Euler and D'Alembert was a complicated, professional one, far different from the personal friendship Euler enjoyed with, for example, Daniel Bernoulli. You can find more on this matter in the paper by Varadaraja V. Raman, "The D'Alembert-Euler Rivalry," *Mathematical Intelligencer* 7, no. 1, 1985, pp. 35–41. This paper, while quite interesting, should be read with some care because the author occasionally overstates his criticism of Euler. For example, he takes Euler to task for not citing (in one of his works) Johann Bernoulli's publication of a derivation of $\zeta(2)$. Well, why *should* Euler have have cited Bernoulli?—it was *Euler*, not Bernoulli, who *first* calculated $\zeta(2)$.

13. Just the thought of an eye operation in the 1770s is probably enough to make most readers squirm. Euler's operation was a procedure called *couching*, which can be traced back to at least 2000 B.C. With the patient's head clamped in the hands of a strong assistant, the doctor would use a sharp needle to perforate(!) the eyeball

and push the cataractous lens aside to let light once again reach the retina (today, the lens is totally removed and replaced with an artificial one). The risk was that lens proteins would be dispersed in the eye, which would cause a severe intraocular inflammation that could result in blindness. That's just what happened to Euler.

14. Ronald Calinger, "Euler's 'Letters to a Princess of Germany' as an Expression of his Mature Scientific Outlook," *Archive for History of Exact Sciences* 15, no. 3, 1976, pp. 211–33.

15. F. Cajori, *A History of Mathematics* (2nd edition), Macmillan, 1919, p. 233.

16. See Dirk J. Struik, "A Story Concerning Euler and Diderot," *Isis*, April 1940, pp. 431–32; Lester Gilbert Krakeur and Raymond Leslie Krueger, "The Mathematical Writings of Diderot," *Isis*, June 1941, pp. 219–32, and B. H. Brown, "The Euler-Diderot Anecdote," *American Mathematical Monthly*, May 1942, pp. 302–3.

17. You can find a photograph of Euler's tomb in Frank den Hollander, "Euler's Tomb," *Mathematical Intelligencer*, Winter 1990, p. 49. Hollander says the tomb is in Leningrad, Russia, which was true in 1990—St. Petersburg had its name changed to Leningrad in 1924 (since 1914 it had been known as Petrograd). But in 1991 the original name was restored. Euler was buried in St. Petersburg, and that is where he is today. Another photograph of his tomb, as well as one of his house at #15 Leytenant Schmidt Embankment, along the Neva River, is on p. 18 of the March 2005 issue of *FOCUS*, the newsletter of the Mathematical Association of America.

18. For much more on the mathematics of Euler, see the really outstanding book by William Dunham, *Euler: The Master of Us All* (Mathematical Association of America, 1999).

Acknowledgments

This book was written while I was still a full-time member of the electrical engineering faculty at the University of New Hampshire, from which I retired in 2004. Through my entire thirty-year teaching career at UNH I received enthusiastic encouragement for my writing efforts (even during my science *fiction* writing days!), and this book was no exception. I am therefore most grateful to my former colleagues in EE, and to the administration at UNH, for that support. UNH academic librarians Professors Deborah Watson and Barbara Lerch helped me in countless ways in obtaining technical information that wasn't available at UNH. After my retirement I continued to receive library support by UNH's enlightened policy of allowing emeriti faculty free Web access, via the JSTOR database, to all the technical journals one could want, as well as free ILL access to the more obscure journals at other libraries but not on JSTOR. Much of the material in this book first served as lecture notes and homework problems in my third-year systems engineering classes (EE633 and EE634) in the academic year 2003–2004. I thank my students in those classes who helped me wring most (*all* would be too much to ask for!) of the errors out of the text. Probably one of the best students I have ever had, Timothy E. Bond, was particularly helpful. Alison Anderson did an outstanding job of copyediting the entire book; she saved me from a number of awkward blunders. As with all university press books, this one received numerous critical pre-publication reviews, and I am particularly in the debt of Desmond J. Higham, Professor of Mathematics at the University of Strathclyde, Scotland, and that of Dr. Garrett R. Love of the Shodor Education Foundation, for their helpful commentary. And finally, as any honest writer will admit, there could simply be no book at all without a receptive editor. I have been so fortunate, indeed, with the editor of this book, Vickie Kearn, who stood by me even when matters at one time looked ominously dark, gloomy and, in fact, downright depressing. In my math book, Vickie is a ten. No, make that an eleven.

Tellico Village (Loudon, Tennessee)
August 2005

Index

Also by Paul J. Nahin

Oliver Heaviside (1988, 2002)

Time Machines (1993, 1999)

The Science of Radio (1996, 2001)

An Imaginary Tale (1998)

Duelling Idiots (2000, 2002)

When Least Is Best (2004)

Chases and Escapes (2007)